SEARCHING THE HEAVENS AND THE EARTH:
THE HISTORY OF JESUIT OBSERVATORIES

ASTROPHYSICS AND SPACE SCIENCE LIBRARY

VOLUME 286

SEARCHING THE HEAVENS AND THE EARTH: THE HISTORY OF JESUIT OBSERVATORIES

by

AUGUSTÍN UDÍAS

Department of Geophysics and Meteorology,
Universidad Complutense, Madrid, Spain

KLUWER ACADEMIC PUBLISHERS

DORDRECHT / BOSTON / LONDON

A C.I.P. Catalogue record for this book is available from the Library of Congress.

ISBN 1-4020-1189-X

Published by Kluwer Academic Publishers,
P.O. Box 17, 3300 AA Dordrecht, The Netherlands.

Sold and distributed in North, Central and South America
by Kluwer Academic Publishers,
101 Philip Drive, Norwell, MA 02061, U.S.A.

In all other countries, sold and distributed
by Kluwer Academic Publishers,
P.O. Box 322, 3300 AH Dordrecht, The Netherlands.

Cover figure: Manuel Figueredo showing the European astronomical tables to the Maharajah Sawai Jai Singh at the Astronomical Observatory of Jaipur, India.

Printed on acid-free paper

TABLE OF CONTENTS

Preface

In 1957, as a young Jesuit student in Havana, I was asked if I would like to be trained for work in Belén Observatory. Though I had joined the Jesuits in Spain I had been sent to Cuba during my novitiate. The same year I was sent to Saint Louis University where I began to study philosophy and geophysics. When I finished my doctoral studies in geophysics, the Colegio de Belén had been confiscated by Castro's government and the observatory closed. I finally ended teaching geophysics in the Universidad Complutense in Madrid, but I have always kept contact with Jesuit observatories, especially Ebro, Cartuja, Saint Louis, La Paz and Bogotá. During the 1970s Jesuit observatories began to close and I felt that their history should not be forgotten. This book tries to achieve this aim.

I have been collecting information since 1985. It was not always easy. Some observatories had already been closed for some time and little information was available. William Stauder of Saint Louis University, who had been my thesis director, gave me great encouragement and help. Together we coauthored three articles on the subject in 1991, 1997 and 1998. Writing a book, as much a complete history as possible, was always in my mind. Little by little the files for each observatory were growing. Besides printed material visits to Jesuit Archives in Rome, Paris, Cologne and London and many contacts by letter gave me important non-published information concerning the observatories, such as letters and reports. Thus, the history of the Jesuit observatories, founded after the restoration of the Society of Jesus in 1814, became clear to me. The history of the observatories in the first period of the Society, from its foundation in 1540 to its suppression in 1773, was a much more difficult task. A complete history of these observatories would require a detailed search in many different archives. I did not have time for this and I decided to limit myself for this part to references in already published work. Therefore, I feel that my history is fairly complete for modern observatories but only an introduction for the earlier ones.

This book is divided into three parts. Part I gives a historical account of the observatories. In this part I have tried to present a narrative that can be easily read. Hence, there are no footnotes. The references can be easily found from the bibliography. Years of birth and death are only given for Jesuits. In some cases dates could not be found for Jesuits who died after the suppression of the Society in 1773. The same happens for Jesuits who later left the Society. In chapter 1, I deal with the scientific tradition of the Society of Jesus and the work of Jesuits in the foundation and maintenance of observatories. I try to link this work with the especial character of Jesuit spirituality. Chapters 2 and 3 present the history of observatories in the 17th and 18th centuries in Europe and mission countries, especially, China and India. With the suppression of the

Society in 1773, these observatories ceased operation. For these chapters the references used are given at the end of the book. I feel that the bibliography I have used is fairly complete. Chapters 4 to 7 give the history of the observatories founded after the restoration of the Society in 1814. Observatories are divided according to their geographical location in Europe (Chapter 4), North America (Chapter 5), South and Central America (Chapter 6) and Asia, Africa and Australia (Chapter 7). In each chapter observatories are given according to chronological order of their foundation. I have discussed first astronomical observatories and afterwards meteorological and seismological observatories. For these chapters bibliography and references used are given with the description of each observatory in Part II and in the short biographic notes of some of their directors in Part III.

Part II, gives a short description of 74 observatories of the modern Society (1814-2000). The order followed is first geographical, as in chapters 4 to 7, and for each region by chronological order of the observatory's year of foundation. I give the years of foundation and closing of each observatory. Finding the year when an observatory closed required, sometimes, considerable research. I consider an observatory closed when it ceased operation or was transferred to a non-Jesuit institution. Observatories that today continue link to a Jesuit institution are considered open, though there is no longer any Jesuit presence. For each observatory I give a short summary of its history, a list of directors and Jesuit collaborators. Since this information comes from Jesuit Province Catalogues, there is, sometimes, an ambiguity of one year. A director can be appointed in one year, but first appear in the Catalogue of the following year. For collaborators this information has not been always found and the lists may not be complete. Buildings are briefly described and main instruments listed with the year of installation. Bulletins and other type of publications of important observatories are listed when available. Finally, a bibliography is given for each observatory; this includes published and unpublished material; I consider each bibliography fairly complete. Each bibliography is set out in chronological order of publication, since this reflects better the chronological stage of the observatory. Some observatories lasted for over one hundred years during which many changes took place. In general there are more references close to times of the observatories' foundation than for their later stages. Naturally, important observatories are given more extensive coverage.

Part III gives short biographical notes of 58 Jesuits scientists, directors or collaborators of observatories of the modern Society of Jesus (1814-2000). The selection includes all the most important directors. One element of selection was the existence of published biographical notes, especially in scientific journals. I am aware that some important names have been omitted. For each person a list of published books and articles is given limited to five or six. Here also the selection may be questionable. Important books are always given.

References are given in chronological order; I have wanted them to be as complete as possible. For Secchi an abundant literature exists and I have listed only a selection of references.

There are many persons who have helped me during the years collecting data and writing the text, to whom I wish to express my gratitude. In first place Larry Drake, last Director of Riverview and La Paz who read, corrected, and made helpful comments on the text. William Stauder, the last Jesuit director of Saint Louis, whom I have already mentioned, also read through some of the chapters. Corrections were also made by Thomas McCoog of the Jesuit Historical Institute in Rome, by Gero McLaughlin from Edinburgh and by Chris Hannan. Elisa Buforn, Universidad Complutense of Madrid helped in many ways, especially, with the figures. Important published material was found at the library of Ebro Observatory and the help of its librarian, María Genescá, is greatly appreciated, together with that of the observatory's last two Jesuit directors, José O. Cardús and Luis Alberca. Directors of several Jesuit archives were kind to look for pertinent material and to make it available to me. Their names would make a long list. Robert Bonfils from the Jesuit Archive in Paris deserves special thanks. Finally I received great encouragement from fellow Jesuits, who were not aware of the extent of the observatory work until I mentioned it to them.

Sources and journal titles

Annales de la Soc. Scient. de Bruxelles: Annales de la Société Scientifique de Bruxelles
APFSJ: Archive Province de France S.J.
Arch. Hist. Soc. Iesu : Archivum Historicum Societatis Iesu
ARSI: Archivum Romanum Societatis Iesu (Roman Archive of the Society of Jesus)
ASJCF-St. Jerome: Archive S.J. Canada Français, St. Jerome
Astrophysical Jour.: Astrophysical Journal
Bol. (Spanish): Boletín
Bol. Soc. Astron. de México: Boletín de la Sociedad Astronómica de México
Boll. (Italian): Bollettino
Bull. (English). Bulletin
Bull. Am. Ass. Jesuit Scientists: Bulletin American Association of Jesuit Scientists
Bull. Am. Meteor. Soc.: Bulletin of the American Meteorological Society
Bull. Am. Ass. Petrol. Geol.: Bulletin of the American Association of Petroleum Geologists.
Bull. Econom. de Madagascar: Bulletin Economique de Madagascar
Bull. Seis. Soc. Am.: Bulletin of the Seismological Society of America.
Bull. Union Geod. et Geophys. Int.: Bulletin de la Union de Geodesie et Geophysique International
Coll. Obs. Publ.: Collected Observatory Publications
EO: Ebro Observatory
Historical Studies in the Phys. and Biol. Scien.: Historical Studies in the Physical and Biological Sciences
Imp. (Spanish): Imprenta
Imp. (French): Imprimerie
J. Brit. Astron. Assoc.: Journal of the British Astronomical Association.
Jour. : Journal
Kaiserl. Akademie der Wissenschaften in Wien, Mathem. Naturw. : Kaiserliche Akademie der Wissenschaften in Wien. Mathematik und Naturwissenschafte
Memoirs of the Roy. Astr. Soc.: Memoirs of the Royal Astronomical Society
Mitt.: Mitteilungen
MO: Manila Observatory
Monthly Notices Roy. Astr. Soc. : Monthly Notices of the Royal Astronomical Society
OSC: Observatorio San Calixto.
Phil. Trans. Royal Soc.: Philosophical Transactions of the Royal Society
Proc.: Proceedings
Proc. Roy. Soc.: Proceedings of the Royal Society
Pub. (German): Publikationen
Pub. (English): Publications

Publ. (Spanish): Publicaciones
Quart. Jour. Roy. Astr. Soc.: Quarterly Journal of the Royal Astronomical Society
Rev. (French) : Revue
Rev. (Spanish): Revista
Rev. de Geogr. Phys. et Geol. Dyn. : Revue de Géographie et Géologie Dynamique
Riv. (Italian): Rivista
Rivista di Fisica, Mat. e Scien. Natur. di Pavia: Rivista di Fisica, Matemática e Scienza Naturale di Pavia.
Roy. Soc. of New South Wales: Royal Society of New South Wales
SJBPA: Society of Jesus British Province Archive
SLU: Saint Louis University
Studies in Hist. and Phil. of Science: Studies in History and Philosophy of Science
Trans. Am. Geophys. Union: Transactions of the American Geophysical Union

PART I. HISTORY

Chapter 1

A JESUIT TRADITION

Jesuits, science and observatories

Jesuit interest in the natural sciences began in the very early days of the foundation of the Society of Jesus in 1540. This work can be related to the development of Jesuit colleges in the sixteenth and seventeenth centuries, coinciding with the birth and progress of modern science. From the very beginning, mathematics, astronomy, and natural sciences were important features of their academic programs. Early scientific work of Jesuits and their contribution to the establishment of modern science have recently received considerable attention, for example, in the works of Heilbron (1982), Dear (1987), Wallace (1989), Harris (1989) and Baldini (1992). Steven J. Harris explicitly links Jesuit scientific tradition with the order's "apostolic spirituality," using the thesis established by Robert K. Merton on the relation between the development of modern science in seventeenth century England and the Puritan ethos. For Harris the values of apostolic spirituality strongly encouraged an activist mentality among Jesuits, which led to the acceptance of the modern sciences, because they were useful means in fulfilling the goals of the Society's apostolates in education and mission work. This became in time an established tradition, which has been continued in different ways until our days. We will treat this point later in this chapter to find what there is specifically in Jesuit spirituality, which constitutes the basis of this scientific tradition.

An important part of the activities of Jesuits in the different fields of science has been the foundation of observatories. The first astronomical observatories were installed in Jesuit colleges in Europe in the seventeenth and eighteenth centuries. In some of them, besides work in astronomy, meteorological observations began to be made. The number of Jesuit observatories in the colleges of the old Society in Europe, in the eighteenth century was, as we will see, about 30. This is about one quarter of all the then existing observatories. Despite this high proportion, Jesuit observatories are seldom mentioned in the histories of astronomy.

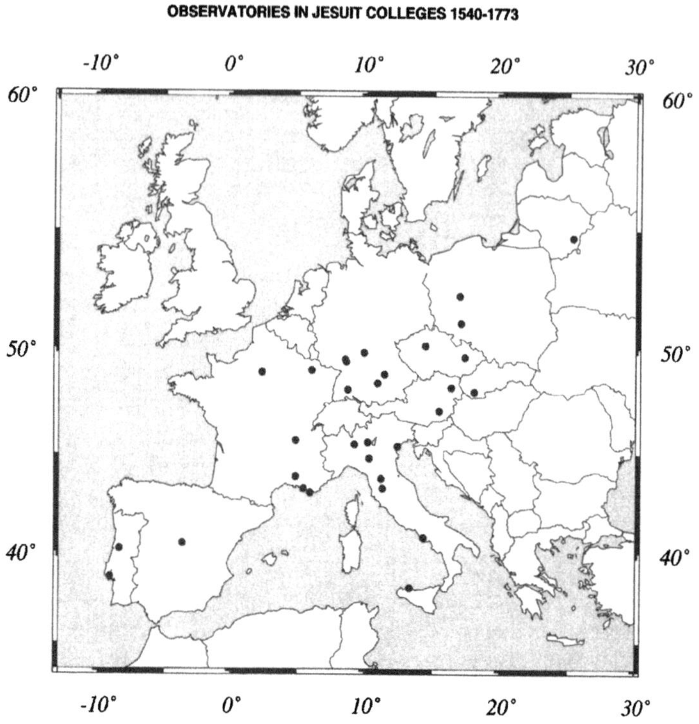

Figure 1. Location of the observatories established in Jesuit colleges, 1540-1773.

These observatories were connected with the chairs of mathematics established in all major colleges. I have found information about 29 observatories: six in France, 13 in Central Europe, eight in Italy, and one in Spain and in Portugal. To these observatories we must add those in mission countries, especially, the Imperial Observatory of Peking, which from 1644 until 1773 was directed by Jesuits, and, after that date to 1803, by former Jesuits. Jesuit astronomers worked also in India, Indochina and in the mission of Paraguay. All these observatories ceased with the suppression of the Society in 1773, or earlier with the expulsion of the Jesuits from Portugal (1759), France (1764) and Spain (1767). The history of most of these early observatories has not been yet written in detail. This work, however, as we will see in Chapters 2 and 3, was very important and laid the foundations of this tradition in the Society of Jesus. It also had a great influence in the founding of the new observatories of the restored Society.

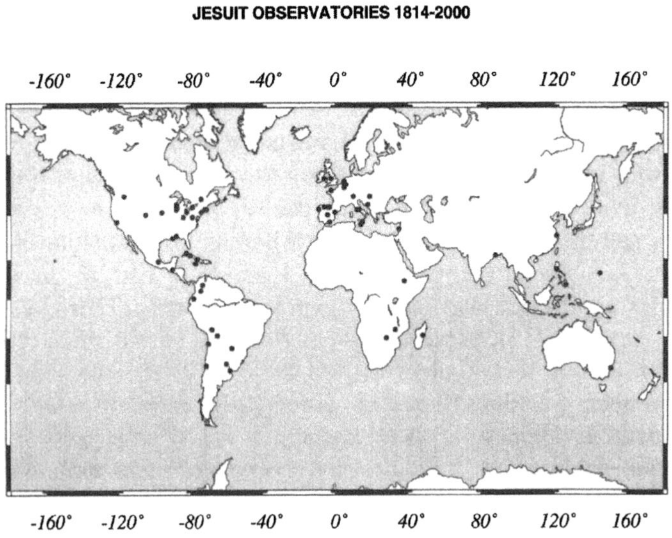

Figure 2. Location of the Jesuit observatories, 1814-2000.

Soon after the restoration of the Society of Jesus in 1814, the foundation of the new observatories began: the first in the Collegio Romano in 1824, followed by those of Stonyhurst College, England, founded in 1838, and in Georgetown University, Washington, founded in 1841. Many others followed on every continent where Jesuits were present. The last observatories to be founded were in Montreal, Canada, in 1952, and in Addis Ababa, Ethiopia, in 1957. The first of these observatories were mainly dedicated to astronomy and meteorology; later they specialized in geomagnetism, solar physics and seismology. Although some specialized in a particular subject, for example, meteorology or seismology, usually they combined different types of observations. Observatories were often attached to secondary schools, to colleges, to faculties of philosophy and to universities, but some were independent institutions, especially in mission countries. Some observatories became important research centers with a worldwide reputation in astronomy, meteorology, and geophysics, while others were small meteorological or seismological stations. In the list of astronomical observatories of André and Rayet of 1878, we find mention of three Jesuit institutions, namely, Stonyhurst, Collegio Romano and Georgetown. Twelve years later, in his list of observatories, Lancaster (1890) mentions nine, namely, Calcutta, Georgetown, Gianicolo (Rome), Kalocsa, Omaha, Puebla, Stonyhurst, Tananarive and Zikawei. A worldwide list of astronomers for 1904 by Stroobart et al. included the names of 15 Jesuits. In 1910, in a worldwide list of seismological and geomagnetic observatories by Merlin and Somville, there appeared 24 Jesuit observatories.

The total number of Jesuit observatories in the modern Society, from 1814 to 2000 is difficult to assess. It depends on the criterion used to determine what constitutes an observatory. At some colleges and faculties of philosophy, meteorological and astronomical observations were made, but not with the formal structure of an observatory. Those recognized as observatories have Directors who appeared as such in the Jesuit Province Catalogues (the Society of Jesus is divided into provinces which publish yearly catalogues with their institutions and the Jesuits assigned to them). With this limitation in mind, I have taken as observatories a total of 74 given in Part II. In this part, observatories are divided into four sections by geographical location (Europe 26, North America 21, Central and South America 18 and Asia, Africa and Australia 9). In each section observatories are given in chronological order of the date of their founding. For each observatory, dates of foundation and closing, main disciplines, a short history, a list of directors and Jesuit collaborators, publications, buildings and instruments, and bibliography are given. Their duration varies from a few years (Quito 3 years and Asunción 5 years) to over 100 years (Stonyhurst 136 years and Manila 135 years).

Biographic and bibliographic information about the observatories and about many of the Jesuits working in them can be found in the reference works of Somervogel (1890-1909), Polgar (1981-1990) and O'Neill and Domínguez (2001). More particular information about observatories of the early Society is very dispersed, as we will see in Chapters 2 and 3. A detailed history of each of these observatories would require much work on unpublished manuscripts preserved in archives. For the modern Society the information is also difficult to find, but there is abundant published literature. For a general history, the work of Pierre de Vregille (1906) gives a very complete description of 25 observatories operating in 1906 and it is the basic reference for an overview of the situation at that time. Short accounts of 18 observatories existing in 1914 are found in Albers (1914). Partial studies have been published about the most important observatories in special publications and in articles in different kinds of journals. Information is also given in the articles about Jesuits' work in astronomy, meteorology and geophysics. For each individual observatory more detailed information is found in its published bulletins and in a variety of publications and journal articles, many of them difficult to find.

Jesuits and scientific societies

Modern science led beginning in the seventeenth century to the creation of scientific societies. Jesuits working in the observatories felt the need to participate in these societies where they could keep an active collaboration with other scientists. The first Jesuit to become a member of a scientific society was Terrenz (Chapter 3), who took an important part in the astronomical work in

China. Not yet a Jesuit, Terrenz was admitted on May 3, 1611 into the Accademia dei Lincei, founded in Rome by Prince Federico Cesi in 1603. Galileo had been admitted only eight days before. In November of the same year Terrenz joined the Jesuits, but continued his relations with the academy. Two of the oldest and most prestigious scientific societies are the Royal Society of London and the Académie des Sciences of Paris. The Royal Society of London was founded in 1661. Four Jesuits have been elected fellows; two in the 18th century, Boscovic, Professor in the Collegio Romano in 1765, and Poczobut, Director of the Vilnius Observatory, in 1771 (Chapter 2) and another two in the nineteenth century, Secchi, Director of the Observatory of the Collegio Romano, in 1856 and Perry, Director of Stonyhurst Observatory, in 1874 (Chapter 4).

Several French Jesuits were members or corresponding members of the Académie Royale des Sciences of Paris founded in 1666. The Academy was reorganized in 1699, with the name of Académie des Sciences, when it was established that no member of a religious order could be admitted to membership, except as honorary or corresponding members. An exception was Gouye, Professor at the Collège Louis-le-Grand in Paris, appointed a member by Louis XIV in 1699, and elected president in 1711. The Jesuit members of the expeditions to China and Siam in 1685 and 1687 were admitted into the academy by a special privilege (Chapter 3). Among Jesuit directors of observatories in the eighteenth century, Beraud (1751, 1762), Lyon, Pezenas (1750), Marseille, Hell (1758, 1762, 1769), Vienna, Ximenes (1751, 1762, 1769), Florence, Belgrado (1762, 1771), Parma, Boscovič (1748, 1771), Rome, and Poczbout (1778), Vilnius were corresponding members (Chapter 2). In the modern Society the following directors of observatories were corresponding members Berloty (1924), Ksara, Colin (1899) and Poisson (1932), Tananarive and Lejay (1935), Zikawei (Lejay was elected non-resident member in 1946) (Chapter 7).

In 1847 Pope Pius IX refunded the old Accademia dei Lincei, which had disappeared in 1651, with the new name Pontificia Accademia dei Nuovi Lincei. In 1936, Pius XI reorganized the Academy with a new name, Pontificia Academia Scientiarum (Pontifical Academy of Sciences). The Popes wanted to select an international body of renowned scientists, at present limited to eighty. Several Jesuits, Directors of observatories, were members of this academy. De Vico, Observatory of the Collegio Romano, was a founding member in 1847 and his successor, Secchi, was also a member (Chapter 4). Other members were Algué, Manila Observatory, Dechevrens, elected in 1887, and Gherzi, Zikawei Observatory (Chapter 7), Fenyi, Kalocsa Observatory, elected in 1916, Hagen, Stein and O'Connell (President from 1968 to 1972), Vatican Observatory (Chapter 4). Coyne, the present Director of the Vatican Observatory, is an honorary member and belongs to its Council.

A number of Jesuit directors of observatories became members of national academies of science. These are selected institutions and election to them represents a measurement of excellence in the scientific circles of each country. Macelwane of Saint Louis University was elected in 1944 a member of the U. S. National Academy of Science (Chapter 5). Romaña and Cardús, Ebro Observatory, were members of the Real Academia de Ciencias Exactas, Físicas y Naturales of Spain. O'Leary, Riverview, and Ingram, Rathfarnham Castle, were members of the Royal Irish Academy. Stein, Vatican Observatory was a member of the Royal Academy of Holland. Fényi, Kalocsa Observatory, was elected in 1916 a corresponding member of the Academy of Science of Budapest (Chapter 4). Ramirez, Instituto Geofísico (Bogotá), was a corresponding member of the Spanish Academy an also member and president of the Academia Colombiana de Ciencias Exactas. Puig and Bussolini, of San Miguel Observatory, were members of the Academia Nacional de Ciencias de Buenos Aires. Viñes, Belén Observatory, was a member of the Real Academia de Ciencias Médicas, Físicas y Naturales of Cuba. Spina, Puebla Observatory, was a member of the Academia Mexicana. Cabré, La Paz Observatory, was elected a member of the Academia Nacional de Ciencias de Bolivia in 1971 (Chapter 6).

Jesuits were active participants in many scientific societies in the fields of astronomy, meteorology and geophysics. In 1820 the Royal Astronomical Society was founded in London. This has become one of the most prestigious astronomical societies. Thirty-one Jesuits from the observatories of Collegio Romano, Vatican, Stonyhurst, Georgetown, Fordham, Rathfarmham Castle, Puebla, Addis Ababa, Bulawayo and Riverview became Fellows of the Society. The first were de Vico, elected in 1841, Weld in 1849 and Secchi in 1853. Almost all the directors of Stonyhurst Observatory were fellows. Among them, Perry in 1873-1878 and in 1887-1890; Sidgreaves in 1892-1897; Cortie in 1904-1905, 1908-1910, 1912-1916 and 1918, served on its council. De Vico, Secchi and Hagen were Associate Fellows. Of the societies for meteorology, the American Meteorological Society and the Royal Meteorological Society of London deserve special mention. Several Jesuits belonged to them. A number of Jesuits in the United States were members of the American Geophysical Union. Macelwane of Saint Louis University was its President in 1953. This society and the American Meteorological Society established in his honor medals awarded each year to young scientists of outstanding ability. Jesuits had a special relation with the Seismological Society of America. One of its thirteen founding members was Ricard, Director of Santa Clara Observatory, who was elected a member of the first Board of Directors. Macelwane and Stauder of Saint Louis University were Presidents in 1928 and 1966 respectively. Seven Jesuits were Chairman of the Eastern Section of the Society founded in 1926, with Macelwane as first Chairman. For many years the Jesuit Seismological Association met jointly with the Eastern Section (Chapter 6).

Jesuits helped to found some scientific societies, for example, Lafont, in 1869, the Indian Association for the Cultivation of Science (Chapter 7) and Secchi in 1872, the Societá degli Spettroscopisti Italiani (Chapter 4).

In 1919 the International Union of Astronomy (IUA) and the International Union of Geodesy and Geophysics (IUGG) were founded. Both of them comprised a number of international associations on different aspects of their sciences. Since 1931 they have depended on the International Council of Scientific Unions (ICSU). These two unions have become the most important scientific institutions in the fields of astronomy and geophysics and their general assemblies periodically convene scientists from all countries. Jesuits have been present at their meetings and have participated in the different commissions and working groups established for particular studies. For example, in the Third General Assembly of the IUA, in 1928, in Leiden, seven Jesuits attended, namely, Lejay (Zikawei), O'Connor and Rowland (Stonyhurst), Angehrn (Kalocsa), Rodes (Ebro), Phillips (Georgetown) and Stein (Vatican). Hagen, O'Connor, Lejay, Rodes and Stein were elected members of different commissions. Lejay was Vice-president of ICSU and President of the Section of Gravimetry of the IUGG. Romañá and Cardús, of Ebro Observatory, were for many years members, and sometimes Cardús was President, of the Finance Committee of the IUGG.

The reasons for scientific work and the foundation of observatories

The reason behind the founding of the Jesuit observatories is rather complex. Harris's work, already mentioned, about the Jesuit scientific tradition put the basis of this tradition on the apostolic spirituality. Though he is examining Jesuits' scientific work of the seventeenth and eighteenth centuries, his analysis can be extended also to the scientific work after the restoration of the Society in the nineteenth and twentieth centuries. For Harris the main elements of the apostolic spirituality are the following: the emphasis placed on Christian service which leads to activities not usually associated with religious life; an active engagement with the world, which leads to a respect for experience or a "testing against experience"; an esteem for learning and what Harris calls the "sanctification of learning," which led to the enormous effort of Jesuits in the field of education. Harris distinguishes three realms in Jesuit activity of the early Society, namely, educational, court and missionary. In the modern Society the first and third remain of great importance.

Accepting the general lines of Harris' analysis I would like to add some considerations which refer to more specifically Jesuit elements. As Harris states, half of all clerics working on science in the seventeenth and eighteenth centuries were Jesuits. For modern times a study has not been made, but there is no doubt that largely this is also still true. Hence, there must be something specific in Jesuit spirituality that explains this fact. Some of Harris' elements of the "apostolic spirituality" are also found in other religious orders and congregations founded after the sixteenth century, but they have not developed a comparable scientific tradition. Hence, we must find some more specifically Jesuit elements in order to explain their especial dedication to science.

First of all, there are historical factors, also mentioned in Harris's analysis. The most important is the coincidence with the birth of modern science of the founding of the Society and the specifically Jesuit emphasis in education. Jesuits have created in the old and modern Society a network of secondary schools, colleges and universities with no comparison in any other religious group. Moreover, the Jesuit Order, founded in the sixteenth century, when it established the first colleges was not hindered by a medieval tradition, present in the older orders, like the Dominicans and Franciscans, who were also active at universities. So Jesuits could more easily enter into the new fields of modern science. A sign of this is the recommendation in the *Ratio Studiorum* of establishing chairs of mathematics in all major colleges, where there were philosophical studies. Clavius, the most influential professor of mathematics in the Collegio Romano, who was responsible for this recommendation, was aware of the growing importance of mathematical sciences in his time and the need of Jesuit colleges to accommodate to this trend. There also was at that time a social demand for this type of studies that Jesuit colleges tried to fulfill. This social demand was recognized by Scheiner, a successor of Clavius, who stated: "It is evident that mathematics are the nets with which one can catch the magnates and nobles and bring them to God's service." The novelty of the dedication to science by the Jesuits in the seventeenth century was manifested by the memorial sent to the Spanish King Philip IV by members of the Universities of Salamanca and Alcalá de Henares on the occasion of transforming the of Madrid into Reales Estudios with university rank. They were opposed to the royal plans and argued that it was not decent for religious to teach such profane matters as mathematics, astrology, navigation and military science. Jesuits answered that these subjects were necessary for philosophy and could be equally well taught by religious.

Besides these historical factors, there are some internal elements. The specific Jesuit spirituality is based on the *Spiritual Exercises*, where Ignatius of Loyola set out his ascetical and mystical experiences. This book presents the guidelines for a month-long retreat, where a person, under the guidance of a director, tries to find what is God's will in his life. Jesuits make this "exercises" twice, once in the novitiate and again at the end of their studies after being ordained priests, and they go over every year the same ideas in an eight-day retreat. Thus the ideas of the *Spiritual Exercises* formed the core of every Jesuit's spirituality and we should find in them the motivation which explains

Jesuit dedication to science. Two meditations at the beginning and at the end of the retreat entitled "Principle and foundation" and "Contemplation to attain love" mark the basic guidelines of the experience. In these meditations the importance of the service of God and how one should strive to "find God in all things" are stated. This last expression was often used by Ignatius himself referring to the Jesuit attitude in the active life. Jerónimo Nadal, a collaborator of Ignatius, expressed this attitude saying that Jesuits must be "contemplatives in action." In this way, for Jesuits, work and prayer are intimately united and there is no activity so profane that cannot be turned into prayer. Teaching mathematics in a university, observing the stars in an observatory or doing cartographic work in unknown lands are activities that a Jesuit finds perfectly compatible with his vocation, and through which he can find God in his life. For other religious groups this can also be a personal attitude, but in the case of Jesuits, this forms part of the very core of their vocation. This explains how, from the very beginning, Jesuits became involved in activities which other religious groups found incompatible with religious life. Their involvement in science is but one example of this mentality. So even today one can find in the same Jesuit community one dedicated to social work with the homeless, a parish pastor, a poet and an astronomer. Even though their activities are so different, they all partake in the same spirituality that moves them to "find God in all things".

Another aspect of this mentality which may explain Jesuits' work in science is their preference for work among those outside the Church and in frontier subjects. Rather than dedicate all their efforts to those already inside the Church in traditional pastoral activities, Jesuits are attracted by work with outsiders in non-conventional activities. This can explain how one can find a Jesuit in a Buddhist monastery in Tibet or taking part in experiments with a particle accelerator. A certain frontier spirit is also considered to be part of the Jesuit spirituality. Other religious groups may find themselves more at their ease in more traditional work with the community of the faithful, but Jesuits are always driven toward the outside, to those places and situations where the presence of the Christian Gospel is not felt. As an example, we will see how Ricci adopted the dress and customs of Chinese scholars to gain access to the court of the emperor (Chapter 3). Science, especially during the nineteenth and early twentieth century, was a field considered alien if not hostile to religious faith. Jesuits felt their responsibility to show that these apparently profane subjects could be also a vehicle to find God. The scientific community became, in this way, the parish of the Jesuit scientist. He found that his activity brought him in to contact with many persons who otherwise would not have had this opportunity, and that being himself a scientist could break many barriers.

Although there is a certain continuity in the establishment of observatories in the old and modern Society of Jesus, there are also differences, since they respond to different historical conditions. In the seventeenth and eighteenth centuries, Jesuit observatories were established in Europe in colleges and universities and were part of their educational programs. As we will see, the *Ratio Studiorum*, which regulated Jesuit teaching, stated that chairs of mathematics should be established in all major colleges and universities. Astronomy was considered at those times an important part of mathematics where rapid progress was being made. Modern science, with its emphasis on experiments and mathematical analysis, was beginning and it was natural that the new Jesuit colleges considered mathematics and astronomy as important subjects, as well as other parts of physics, such as optics, mechanics, electricity and magnetism. Teaching of astronomy led soon to the founding of observatories. They had an academic character, but also functioned as true research facilities. This happened in the second part of the seventeenth and the first half of the eighteenth century, when the experimental and observational aspects of modern science were emphasized. The same happened with the work of Jesuit astronomers in China and India. There astronomy was considered an important tool in the missionary work. Interest of Chinese and Indian rulers in astronomy made Jesuit astronomers welcome to their courts. Jesuits were asked to direct observatories or help local astronomers. They were for some time the only Western astronomers available and they introduced modern astronomy into these countries. The suppression of the Society in 1773 cut short all these developments. We can never know what would have been the future of the Jesuit observatories if they had been allowed to continue their work.

In the observatories founded after the restoration of the Society in 1814, there was, in the first place, a sense of continuity with the work done in the old Society. For example, the observatory of Zikawei in China was considered a continuation of the work of the Jesuits in the Imperial Observatory of Peking in the seventeenth and eighteenth centuries. But the historical situation had changed. In the nineteenth century modern science was firmly established and a rationalistic mentality was spreading that led in some instances to the presentation of opposition between science and religion. This mentality was clearly expressed, for example, in John W. Draper's book published in 1874 where he said: "Then has it in truth come to this, that Roman Christianity and Science are recognized by their respective adherents as being absolutely incompatible; they cannot exist together; one must yield to the other; mankind must make its choice - it cannot have both". The existence of this mentality was also felt inside the Church. Thus, in the opening paragraph of the document on the occasion of the founding of the Vatican Observatory in 1891, Pope Leo XIII refuted these ideas: "Those borne of darkness are accustomed to calumniate her (the Church) to unlearned people and they call her the friend of obscurantism, one who nurtures ignorance, an enemy of science and of progress, all of these accusations being completely contrary to what in word and deed is essentially the case. Right from its beginnings all that the Church has done and taught is an adequate refutation of these impudent and sinister lies." The Pope considered the founding of the observatory an answer to these false accusations: "This plan is simply that everyone might see clearly that the Church and her Pastors are not opposed to true and solid science, whether human or divine, but that they embrace it, encourage it, and promote it with the fullest possible dedication."

Against this mentality Jesuit scientific work was a practical way to show that the opposition between science and religion does not exist and furthermore that a harmony between them is possible. These ideas were present in Secchi's report in 1877 about the work in the observatory of the Collegio Romano. Cortie, Director of Stonyhurst Observatory, writing about Secchi in 1929 affirmed: "The enemies of Holy Church have made such unwarranted use of science as a weapon of attack against her most fundamental truths, that an impression has sometimes been produced among many of her children that the pursuit of science is damaging and dangerous to faith... Father Secchi is a striking example of one who knew how to unite religion and science". Thus, the presence of Jesuits in science, through their own scientific institutions, was considered a clear argument against these false accusations and was presented as an example of the compatibility between Christian faith and science.

In mission countries, where a number of observatories were installed, the help afforded to missionary work by the scientific prestige of these institutions was added to the previous arguments. Berloty, founder of the Ksara Observatory in Lebanon, wrote in 1912: "Missionaries (working in the observatories), helping scientific development, perform work useful to the countries where they work and show, once more, that the Catholic religion, working mainly for the salvation of souls, has never neglected true science which adorns the human spirit". This has been also the spirit of the great missionaries of the old Society in China, as expressed by Verbiest, who used to repeat that Christian religion had entered China by the hand of astronomy. There were also other reasons, such as the absence of scientific institutions that made the pioneer work of Jesuits of great value and the fact that they often were the only scientific institutions in those countries. In fact Jesuits carried out some of the earliest astronomical, meteorological and seismological observations in Africa, Asia, and Central and South America. In countries where earthquakes and hurricanes are frequent occurrences, the human interest in the mitigation of damage owing to these natural disasters was united to the scientific interest of studying their causes. This explains the work of seismological stations in South America and meteorological observatories dedicated to the study of tropical hurricanes in the Caribbean and the Far East. As we will see, Jesuits were among the first to forecast successfully the path of hurricanes and the time of their arrival at populated cities and the first to install a seismographic network on a continental scale. Other services offered by

Jesuit observatories in mission lands were the establishment of exact determinations of geographical coordinates, which served as support for cartographic work, and the provision of an accurate time keeping service by means of astronomical observations.

The observatories and the Society of Jesus

Despite their large number, the foundation of observatories in the modern Society did not respond to a unified program explicitly directed by the Jesuit central authorities. The only reference in official documents in that sense, that I have found is a recommendation in 1934 of establishing observatories and laboratories in the Jesuit faculties of philosophy, as an aid to students in their training in science. In general, observatories were created by local initiatives, although they could count on the approval of the superiors in Rome. These observatories were not the work of amateur scientists. In most cases, Jesuits dedicated to the observatories were well trained with licentiate and doctoral degrees in science from prestigious universities. In Chapters 4 to 7 and in the short biographies in Part III, the universities where Jesuits Directors studied and obtained their degrees are mentioned. Jesuits working at the observatories kept up an active participation in scientific meetings and congresses and had close ties with other scientists. Many of them were active members of scientific societies, as we have seen. Jesuit observatories were informally linked among themselves, often with exchange of personnel and publications. For example, several Jesuits, who later worked at different observatories spent some time at Stonyhurst Observatory or studied at Georgetown University or Saint Louis University. A large collection of publications from many Jesuits observatories is today preserved in the library of Ebro Observatory, Spain.

Jesuits were proud of their observatories and the Fathers General praised their work. For example, Father General Luis Martín in a letter to Miguel Saderra-Masó of the Manila Observatory wrote in 1893: "The studies of our brothers in the observatory result not only in the advance of science and the honor of our Society but also are useful to scholars and contribute to the help to the souls to which our studies and work must be directed according to our Institute". Another Father General, Wladimir Ledokowski, wrote in 1918 to Algué, Director of Manila Observatory: "I consider very adequate to promote the greatest glory of God the scientific prestige that the observatory has obtained among people, especially mariners, of the eastern countries". Superiors recognized the importance of this work and dedicated the necessary funds for the construction of buildings, acquisition of instruments and their maintenance. This amounted sometimes to large financial commitments. Superiors also dedicated young Jesuits with an interest in science to the work in the observatories and financed their scientific formation at the best universities.

However, the work in science and in the observatories was not always well regarded, or even understood, by some fellow Jesuits, who sometimes regretted the small involvement of Jesuit scientists in direct apostolic activities. Already in the early times, Scheiner complained: "Our philosophers denigrate mathematics orally, in writing and even in their publications... It is unacceptable that Jesuits write against this science not knowing anything about it". In modern times complaints against the observatories were more often expressed in mission countries where the needs of direct evangelization were more urgently felt and this was the source of occasional personal friction. For example, Lejay, Director of Zikawei Observatory, complained in 1933 in a letter to Father General that the importance of the intellectual work of the observatory was not sufficiently recognized in the mission and that many considered a lost time that dedicated to scientific studies. Father Ledokowski answered saying that he had always protested against this unjust judgement and expressed the great importance of this work and the chosen place given to it by the Church and the missionary tradition. He concluded that the two modes of apostolate (pastoral and scientific) are complementary and both are essential.

Life and death of the observatories

In general, observatories flourished during some time with Directors with high scientific interest and strong personalities, which imposed high standards in observation and research. This work was not always kept up by their successors and observatories fell into a slow decay. Since observatories were local institutions, finding qualified and interested people for their direction was not always possible. However, a very large number of Jesuits with different degrees of competence during four centuries dedicated great efforts to the scientific work of the observatories. This work as a whole is beyond comparison with that done by any other single institution inside or outside the Church.

We have already mentioned that all observatories in the old Society of Jesus ended with the suppression in 1773. In the observatories of the modern Society their duration in operation varies widely (see dates in Part II). Their closing depended on many circumstances, internal and external to the Society. Some ceased operating or became government centers due to political changes, for example, when communist governments came into power (Belen, Kalocsa and Zikawei). Other were transferred to state universities or simply closed due to the high cost of operation and lack of personnel. This evolution took place in many cases between 1960 and 1980, but the crisis started earlier after the Second World War about 1950. Owing to the high costs in upgrading astronomical instrumentation, these sections were the first to decay and close, once the original instruments became obsolete. Geophysical observations with minor costs were kept for a longer time. Sometimes growth of governmental services, especially in meteorology and seismology made the work of Jesuit observatories no longer necessary.

Since apologetic motives in the science versus religion controversy, as we have seen, was an important factor in the foundation of observatories in the nineteenth and early twentieth centuries, the abatement of this controversy late in the twentieth century also had a great influence in their closing. Inside the Catholic Church an important change had happened with the Second Vatican Council, completed in 1965, where a new and more open attitude of the Church with respect to the modern world was established. In the document about the relation between the Church and the modern world (*Gaudium et Spes*), a new mentality regarding science was presented, based on the acceptance of the mutual autonomy of faith and science. The attacks against the Church using the excuse of science had also to great extent disappeared. The apologetic mentality was no longer urgently felt and apostolic priorities in the Jesuit Order moved toward other activities. The 32nd General Congregation in 1975 established work for faith and justice as the leading motivation of Jesuits. Its fourth decree insisted on the priority which was to be given to social justice. The scientific apostolate, which in the previous 31st General Congregation, in 1965, was still given an important role, was not even mentioned. These and other motives, together with the decrease of vocations to the Society, have motivated the closing of practically all Jesuit observatories in the last thirty years. Furthermore, the presence of Jesuits in the natural sciences has also experienced in recent times a sharp decline. Only a few observatories, namely, Manila, Ebro, Bogotá, Buffalo, St. Louis, Weston and Montreal, are still today in operation, but with a minimal Jesuit presence. The Vatican Observatory, which is a Pontifical institution entrusted to the Society, is an exception, where there is still today a strong number of Jesuits actively working in astronomy and astrophysics.

Chapter 2.

OBSERVATORIES IN JESUIT COLLEGES AND UNIVERSITIES IN EUROPE (1540-1773)

Mathematics and astronomy in Jesuit colleges

In 1548, eight years after its foundation by Ignatius of Loyola (1491-1556), the Society of Jesus founded its first college in Messina, Sicily and three years later in 1551, the Collegio Romano in Rome. The Collegio Romano became a university in 1553, with the right to award doctoral degrees in philosophy and theology. Throughout the first centuries of the Society, the Collegio Romano served as a model for the approximately 625 Jesuit colleges and universities founded in Europe before the Society's suppression. Colleges were of two types: minor colleges, equivalent to primary and secondary schools today, with teaching of grammar, oratory and poetics, and major colleges, with faculties of philosophy and theology. Major colleges provided higher education and some of them were actually called universities. In 1710 there were in Europe 24 universities run by Jesuits, but many of the major colleges had a similar level of higher education. For example in France, 46 out of the 89 colleges had faculties of theology. The method of teaching in these colleges was based on the system used in the University of Paris, where Ignatius and his first companions had studied. Modified through the teaching experience in the new Jesuits colleges, the method was formalized in the *Ratio Studiorum*, first published in 1586 and in its definite form in 1599. The *Ratio Studiorum* specified the programs and methods of teaching which were to be followed in all Jesuit colleges. Among the disciplines in the curriculum of the faculty of philosophy, an important place was given to mathematics, which included at that time astronomy, mechanics and optics. Mathematics was established as an obligatory subject in all Jesuit major colleges which had studies of philosophy. Mathematics was generally taught in the second or third year of the philosophical studies. However, not all philosophy students were obliged to attend the courses of mathematics. For example, according to the Jesuit historian of education, François de Dainville (1978), in 1627 in the colleges of Paris and La Flèche in France out of the 873 students following the philosophy courses, only 64 took the courses in mathematics. As we will see not all major colleges had stable chairs of mathematics.

The establishment of Jesuit colleges and universities throughout Europe coincided with the beginnings of modern science. In 1543 Nicolaus Copernicus published *De revolutionibus*, with the first presentation of the heliocentric system. From 1576 to 1601 Tycho Brahe carried out his accurate astronomical observations, which contributed to the dismantling of medieval cosmology. Based on these observations Johann Kepler proposed his three laws of planetary motion and published in 1609 his treatise on Copernican astronomy *Astronomia Nova*. From 1610 to 1638 Galileo Galilei published his most important works, which established the modern views in mechanics and astronomy. In 1637 and 1641 René Descartes published his most important philosophical and scientific works. Finally, in 1687 and 1704 Isaac Newton published his two main works *Principia Mathematica* and *Optics,* where definitely the foundations of the modern sciences of mechanics, astronomy and optics were established. Thus, from 1543 to 1707, the old medieval science based on Aristotelian physics and the geocentric cosmology was abandoned, and the modern scientific method, founded on observations and experiments and their mathematical analysis, was introduced.

An important element in the establishment of modern science was the change of the cosmological model, from the medieval finite geocentric model of solid spheres, based on Ptolomaic astronomy and Aristotelian physics, to the new heliocentric infinite model, based on the ideas of Copernicus, Kepler and Galileo, and finally explained through the gravitational law of Newton. From its initial proposal by Copernicus, the heliocentric system encountered serious difficulties. Motion of the earth was against common experience, the accepted geocentric cosmological model, as well as the principles of Aristotelian physics. For this reason, the Danish astronomer Tycho Brahe proposed a mixed system in which the sun orbited the earth with the planets orbiting the sun. In this way he kept the earth stable at the centre of the universe. In 1609 Galileo turned his newly developed and still rudimentary telescope to the sky and observed a number of new phenomena, such as the satellites of Jupiter, the phases of Venus, new stars and the mountains of the moon, which for him constituted the clear proof of the Copernican astronomy. About 1611 Galileo's public teaching of the Copernican doctrine disturbed the Roman ecclesiastical circles. The Holy Office saw an incompatibility between the literal interpretation of certain passages of the Holy Scriptures and the motion of the earth. This resulted in the condemnation of the propositions which held the motion of the earth in 1616 and Copernicus' book being included in the Index of forbidden books until it was corrected. Later, in 1633, the condemnation of Galileo reinforced the idea that all considerations of the motion of the earth were against the accepted interpretation of the Bible. The ecclesiastical position was that, in the absence of a definite proof of the motion of the earth, the traditional geocentric doctrine, consistent with the literal interpretation of the Scriptures, should be held. This was a serious obstacle for the acceptance of the Copernican system by Jesuit astronomers. However, the Copernican system could be taught as a useful hypothesis in the determination of the positions of the planets. Jesuit mathematicians accepted this position and taught the three astronomical systems, namely, the Ptolomaic geocentric system, the

Copernican system and the mixed system of Tycho Brahe, as hypotheses for astronomical calculations. For the physical situation, however, from about 1630, a modification of Tycho's system was generally adopted by Jesuit astronomers. Thus, they kept the earth stable at the centre of the planetary system, as required by the ecclesiastical norms, and explained the new discoveries by letting the other planets orbit the sun. The ecclesiastical prohibition on heliocentrism was lifted in 1757 and, from the middle of the eighteenth century, Newtonian physics was taught in Jesuit colleges, so that Jesuit professors and teachers were incorporated into the mainstream of modern science.

Another problem for Jesuits teaching sciences was the complete rejection of Aristotelian physics by the emerging modern science. In 1594 the Fifth Jesuit General Congregation established that Aristotelian doctrine should be followed in all Jesuit teaching. Still in 1751 the Seventeenth General Congregation insisted that the natural sciences be taught according to the Aristotelian doctrine. For science professors and teachers this led to a compromise solution with the separation of the teaching of physics according to Aristotelian doctrine (we would call it today philosophy of nature) and of mathematics, which included also the modern disciplines of astronomy, mechanics, optics and other parts of physics, according to the modern mathematical approach. Thus, if we want to find the Jesuits' contribution to modern science, from the sixteenth century to the eighteenth century, we have to look to those who held the chairs of mathematics in their colleges and universities. Moreover, Jesuit mathematicians strove to show that the mathematical sciences were true sciences even in the strict Aristotelian sense. Thus, they developed the first discussions about the validity of using mathematics in physics and about the form in which the statements of mathematical physics were justified by experiment and observation. Some of them even insisted that the demonstrative value of mathematics applied to physics was superior to that of the traditional Aristotelian physics. In this they clashed sometimes with their fellow professors and teachers of physics. By discussing the validity of mathematical sciences, they were, in some sense, pioneers of the modern philosophy of science.

From their establishment, the first two Jesuit colleges of Messina and Rome began teaching mathematics. Teaching was formalized with the creation of chairs of mathematics, one of the first chairs in the Collegio Romano in 1553. A detailed account of the establishment and early development of the chairs of mathematics in Jesuit colleges can be found in the work of Antonella Romano (1999). As has already been mentioned chairs of mathematics included teaching the strict mathematical disciplines of arithmetic, geometry, algebra and calculus and those under the title of applied mathematics ("mathesis applicata"), namely, astronomy, mechanics, optics, acoustics, hydraulics, cartography and military engineering. Toward the end of the eighteenth century, there were in Jesuit colleges and universities of Europe about 95 chairs of mathematics. The total number of Jesuit professors and teachers of mathematics is difficult to calculate. Karl A. Fischer (1978, 1983) has published the lists with the names of the professors of mathematics and the years they taught for the colleges of France, Italy and central Europe. In France, from 1600 to 1762, in 30 colleges out of a total of 89, there were about 620 professors and teachers of mathematics, but some of them only for a short time. In Germany, Austria, Holland, Belgium and Bohemia, of 163 colleges, there were chairs of mathematics in 42 of them, with about 900 professors or teachers of mathematics in the period 1550 to 1773. In the same period in Italy there were chairs of mathematics in 18 colleges out of 183, with about 273 professors or teachers. In Spain, a chair of mathematics existed in the Colegio Imperial of Madrid, where there were 27 professors from 1627 to 1767.

The key figure in the beginning of Jesuit teaching of mathematics was Christopher Clavius (1537-1612), Professor of Mathematics in the Collegio Romano from 1564 until his death. During his 48 years as professor he had a profound influence on the teaching of mathematics and science in all Jesuit colleges. Clavius contributed to the writing of the *Ratio Studiorum* and was responsible for the importance given there to mathematics. He insisted that the same importance should be given to mathematics as to the other branches of study. Clavius was aware of the importance of mathematics in the trend of the studies of his times and commented that it would be sad if Jesuit colleges were to be left behind. He insisted that there must be two professors of mathematics at least in each college, and that mathematical academies should be established for the better students. Clavius participated in the reform of the calendar, promoted in 1582 by Pope Gregory XIII, and was considered to be the Euclid of his time. He was held in high esteem by Galileo, who sought his approval and support. Among his many mathematical books are his commentary to the books of Euclid, a textbook of algebra and a commentary on *The Sphere* by John Sacrobosco (Hollywood), a medieval textbook of astronomy. His books were for many years the standard textbooks on mathematics and astronomy in all Jesuit colleges. The book on the sphere, *In Sphaeram Ioannis de Sacrobosco, Commentarius*, first published in 1570, went through six editions during Clavius' life. In each edition, he introduced the most recent astronomical developments. In the last edition of 1611, Clavius wrote, after commenting on the new discoveries, among them those of Galileo, that a reform of the astronomical system was necessary.

Clavius was the first of a long tradition of Jesuit mathematicians and astronomers. I will mention here only a few of them, not directly related with observatories. Firstly we have Clavius' immediate successors in the chair of

Figure 3. Christopher Clavius, Professor of Mathematics at the Collegio Romano, 1564-1612, the most influential figure in the beginning and early development of the Jesuit scientific tradition (ARSI).

mathematics in the Collegio Romano, Christopher Grienberger (1564-1636) and Orazio Grassi (1583-1654). Grienberger was a more experimental astronomer than Clavius. He was one of the first to introduce the equatorial mounting of telescopes and began a program of determining the positions of the stars with the telescope. Grassi is best known for his observation of comets, about whose nature he entered into a controversy with Galileo. Christopher Scheiner (1573-1650) was professor, first in Ingolstadt, Germany, and later in Rome. He was one of the first to observe and study sunspots and engaged in a controversy with Galileo about this discovery. In 1630 Scheiner published his *Rosa Ursina*, the first complete study of the sun with observations obtained from the use of the telescope. We will see how the study of the sun was to become a favourite subject of later Jesuit astronomers. Scheiner's successor in Ingolstadt was Johann B. Cysat (1588-1667). In 1618 he was the first to study a comet by means of the telescope, determining its trajectory and possible composition. He also discovered the Orion Nebula and in 1631 he was one of the first to observe the transit of Mercury across the sun.

Giovanni Battista Riccioli (1598-1671), Professor in Parma and Bologna quickly acquired a reputation as an astronomer. In 1651 with the help of his disciple, Francesco Maria Grimaldi (1613-1663) he defended the geocentric system in *Almagestum Novum*. In this work, a true astronomical encyclopaedia, he explained, at great length, the foundations of all the astronomical theories held in his time. He considered the Copernican system as the most simple, elegant and best constructed, but could not accept it because of the ecclesiastical condemnation of the motion of the earth and the preference for Aristotelian doctrine. He proposed a modification of the Tychonian system with only Mercury, Venus and Mars orbiting the sun. The earth was stable at the center, around which orbit the moon, the sun and, much further out, Jupiter and Saturn. Riccioli and Grimaldi drew one of the first detailed maps of the surface of the moon giving names to its topographical features which are still in use. His astronomical works were widely accepted and were used in place of Clavius' textbook in Jesuit colleges until the end of the 18th century. From 1638 to 1680, Athanasius Kircher (1601-1680) was professor at the Collegio Romano and occupied the chair of mathematics in 1639-1630 and 1644-1646. He was a prolific author who touched upon many subjects, from Egyptian hieroglyphics to optics and magnetism. In 1664 he wrote *Mundus Subterraneum*, a book in which he speculated about the nature of the interior of the earth. In this book he discussed the origin of earthquakes and volcanoes, as originating from the action of internal conduits of fire, connected with a fire at the earth's centre. This work was very influential in the ideas about the earth's interior in the 17th and 18th centuries.

The last professors of mathematics in the Collegio Romano were Giuseppe Asclepi (1707-1775) and Roger Josep Boscovič (1711-1787). Boscovič, a Croatian, was one of the first Jesuits to accept Newtonian physics. In his most important book, *Philosophiae Naturalis Theoria*, published in 1758, he presented an atomic theory of matter in which atoms were center points of forces without dimensions. Near the atoms, forces alternated as repulsive and attractive and, far away, only the attractive forces were active, following the Newtonian law of the inverse of the square of distance. In 1751, together with Christopher Maire (1697-1767) Boscovič carried out the measurement of a degree of latitude from Rome to Rimini and published a book about the shape of the earth. In 1765 he was elected a fellow of the Royal Society of London, the first Jesuit to merit this distinction. Dmitri I. Mendelevev, the author of the periodic table of elements, said that Boscovič , together with Copernicus, was the pride of the Slavic peoples and that he can be considered as the founder of modern atomism.

Other professors of mathematics and astronomy are Gregoire de Saint-Vincent (1584-1667), author of a series of textbooks, who claimed to have solved the insoluble problem of the quadrature of the circle, and Claude Richard (1588-1664), Professor at the Colegio Imperial of Madrid who published commentaries on the works of Euclid and Apollonius. Among Jesuit mathematicians, Girolamo Sacheri (1677-1733), Professor in the college of Pavia and the university of Milan, a precursor of the non-Euclidian geometries deserves a special mention, together with Vincenzo Ricatti (1707-1775) Professor at Bologna.

The first Jesuit European observatories

The development of astronomy required the establishment of fixed places of observation and thus resulted in the establishment of the first observatories. Since Galileo first directed his telescope to observe the heavens, this instrument has become the standard means for astronomical observations. Because instruments were small and portable, stable observatories were not necessary. As telescopes grew in length from some centimetres to several metres, it was necessary to have observatories to keep these instruments permanently installed. One of the oldest observatories in Europe was installed by Tycho Brahe in 1580 in Uraniborg, Denmark. Two important ones were established in 1667 at Paris and in 1676 at Greenwich. Jesuits began astronomical observations in the Collegio Romano in the time of Clavius and Grienberger. As soon as Galileo reported his astronomical observations with the telescope, Jesuits made the same type of instrument and began observing the planets and the stars. In the Collegio Romano, despite the active astronomical observations carried out by the professors of astronomy, a true observatory was not established until much later (Chapter 4). In other colleges true observatories were established during the seventeenth and eighteenth centuries, but they only lasted until the suppression of the Society of Jesus in 1773. Information about these observatories is difficult to find. Joseph Jerome de Lalande (1803), director of the Paris Observatory from 1795 to 1807, included information about observatories in his edition of Jean E. Montucla's history of mathematics. He discussed observatories at the Jesuit colleges of Lyon, Marseille, Avignon (France), Mannheim (Germany), Vienna, Graz (Austria), Prague, Tyrnau (Trnava, Czech Republic), Vilnius (Lithuania), Rome and Milan (Italy). Even though he wrote some years after the suppression of the Society, he provided a short notice about each of these observatories with the names of Jesuits who worked in them and the type of work they carried out. A second list was given by Johann Schreiber (1903), a Jesuit astronomer from Kalocsa Observatory (Hungary), which included, besides those mentioned by Lalande, the observatories at the colleges of Schwetzingen (Germany), Lemberg, Pont-à-Mousson (France), Lisbon (Portugal), Florence and Parma

(Italy). Other less important observatories were mentioned by Julien Thirion (1880) in his work about Jesuit astronomers in response to Franz Xavier Baron von Zach, and by M. Maynard (1853) writing about Jesuit education between 1750 and 1773. These are the observatories established at the colleges of Augsburg, Würzburg, Ingolstadt, (Germany), Olmütz (Olomouc, Czech Republic), Breslau (Wroclaw), Posen (Poznan) (Poland), Naples, Siena, Brescia, Palermo (Italy), Coimbra (Portugal). Of this last group which were only mentioned by name there is very little information. Thirion, mentioned that of the about 130 observatories existing in the world at the beginning of the nineteenth century, 32 had been founded or directed by Jesuits.

We know the detailed history of only a few of the observatories. Astronomical observations with small telescopes were carried out in many Jesuit colleges with no formal observatories, since astronomy was part of the programs of mathematics. The establishment of a formal observatory implied that a site was dedicated for this purpose, instruments were installed in a permanent form and a Jesuit was appointed as director, generally, a professor or teacher of mathematics. The instruments used were quadrants, sextants and refracting telescopes, generally built at the same observatories. Though the principal work of the observatories was astronomy, meteorological observations were also made in many of them. These included more or less regular instrumental measurements of temperature, atmospheric pressure, humidity and rainfall. The basic instruments, such as thermometers, barometers, hygrometers and pluviometers were developed during the seventeenth century. In some observatories, for example, in those of Lyon, Marseille, Prague, Milan and Vienna, meteorological observations were published. Two of the earliest observatories were those established in the College of Avignon in 1632 and in the University of Ingolstadt about 1635. Most observatories were established in the middle of the eighteenth century. With the expulsion of the Jesuits from Portugal (1759), France (1764) and Spain (1767), and their suppression in 1773, Jesuit work in these observatories came to an end. Thus, when they were closed, many of the observatories had had only a few years of existence. We will never know what the development of these institutions would have been had they been allowed to continue.

France

Some of the earliest Jesuit observatories were established in France. We owe to Pierre de Vregille (1906) for the detailed history of two of them, those founded in Lyon in 1701 and in Marseille in 1702. In 1565 the Collège de la Trinité in Lyon, founded in 1527, was entrusted to the Jesuits. Forced to leave Lyon in 1594, Jesuits returned to the college in 1604. The college grew rapidly. New buildings were added in 1607 and a church in 1617. Building continued until 1660. The college founded a Chair of Mathematics in 1605 in which a number of Jesuit mathematicians taught. Honoré Fabri (1607-1688), who taught in Lyon from 1640 to 1647, was one of the first Jesuits to occupy the chair of mathematics. Afterwards he went to Avignon and to Rome where he published most of his scientific works. In his textbook of physics *Physica, Scientia Rerum Corporearum* (1669), he departed from the strict Aristotelian doctrine and adopted a type of corpuscularism, similar to that of Pierre Gassendi and Descartes, with whom he corresponded. Other correspondents were Christian Huygens and Gottfried Wilhem Leibniz. Two of the seven founding members of the Académie du Lyon in 1700 were Jesuit professors: Jean de Saint-Bonnet (1652-1702) and Pierre Thaillandier (1676-1713). The former drafted plans for an observatory and broke ground for its construction in 1701. Unfortunately, as he was overseeing construction, he died from a fall. The observatory building, a tower of eight stories was built on top of the church of the college. Thaillander was the first Director. In 1707 he departed to India and made astronomical observations in Pondicherry, as we will see (Chapter 3).

In 1740 Laurent Béraud (1702-1777) was appointed Professor of Mathematics in the college and Director of the Observatory. Previously he had taught in the colleges of Vienne, Avignon and Aix. He was an excellent professor and among his students in Lyon were the French astronomers, Montucla and Lalande. Lalande wrote in 1780: "The lectures of Father Béraud aroused my interest in astronomy and I have devoted my life to its study." Béraud, a very active astronomer, diligently observed and studied comets, eclipses, sunspots and auroras. In 1753 he observed two solar transits of Mercury and used these observations to determine the diameter of the planet, its inclination with respect to the plane of the ecliptic and the descending node. He also observed the solar transit of Venus of 6 June 1761. Transits of Venus are rare astronomical phenomena which occur approximately every 120 years, in pairs with an interval of eight years between them. Their observation allowed at that time accurate determinations of the distance between the earth and the sun. As we will see, these phenomena attracted the interest of Jesuit astronomers. Béraud sent reports of his astronomical and meteorological observations to the Académie du Lyon and the Académie des Sciences of Paris; he was a member of both. Beraud's scientific interests were very broad; besides astronomy, he studied meteorology, physics, chemistry and natural sciences. His numerous unpublished manuscripts are preserved in Lyon. In 1662 the Jesuits were expelled from Lyon. After the expulsion of the Jesuits, the college and the observatory were entrusted to the Oratorians until 1797, when, during the revolution, both were closed. Béraud returned privately to Lyon in 1768 and made observations of the second solar transit of Venus on 3 June 1769 on the hill of Fourvière.

The Sainte-Croix Observatory was established in 1702 in Marseille. This observatory was not in a Jesuit college, but it was part of the Royal School of Hydrography, founded in 1669, where there were Jesuit professors of mathematics and other subjects. One of them was Jean Bofa (1638-1724), who had been a professor of mathematics in the college of Avignon and arrived at Marseille in 1680. He returned to Avignon three years later. The observatory was built in 1701, a solid construction of four stories and was equipped with a number of telescopes of different sizes, clocks and other astronomical instruments. It had also meteorological instruments such as barometers and thermometers. The first director was Antoine de Laval (1664-1728), Professor of the School of Hydrography from 1696. He was very active in astronomical observations: for example, he observed six solar eclipses and 12 lunar eclipses. Moreover, he was very interested in the problem of the determination of geographical longitudes. Laval left Marseille in 1718 to become Royal Professor of Hydrography in Toulon. His successor was Jean Baptiste Thioly (1676-1720), who died in the plague which ravaged the city in 1720. After this year the observatory decayed and was practically abandoned.

In 1728 Esprit Pézenas (1692-1776) arrived at Marseille as Professor of Hydrography in charge of the observatory, where he installed new instruments. In 1749 the school of hydrography was closed but the observatory acquired a new importance; King Louis XV gave it the title of "Observatoire Royal de la Marine" and bestowed generous funding. Pézenas dedicated himself completely to the observatory work, reorganising it and installing new instruments, such as a telescope of 1.8 meters length made in England by Short, and a large quadrant of 3.6 meters radius. Pézenas was very active and published his observations in the Jesuit journal *Mémoires de Trévoux* and in the *Mémoires de l'Académie* of the Académie des Sciences of Paris, of which he was a corresponding member. He made a very detailed study of sunspots, their time of revolution and the position of their axis of rotation and worked on the problem of the determination of longitude. Pézenas trained in the observatory many astronomers among them the Jesuits Lagrange who would become Director of the observatory in Milan, and Poczobut who was to be Director of the observatory of Vilnius. Pézenas translated English scientific books into French, such as the algebra and calculus of Colin Maclaurin. After the expulsion of Jesuits from France in 1763, he retired to Avignon where he continued his research, publishing two books about the problem of the determination of longitudes, and further translations of English authors. The observatory continued in its original location until 1863 when it was moved to Longchamp.

Other observatories were established in the colleges of Pont-à-Mousson, Avignon and Toulon. The college of Pont-à-Mousson, situated at approximately equal distance from the cities of Metz, Nancy and Toulon, was founded in 1575. The observatory was established in 1720 in a special building on the gardens of the college. It had a telescope and several astronomical instruments. Jean Geiger (1715-1785), a professor of physics from 1749-1757, built several instruments to measure atmospheric pressure and a sort of planetarium for teaching the motion of the planets. Though Jesuits were prevented from accepting the heliocentric system because of ecclesiastical intervention, this situation changed toward the middle of the eighteenth century. In 1759 a thesis was defended in Pont-à-Mousson with the title: "The hypothesis of Copernicus explains perfectly the motion of the celestial bodies and it is in accord with the observations; the system of Ptolemy is against the observations and that of Tycho Brahe is too complicated." In those years Newtonian physics and the theory of gravitation were also taught in the college. Jesuits were forced to leave the college and the observatory in 1762.

The observatory of the College of Avignon was founded by Athanasius Kircher, who was professor in Avignon from 1631 to 1633. Thus, this may have been one of the earliest Jesuit observatories. Kircher installed the observatory in the tower of La Motte. He also built a kind of planetarium, where, by means of a system of mirrors, the positions of the sun, moon and planets were projected on the walls and which became a popular attraction in the city. Kircher published in 1635 a book about mirrors, *Primitiae Gnonomicae Catoptricae,* where he described the system used in the planetarium. One of his successors was Bonfa, who spent some years as professor in Marseille and returned to Avignon in 1683. Bonfa observed in Avignon the solar eclipses of 1684 and 1699 and the lunar eclipses of 1678, 1679, 1686, 1701 and 1703 and published some of his observations. The observatory was renovated in the eighteenth century by Jean Claude Ignace Morand (1707-1780) and continued until the expulsion of Jesuits in 1762.

Founded in 1685, the College of Toulon was a royal foundation for the formation of chaplains and officers of the French Navy. For the training of the marine officers an important place was given to instruction in mathematics, astronomy and hydrography. Laval arrived at Toulon in 1718 from Marseille and the following year founded an observatory. His nephew Jean Jacques du Chatelard (1693-1757), Professor from 1728, carried out an active program of astronomical observations which he published in the Jesuit journal *Mémoires de Trévoux*. Among them were observations of four solar eclipses, in 1737, 1738, 1748 and 1750, ten lunar eclipses, the solar transits of Mercury in 1753 and 1743, and observations of comets and of the occultation of Jupiter by the moon. He also published meteorological observations between 1737 and 1748 and observations of the aurora borealis. The observatory was continued by the Navy until the end of the nineteenth century. This observatory is not named in the lists of Lalande and Schreiber, but it is mentioned by Pierre Delattre (1953) and Dainville (1978).

One of the most important Jesuit colleges in France was that established in Paris in 1563. First it received the name of Collège de Clermont because it was founded by Guillaume Dupart, Bishop of Clermont. In 1594 the college was closed and Jesuits expelled from Paris. In 1618 the college was restored and received the support of kings Henry IV and Louis XIII. In 1682 the college was refounded by Louis XIV and its name changed to "Collège Louis-le-Grand." The name of this college does not appear in the lists of observatories we have mentioned above. However, as stated by Bigourdan (1918), the college had a small observatory in which a large number of astronomical observations were made. The college is important because it was the scientific center for many missionary astronomers, especially, for two expeditions to Siam and China, as we will see in the next chapter. Before the foundation of the Académie des Sciences the college was the meeting place of scientists and the aristocracy of Paris, where scientific topics were discussed.

It seems that there was not a particular site for astronomical observations until 1674 when a small place was used as observatory on a building called the "belvédère" or the "guérite". The observatory had two stories and a platform where instruments were installed. Pierre Bourdin (1595-1653) made the first observations at the college: lunar eclipses in 1645, 1646 and 1653 and a solar eclipse in 1652. One of the most important professors of mathematics was Ignace Gaston Pardies (1636-1673), author of a popular textbook of geometry and an unfinished treatise on optics. He corresponded with Newton on optical problems. Pardies observed a comet from the observatory in March 1672. Jean de Fontaney (1643-1703), who was to be the head of the expedition to China, made in Paris observations of a solar eclipse in 1684, lunar eclipses in 1678 and 1682 using two telescopes of 90 cm and 3.6 m length and comets in 1680 and 1681 which he discussed in a small book. His successor in the chair of mathematics Thomas Gouye (1650-1725), a member and president of the Académie des Sciences, published in 1688 and 1692 the observations made by the French Jesuits in Siam and China. There is no mention of observations made from 1684 until those by Jean Nicholas Cairon de Merville (1714-1768) of the transit of Mercury in 6 May 1753 with a telescope of 4.8 m length, and of the transit of Venus on 6 June 1761 with a Newtonian reflector of 1.3 m length. After the expulsion of the Jesuits, the observatory was still used by the astronomers Charles Messier and Joseph Lalande, at least until 1782.

Central Europe

Several observatories were established at Jesuit colleges in some of the cities of central Europe, which today belong to different countries. One of the first was established at the University of Ingolstadt, Germany, sometime about 1635. According to Benhard Durh (1921) the observatory was first installed in a tower of the Kreuzkirche. A wooden tower was afterwards built, called "turris mathematica," with four windows in the directions of the cardinal points. In 1650 a platform for astronomical observations was built in the professors' lodge which was called the "spekula." The students continued using the wooden tower until it burned down in 1684. Nothing has been found about these early observations. Scheiner was professor of mathematics from 1605 to 1616 and made his first observations of sunspots in 1612, but this was before the observatory was founded. From 1722 to 1726, Nicasius Grammatici (1684-1736) made astronomical observations and published in 1726 a table of lunar positions using Newton's theory. Joseph Schreier (1681-1754), professor from 1726 to 1730 published in 1728 a book on the elliptical orbits of the planets.

The college in Vienna, Austria, was founded in 1551 and from 1568 enjoyed the protection of Emperors Maximilian II, Ferdinand II and Leopold I. The college had Chairs of Mathematics and Physics. From 1623 the college was part of the University of Vienna, but with considerable independence. The observatory was founded for the first time in 1734 by Josef Franz (1704-1776), an astronomer in charge of the philosophical studies, who was director from 1738. The observatory was located on the roof of the college. In 1755 Empress Maria Theresa refounded the observatory and gave it her support and the title of Imperial Observatory. In 1756 Maximilian Hell (1720-1792), born in Schemnitz (Selmecz), Hungary, was named Professor of Mathematics and Director of the Observatory. In 1745 Hell had worked in the observatory and in 1751 he founded an observatory in Tyrnau, then Hungary (Trnava, Czech Republic). He began working at the observatory and also carrying out research in other fields besides astronomy, such as magnetism, electricity and mathematics. Hell's main work at the observatory was to publish each year, *Ephemerides Astronomicae* (astronomical ephemerides), that is astronomical almanacs with the precise times for every day of the year of sunrise and sunset, positions of the moon and planets and other astronomical data. This type of publication, which was to become for a long time a common procedure of observatories, had only begun to be published in the Observatory of Paris a few years before. The first volume of 250 pages, written in Latin, appeared in 1757; it also included other astronomical observations and scientific results. This was the first publication of this type in German-speaking countries. Hell published 37 volumes of ephemerides between 1757 and 1792. In this work he was helped by his coworkers and students, among them, Franz Triesnecker (1745-1819) and Anton Pilgram (1730-1793).

Figure 4. Maximilian Hell, Director of the Vienna Observatory, Austria, 1756-1792, during his observation of the transit of Venus in 1769 in Vardö, Lapland (Norway) (ARSI).

Hell's recognition as an astronomer resulted in his being invited by Christian VII, King of Denmark and Norway, to observe the solar transit of Venus of 3 June 1769 and the solar eclipse of the following day in the northern part of Norway. This was a remarkable invitation by a Protestant king, who did not allow Jesuits in his territory. Hell was accompanied by Johann Sajnovics (1733-1785), a Hungarian Jesuit who had been Subdirector of the observatory in Tyrnau. Hell and Sajnovics arrived in October 1768 at the island of Vardö in Lapland, at a latitude of 70 degrees north, where they remained for over a year. Besides the astronomical observations, they carried out many other experiments and observations, of auroras, magnetic declination and tides. They also studied the culture, religion and customs of the Lapps, and compared their language with Hungarian. Sajnovics is credited with being the first to point out the relationship between the Finnish and Hungarian languages. Hell's results, among them, a determination of the distance between the earth and the sun, which he first published in Copenhagen in 1770 and later in Vienna, differed from those of other observers and were questioned by some subsequent astronomers, especially by Carl Ludwig Littrow in 1835 who claimed that some of the observations had been either manufactured or falsified. In 1883 the American astronomer Simon Newcomb, examining Hell's original notes, demonstrated that this was not true. He showed that Hell's observations were more exact than those of other contemporary observers. Hell's complete work with all his observations and results during his journey to Lapland *Expeditio Litteraria ad Polum* was never published. During his absence he was replaced at the observatory by Pilgram, who published the ephemerides for those years. After the suppression of the Society in 1773, Hell continued as director of the observatory until his death.

Joseph Liesganig (1719-1799) was professor of mathematics in Vienna from 1752 until his death. He replaced Hell as Director of Philosophical Studies when Hell was named Director of the Observatory in 1756. Liesganig's main work was geodetic and cartographic. From 1762 to 1766, commissioned by the Empress Maria Theresa, he carried out the measurement of nearly three degrees (2 degrees 56 minutes 35 seconds) of latitude from Oslavanik (Czech Republik) through Vienna to Varazdin (Croatia), published in *Dimensio Graduum Meridiani Viennensis et Hungarici* (1770). This was the first of this type of measurement done in central Europe. In 1762 he measured personally the base line (12 158 meters) near Vienna. When this measurement was repeated in 1806 it was found that it had only an error of 7 mm per kilometre. In these measurements he was helped by Triesnecker and also by the Jesuit mathematicians Karl Scherflers (1716-1783), who did most of the triangulation work, and Georg Ignaz Metzburg (1735-1798). Liesganig and Scherflers did very extensive cartographic work in Austria, for example, the map of Ostgalizien in 42 sheets. Hell's successor as Director of the Observatory was Triesnecker, his former collaborator who continued the publication of the ephemerides and worked on geodetic measurements in Austria. With his death in 1819 the presence of Jesuits (or former Jesuits) in the Vienna observatory ended.

In 1556 Emperor Ferdinand I founded a Jesuit college in Prague, which became known as the Collegium Clementinum, after the neighboring church of Saint Clement. The college was attached to the Caroline University of Prague and from 1654 the university was named "Universitas Carolo-Ferdinandea". The college had a chair of mathematics from its foundation and Jesuits professors had friendly relations with Kepler during his stay in Prague. In 1721 began the construction of the tower for astronomical observations which was finished in 1723. Its decoration included a statue of Atlas carrying the globe. At the same time a mathematical and physical museum was founded with a variety of instruments. There is no record of observations being done at that early time. The real beginning of the observatory is connected with Josef Stepling (1716-1778), Professor of Mathematics from 1748, who became Director of the observatory in 1751. He was asked to make observations of solar and lunar eclipses in Prague by the Prussian Academy of Sciences. In 1751 Stepling reconstructed the observatory tower and installed the first astronomical and meteorological instruments. He carried out an active program of astronomical and meteorological observations. In 1752 Stepling published a series of observations of atmospheric pressure, temperature and air humidity. In 1753 he became Director of Mathematical and Physical Studies and introduced the teaching of modern Newtonian physics in the college. In 1761 Stepling used an inheritance from his mother in the observatory and established a mathematical library with more than 600 books. With the suppression of the Society of Jesus, the college and the observatory became state property. Stepling was one of the few Jesuits who stayed in his Chair of Mathematics in the university and continued until his death as Director of the observatory. Antonín Strnad (1747-1799), Stepling's assistant from 1771, became director in 1781. Strnad married in 1784 and became in 1795 Rector of the University of Prague.

Information about other Jesuit observatories in Germany is very scarce. The Jesuits established a college in Würzburg in 1567. After the establishment of the University of Würzburg in 1582, Jesuits taught on its faculties. Kircher was a professor from 1629 to 1631, when he left the city due to the advance of the army of King Gustavus Adolphus of Sweden. After a short stay in Avignon, he went to Rome where he became Professor of Mathematics in the Collegio Romano. His companion Kaspar Schott (1608-1666) returned to Würzburg in 1655 as Professor of Mathematics. A century later Franz Huberti (1715-1789), Professor of Mathematics from 1754, founded an observatory in 1757, with the support of the Bishop of Würzburg. The observatory was located on the tower of the Neubaukirche, where Huberti installed, among other instruments, a reflecting telescope. He carried out astronomical and meteorological observations and he published his thermometric observations between 1765 and 1770. In 1761 Huberti participated in the observation of the solar transit of Venus. From his observations he calculated the solar parallax and the distance between the earth and the sun. Huberti held a scientific correspondence with Stepling (in Prague) and published textbooks of arithmetic, geometry and algebra. In 1773 the Jesuits, Franz Xavier Trentel (1730-1804) and Nikolaus Burkhäuser (1733-1809), were appointed professors of mathematics. The suppression of the Jesuits in the same year did not affect the situation. Huberti, Trentel and Burkhäuser remained as professors in the University. Trentel, who had studied with Hell in Vienna, wrote textbooks on geometry, algebra and conics and Burkhäuser published a book on Boscovič's corpuscular theory in 1770. After Huberti's death, Trentel and Burkhäuser took charge of the observatory until the latter's death in 1809.

The Elector Prince Charles Theodore von der Pfalz founded an observatory in 1764 in Schwetzingen and another in 1772 in Mannheim in the Palatinate, Germany. In Mannheim a tower of 32 meters was built where many astronomical instruments were installed, among them, a meridian telescope by Ramsden. Christian Meyer (1719-1783) and Johann Metzger (1735-?) made astronomical observations. In Graz, Austria an observatory was established in the Jesuit college in 1745 on a tower built for this purpose. The first director was Peter Halloy (1707-1789) who held this post five times between 1745 and 1755. From 1765 to 1771 Karl Tiernberger (1731-1780) directed the observatory. He observed a comet in 1769 and worked mainly on mechanics and hydraulics. Other Jesuits working at the observatory were Nikolaus Boda (1723-1798) and Leopold Biwald (1731-1805), who published two commentaries on Newton's optics and one on Boscovič's natural philosophy. Activity ceased after the suppression of the Society in 1773 and the tower of the observatory was demolished in 1787. In 1770 an observatory was founded in the College of Dillingen, Germany, whose director was Caesareus Amman (1727-1792). According to Zach, as reported by Schreiber (1903), an observatory was established in the college of Lemberg, Alsace (France) in a tower where several astronomical instruments were installed. It is probable that it was founded by Liesganig. After the suppression of the Jesuits the observatory was closed and the tower itself demolished. Observatories were also established in the colleges Olmütz (Olomouc, Czech Republic) where a Chair of Mathematics existed from 1590, and Breslau, Silesia (Wroclaw, Poland) with a Chair of Mathematics from 1641. No information has been found about them, except that they appear in the lists given by Lalande (1803) and Thirion (1880).

An observatory was founded by Hell in 1751 in the large Jesuit college in Tyrnau, then in Hungary (Trnava, Slovakia). The observatory was established in a special building 36 m high with an added observational tower of 6 m, built between 1753 and 1755. The observatory had astronomical and meteorological instruments. An underground vault was used to compare the measurements of thermometers and barometers. The first Director was Franz Weiss (1717-1785), Professor of Mathematics and Astronomy, with Sajnovics and Franz Taucher (1738-?) as subdirectors. After the suppression of the Society of Jesus, Weiss continued as Director of the observatory and Professor of Astronomy in the university. Weiss was offered the direction of the observatory in Mannheim which he declined. In 1773 the observatory was moved to Buda (today part of Budapest), where Taucher occupied the post of Director.

In 1578 a small college was founded by the Jesuits in Vilnius, Lithuania, known as "Accademia Vilnensis." In 1639, with the support of Wladislaw IV Vasa, Duke of Lithuania and King of Poland, it became a large major college. The observatory was established in 1753 with the funding of the Polish Countess Puzynina, who had a great interest in astronomy. The observatory was added on the third floor of the college with two more floors and two observational towers. The first floor added, which is still preserved today, has six windows with a decoration of astronomical symbols between them. The first Director was Tomasz Zebrowski (1714-1758) who had studied in Vienna and with Stepling in Prague. In 1758 he was succeeded by Martin Odlanicki Poczobut (1728-1810), who had studied astronomy in Italy and in Lyon with Pézenas. He installed several instruments and made observations of the orbit of Mercury which were used by Lalande. In 1769 he was elected a fellow of the Royal Society of London and in 1776 a corresponding member of the Académie des Sciences of Paris. After the suppression of the Society of Jesus, Countess Puzynina came to the rescue of the observatory with more funding. Poczobut continued as Director of the Observatory and Professor of Mathematics in the college, which had become a state university. He received the support of King Stanislas II of Poland who gave the observatory the title of Royal Observatory. Poczobut prepared 34 volumes with his daily observations from 1771 to 1806. He became Rector of the university from 1780 to 1807 where he carried out a reform of the studies. In 1808 he joined the small group of Jesuits who had remained under the protection of Catherine II in White Russia (Belarus). Other observatories were founded in Poland at Poznan by Josef Rogalinski (1728-1802) in 1762 and at Lvov by Sebastian Sierakowski (1742-1824) in 1770. After the suppression of the Society from 1779 to 1799 two ex-Jesuits Estefan Luskina (1725-1793) and Jowin Bystrzycki (1737-1821) established an observatory in Warsaw.

Italy, Portugal and Spain

Observatories were also established in Jesuit colleges in Italy, Portugal and Spain. We have seen that astronomical observations began early in the Collegio Romano, where a Chair of Mathematics existed from 1553. However a formal observatory was never established there. Boscovic and Orazio Borgondio (1675-1741), two of the last professors of mathematics, drafted the first plans for an observatory about 1744, on the occasion of the observations of a comet. The observatory was to be installed on the roof of the church of Saint Ignatius, next to the college. The project, conceived by Boscovic on a grand scale, would occupy most of the roof of the church. Although the project was approved by the Pope Benedict XIV, it was never carried out. However, some instruments were actually installed by Boscovic, for a short time, in another part of the college, the Kircher Museum. Sometimes Asclepi, the last professor of mathematics from 1763 to 1773, is credited with founding the observatory, but there is no evidence of this.

The Brera College (Collegium Breranum) in Milan was one of the first colleges in Italy to establish an observatory. In 1760 Pascal Bovio (1721-?) installed a telescope and other instruments for astronomical observations. In 1762 Louis Lagrange (La Grange; 1711-1783), who had been trained with

Pézenas and became a professor of mathematics and astronomy in Marseille, was called to Milan to direct the observatory. Boscovič, after a long journey, first to Constantinople in 1761, where he failed to observe the transit of Venus, and later to Saint Petersburg, became professor in Pavia in 1764. He came often to the Milan observatory, where he calibrated the instruments and made observations. Boscovič was interested in the more scientific aspects; he left Milan in 1772. Lagrange, who was during all this time the Director, continued in the post also after the suppression of the Society. He formed a good group of Italian astronomers, among them the Jesuits Giovanni Angelo de Cesaris (1749-1832) and Francesco Reggio (1743-1804), who remained at the observatory after the Society's suppression when the observatory continued as a state institution.

In 1756 Leonardo Ximenes (1716-1786) founded an observatory in the Collegio di S. Giovannino in Florence. He made observations of the solar eclipse of 1748 and the lunar eclipses of 1750 and 1755. In 1761 Ximenes was appointed geographer and mathematician of the Grand Duke of Tuscany and professor at the University of Florence. After the suppression of the Society, the college was given to the members of the order of the Priorists, with whom Ximenes continued working at the observatory. The observatory exists today with the name of Osservatorio Ximeniano. In 1757 Giacobo Belgrado (1704-1789) established an observatory in one of the two towers of the college of Parma. He published works on physics and meteorology, for example, one about the thickness of the atmosphere. Francesco Tortosa (1717-1800), a professor of mathematics, collaborated with Belgrado and made observations of the moon from the observatory they established in Parma. He was also the author of several books on physics. Thirion (1880) and Maynard (1853) mentioned also observatories in Brescia, founded about 1760 by Francesco Cavalli (1719-?); in Venice founded about 1756 by Bartolomeo Panigi (1720-?); in Palermo where a chair of mathematics existed from 1609, and in Naples, where the chair of mathematics was established in 1594. No information has been found about these observatories.

An observatory was founded in 1722 by Giovanni Battista Carbone (1694-1750), a Neapolitan, in the Real Colégio de Santo Antão-o-Novo in Lisbon, Portugal. The college had been founded in 1553 in a different location, the new college was inaugurated in 1593. In 1722 Carbone, accompanied by Dominico Capassi (1694-1736) arrived at Lisbon from Rome on their way to the Brazil Mission, but they were retained in Lisbon by King John V, who in 1726 appointed Carbone Royal Mathematician. Carbone and Capassi had met in Rome with Manuel de Campos (1681-1758), a professor of mathematics in the Lisbon college, and author of several books on geometry and trigonometry. Campos recommended both to the Jesuits in Lisbon as good mathematicians. Carbone and Capassi observed solar eclipses in 1725 and 1729 and lunar eclipses in 1724, 1726 and 1730. Carbone published the observations of the 1730 eclipse in the *Philosophical Transactions* of the Royal Society of London. He was named rector of the college in 1749. He enjoyed the favour of the king, who not only appointed him Royal Astronomer but also funded the acquisition of astronomical instruments for the observatory. In 1730 Capassi went to Brazil, where he did extensive cartographic work together with Diego Soares (1684-1748), who had been also a professor of mathematics in S. Antão.

In 1749 Eusebio da Veiga (1717-1798) succeeded Carbone as Professor of Mathematics in the Lisbon college and reorganized the mathematical and astronomical studies in the so-called "Aula da Esfera". In 1756 he published the astronomical ephemerides for 1757, the first in Portugal, under the title *Planetário Lusitano,* and, in 1758, the ephemerides for the years 1759 to 1761. He observed the solar eclipse of 1753 and the lunar eclipses of 1755 and 1757. This last lunar eclipse was also observed by Bernardo de Oliveira (1711-1796) and Jose Texeira (1729-1799) in Coimbra and by Dionisio Franco in Evora. After the Jesuits' expulsion from Portugal in 1758, da Veiga went to Rome, where he was director of the Specola Gaetani, an observatory created by the Duke of Sermoneta. The state confiscated the college after the Society's expulsion from Portugal in 1758 and in 1770 the building was transformed into a hospital (Hospital de S. José), which still exists. An observatory is mentioned by Thirion (1880) in Coimbra, but no other information has been found, except for the existence of a chair of mathematics. The most important professor of mathematics in Coimbra was Inácio Monteiro (1724-1812), who published in 1754 and 1756, *Compendio dos Elementos de Matemática*, a two-volume work where he introduced the new currents of mathematics to Portugal.

In Spain the Jesuit college where mathematics was given the greatest importance was the Colegio Imperial of Madrid. Though founded in 1572, its real importance began in 1609, when it was refounded through a donation of the Empress Maria of Austria, daughter of Charles V and wife of Maximilian II. In 1625 it was given the category of "Reales Estudios" by King Philip IV, somewhat short of being a full university. It was endowed with two chairs of mathematics; astronomy was one of the main subjects. Professors of mathematics were also "Cosmógrafos Reales" (Royal Cosmographers) and carried out occasional astronomical observations, mainly of comets and solar and lunar eclipses. José Zaragoza (1627-1679), professor from 1670 to 1679,was the author of various works on mathematics and astronomy, among them a textbook of astronomy and geophysics, *Esfera en comun, celeste y terraquea,* published in 1675. He made observations of the comets of 1664, 1665 and 1669. Jacobo Kresa (1645-1715), professor from 1678 to 1696, observed a lunar eclipse in 1678. José Cassani (1673-1715), professor from 1705 to 1749, observed the lunar eclipse of 1701 and the solar eclipse of 1706 and sent the observations to the Académie des Sciences of Paris. He observed a comet in 1737 and published in the same year a book on the nature of comets.

In 1752, thanks to the influence of Francisco de Rávago (1685-1763), Jesuit confessor of King Ferdinand VI, an observatory was established in a building near the college, where instruments acquired in England were installed. Not much is known about this observatory, which was the first to be established in Spain, a year before the Naval Observatory of Cadiz. The observatory had several telescopes one of 2.44 meters length and other instruments, such as quadrants and sextants. Johann Wendlingen (1715-1790), a native of Prague, and Professor of Mathematics in Madrid from 1757 to 1767, observed the lunar eclipses of 1757 and 1758 which he published in the *Philosophical Transactions* of the Royal Society of London. Wedlingen used a telescope of 91.4 cm with a blue lens at the ocular. Christian Rieger (1714-1780), who had been for 13 years a professor in Vienna, came to Madrid in 1760 as Professor of Mathematics. He made observations of the transit of Venus of 1761, mentioning that they were made at the observatory of the Colegio Imperial. For these observations he used a quadrant of 61 cm with a sighting of 91.6 cm and two telescopes of 90 cm and of 2.44 meters. The last professor to be incorporated in 1765 was Tomás Cerdá (1715-1791), one of the first to introduce Newtonian physics and differential and integral calculus to Spain. Jesuits were expelled from Spain in 1767 and nothing is known about what happened to the observatory.

Chapter 3.

JESUIT ASTRONOMERS IN CHINA, INDIA AND OTHER MISSIONS (1540-1773)

Astronomy in the Jesuit missions

The interest in astronomy that we have seen in the Jesuit colleges of Europe was carried into the mission countries. In 1552, at the end of his second journey to Japan, Francis Xavier (1506-1552), the first Jesuit missionary to the East, wrote to Rome asking that Jesuits sent to Japan should have some knowledge of astronomy since the Japanese were a very curious people interested in the motion of the heavens, solar and lunar eclipses, lunar phases, the origin of rain, snow, thunder and lighting and other natural phenomena. However, it was not in Japan, but in China that Jesuit missionaries found astronomy a great help in their work of spreading the Christian faith. As Ferdinand Verbiest, one of the Jesuit Directors of the Beijing Observatory, expressed it in his book *Astronomia Europaea:* "Holy Religion makes her official entry (in China) as a very beautiful queen, leaning on the arms of Astronomy and she easily attract the looks of all the heathens. What is more, often dressed in a starry robe, she easily obtains access to the rulers and prefects of the provinces." At the end of the book he repeats the same idea in a more explicit form, "Christian Religion in China is justly represented as a most august queen who appears publicly with her arm leaning on Astronomy ... because she was first introduced in China through Astronomy, because she was left untouched thanks to Astronomy and because after having been banished several times, she was each time called back and successfully restored to her former dignity by Astronomy." Truly, it was through their interest in astronomical knowledge that the first Jesuit missionaries were able to enter China and influence Chinese society, which was closed in the sixteenth century to all foreigners. Even at times when other missionaries were expelled from China, Jesuit astronomers remained in their posts. Another country to which Jesuit missionaries carried their astronomical observations was India. Some Indian princes were interested in astronomy and the Jesuits collaborated with them, but this collaboration was not as extensive as in China. Finally, in a completely different scenario, Jesuits also founded the first astronomical observatory of the western hemisphere in the mission of Paraguay.

Chinese astronomy and Matteo Ricci

The development of astronomy in China is a vast subject; for a detailed account the reader may consult Needham's work (1959). Only a very short description is given here, that may help the reader to understand the work of the Jesuits who used astronomy as a means to enter China to influence its learned circles and to explain the Christian religion. First of all, we must be aware of the official character of astronomy, because of its use in the preparation of the calendar, which each year was promulgated by the Emperor and was used by the whole empire and vassal countries. Though astronomy in China can be traced back to very early times, it was during the Earlier Han dynasty (206 BC to AD 9) that it underwent a period of great development. At this time, the Board of Astronomy was founded as an official department of the Imperial government, and thus astronomy acquired its official status. Simultaneously, important developments in astronomy occurred in the East, centuries before comparable advances in the West. Some of these were: the concept of heavenly bodies floating in an empty infinite space, the use of the equatorial-polar reference system (not used in the West until 1585 by Tycho Brahe), the drawing of star maps and the construction of astronomical instruments, such as armillary spheres and celestial globes. Since the fifth century BC in China, and probably some centuries earlier, celestial phenomena, such as solar and lunar eclipses, novae and supernovae, comets and sunspots were carefully observed and recorded. As a matter of fact, Chinese astronomers were more interested in unusual and extraordinary phenomena than in the regular motion of heavenly bodies. During the Mongolian Yuan dynasty (1279-1368), a fruitful collaboration was established with Arabian and Persian astronomers and they came to serve in the imperial court. During this period, a Muslim school of astronomy was founded in Beijing, which continued after the Mongols were stripped of power. During the Ming dynasty (1368-1644), which succeeded the Mongolian domination, astronomy, mathematics and all the sciences in general suffered a considerable decline. It is at the end of this dynasty that the first Jesuits arrived in Beijing. They found astronomy and mathematics in a feeble state in which the brilliant findings of the past had been practically forgotten. Astronomical predictions of solar and lunar eclipses, for example, manifested notable errors. This situation allowed the Jesuits to demonstrate the superiority of western astronomy and led to their being put in charge of the reform of the calendar and later of the Board of Astronomy.

In 1557, Portuguese merchants had established a permanent commercial post in Macao, from where they were allowed to trade with China through Canton, but the rest of the Empire remained firmly closed to all foreigners. The last missionaries to enter China had been Franciscans during the reign of the Mongolian Emperor Kublai Khan in the fourteenth century. In 1563, there was a small Jesuit community in Macao, which had tried in vain to enter China. An important change in the Jesuit activities took place in 1578 with the arrival of Alessandro Valignano (1539-1606) as Visitor to the missions in the East Indies. He had a new concept of mission work, based mainly on adaptation and cultural dialogue: Christianity should not be advanced by imposing European ways, but by entering quietly into the body of Chinese culture and trying to transform it from within.

Matteo Ricci (1552-1610), born in Macerata, Italy, was the first Jesuit to enter the Chinese Empire. Possessed of a keen mind and an extraordinary memory, together with a charming personality, he was the perfect man to carry out Valignano's projects for the penetration of China. Ricci, who had arrived at Macao in 1582, used his knowledge of mathematics, which he had acquired in the Collegio Romano under Clavius, to gain the goodwill of Chinese scholars. In 1595 he adopted the dress of the Chinese scholars and presented himself as a scholar from the West. In 1598, after several years spent living in the south of China, Ricci received a visit of Wang tso, an important officer in the Tribunal of Rites in Nanjing, who invited him to go to Beijing with the idea of working on the reform of the calendar, the deficiencies of which were at that time notorious. Accompanied by the Spanish Jesuit Diego de Pantoja (1571-1618), Ricci arrived in Beijing in 1600. In January of the following year, both entered the Forbidden City, the imperial palaces of difficult access and where the emperor Wan-li lived a secluded life. In the memorial Ricci presented to the emperor, he presented himself as a western scholar who having studied astronomy, geography and mathematics wished to enter the emperor's service. The two Jesuits brought several gifts among them a clavichord and two pendulum clocks. The need to adjust these clocks gave them an occasion to visit periodically the imperial palace and to confer with the members of the court. Though Ricci never met the emperor personally, he obtained his permission to establish himself in Beijing.

During his stay in Nanjing, Ricci had visited the astronomical observatory, and admired the armillary spheres, celestial globes, gnomes, and other astronomical instruments, all admirably worked in bronze. He discovered that the Chinese astronomers themselves were not very familiar with these ancient instruments. These instruments were, in fact, the work of the astronomer Guo Shoujing (Kuo Shou Ching), who lived during the reign of the Mongolian emperor Kublai Khan at the end of the thirteenth century. Two copies of these instruments had been made and had been installed in the new capital of Beijing and in Ping-yang. Those of Ping-yang were transferred later to Nanjing. Ricci, who had been taking measurements of the latitude, realized that the orientation of the instruments did not fit with their location in Nanjing. This was one sign of the many shortcomings of the contemporary Chinese astronomers. The deficiencies in astronomical knowledge of the astronomers Ricci met led him to conceive of the idea of using his knowledge of European astronomy to gain goodwill in scholarly circles. Conspicuous consequences of these deficiencies were the many errors which were acknowledged to be contained in the actual calendar. Not being himself specially trained in astronomy, Ricci, in a letter to

Rome in 1605, asked for a good Jesuit astronomer to be sent to China in order to carry out the reform of the calendar. In this way he expected, to gain access to and establish good and influential relations with Chinese intellectuals and the imperial court.

In 1607, Ricci by now very fluent in the Chinese language with the collaboration of the Chinese scholar Xu Guangqi (Hsü Kuang ch'i), a Christian convert, published a Chinese translation of the first six books of Euclid; he used Clavius' edition. He published several other of Clavius' mathematical works and wrote some short texts on geometry. At the same time, Ricci drew the first world map, *The great map of ten thousand countries,* giving the correct position of China in relation to other countries and to the newly discovered lands of America, translating all geographical names into Chinese. This chart, which soon became very popular, contrasted sharply with the traditional Chinese maps that reduced the world to the Chinese Empire surrounded by a few small countries. For the first time, the Chinese could see in this map the relation of their land to the rest of the world. In 1610 Ricci insisted again in a letter to Rome on the need to send some Jesuit astronomers to China "to continue the work I have begun with my little forces, my few books and my little knowledge".

The reform of the Chinese calendar

After his death in 1610, Ricci's work was continued by his Jesuit companions and Chinese collaborators. On the 15th of December 1610, the astronomers of the imperial observatory of Beijing were responsible for an important mistake in the prediction of a solar eclipse. The eclipse was correctly predicted by Sabatino de Ursis (1575-1659), an Italian Jesuit, who had arrived at Beijing in 1607 to help Ricci in his astronomical work. This was the first prediction of a solar eclipse made by a Jesuit astronomer in China. One of the earliest observations of a lunar eclipse was that of by Giulio Aleni (1582-1649) in Macao in 1612. Correct prediction of eclipses was very important because of the social consequences attached to this type of phenomenon. Jesuits established the superiority of western astronomy through the precision of their predictions of solar and lunar eclipses. De Ursis' correct prediction was an strong argument used by Xu Guangqi, then a member of the Tribunal of Rites, to persuade the other members, and finally the emperor Wan-li, about trusting the Jesuits with the reform of the calendar. De Ursis and de Pantoja began work on the calendar, but the project had to be abandoned because of the opposition of Chinese astronomers. In 1614 the Portuguese Manuel Dias (1574-1659) published a compendium in Chinese of the Ptolemaic astronomical system which was later widely used by Chinese astronomers. In this book Dias published the first exposition in China of the discoveries of Galileo and of the use of the telescope in astronomy. In 1616, due to the insidious plots of Shen Ch'üeh, an influential person in the court, who resented the presence of the Jesuits, a strong persecution of Christians began. De Ursis was expelled to Macao where he died in 1620. The same year, the emperor Wan-li died. After the short reign of T'ien-ch'i, his brother Chongzhen (Ch'ung chen) became emperor. In 1627 he appointed Xu Guangqi Vice-president of the Tribunal of Rites and later a member of the Imperial Council. From this influential position, Xu Guangqi again proposed the calendar reform. Finally, in 1629 through an imperial edict, the Jesuits were formally put in charge of the revision.

The Chinese calendar was a lunar-solar calendar consisting of 12 lunar months of 29 or 30 days with an intercalary month added seven times every 19 years, so that those years had thirteen months. In certain months some intercalary days were also added. Each month began with the new moon. The first month of the year began when the sun entered the zodiacal constellation of Pisces; that is, between 21 January and 20 February. For a long time many Chinese scholars had tried to solve, in different ways and not always successfully, the difficult problem of synchronizing the discrepancies between the lunations and the solar year. It is difficult nowadays for a western mentality, to comprehend the importance of the calendar in the life of ancient Chinese society. According to Confucius' doctrine, which permeated the whole of Chinese life at that time, the ideal of human life consisted in the harmony of men among themselves and of men with nature and the phases of the universe. From this view there followed the need to adjust society and private lives to the rhythm of the heavens. Thus the position of the heavenly bodies would determine the propitiousness or suitability of certain days for social and religious celebrations, for ceremonies and even for the small details of everyday life. High affairs of state and simple affairs of family life had to be decided with an eye upon the calendar. This explains the need for accurate prediction of astronomical phenomena. The occurrence of unannounced phenomena could mean serious setbacks and bring severe disgrace with it. Each year, a special department of the Board of Astronomy prepared with great care a calendar that was solemnly proclaimed by the Emperor and used by the whole empire. The calendar contained the astronomical ephemerides of the sun, moon and planets; and it included information about and prediction of the weather and of the occurrences of extraordinary phenomena such as severe storms, earthquakes, solar eclipses and the appearing of comets. It also fixed which days were propitious for different activities, according to the position of the planets and stars, and other astrological considerations.

In 1614 Nicolas Trigault (1577-1628) arrived at Rome, the first Jesuit who returned from China. He insisted with Jesuit superiors on the importance of sending to China astronomers, as well as astronomical books and instruments.

In 1619 he returned to China accompanied by two other Jesuits who were to play an important role in the reform of the calendar. The first was Johann Schreck (most commonly known by the latinized form of his name Terrentius or Terrenz; 1576-1630) who like Ricci had been a student of Clavius at Rome. He was already recognized for his knowledge of medicine, philosophy and mathematics and in 1611 was admitted to the Accademia dei Lincei, immediately after Galileo. The same year Terrenz joined the Jesuits. His work on geography and natural science merited him the name of Plinius Indicus. In 1621 Terrenz introduced the first telescope in China, given as a gift to the Emperor. He wrote to Galileo trying without success to get him interested in the astronomical work in China and asking about the determination of eclipses. When his attempt to interest Galileo failed, Terrenz turned to Kepler who answered his questions in a booklet published in Europe in 1629. On 21 June 1629, a solar eclipse took place which was calculated by the two schools of Chinese astronomers, the traditional and the Muslim, and by Terrenz whose prediction was the only correct one. He insisted that this was not an error of the astronomers, but a failure of the system, pointing to the necessity of adopting the European astronomy. This prediction moved the Emperor to follow the counsel of Xu Guangqi. He decided to entrust the Jesuits with the reform of the calendar. In 1629 Terrenz himself took charge of the work, helped by Niccolò Longobardo (1565-1655), but he died the following year.

The second Jesuit important in our story to arrive at Macao in 1619, was Johann Adam Schall von Bell (1592-1666), from a noble family of Cologne, Germany. He also had been a student of Clavius in the Collegio Romano and probably was present at the warm reception that was given there to Galileo in May 1611. After his arrival in China, Schall finally established himself in Beijing in 1630 to work on the calendar. Before coming to Beijing, he had already been active in astronomical work, predicting with precision two lunar eclipses on 8 October 1623 and 9 September 1624. He had also written a short book in Chinese on lunar eclipses which was officially published by the Board of Civil Office. Schall and Giacomo Rho (1592-1638) succeeded Terrenz in the revision of the calendar after Terrenz died in 1630. After Rho's early death in 1638, Schall assumed complete responsibility for this work. From 1630 onwards, Schall worked strenuously on the reform of the calendar and on the translation into Chinese of diverse material on western astronomy and mathematics.

Figure 5. Johann Adam Schall von Bell, the first Jesuit Director of the Imperial Observatory of Beijing, China, 1644-1664, dressed as a Mandarin of the First Class.

After the death of Xu Guangqi in 1633, Li Tianjing (Li Tien king), also a Christian, was appointed Director of the Astronomical Observatory. Schall continued his work, without holding any official position, and only received direct recognition from the Emperor Chongzhen in 1638. In 1641, after he had made a new exact prediction of a solar eclipse, observed by the Emperor himself, Schall was ready to introduce the new calendar which had been finished earlier. At the time, the political situation was deteriorating with frequent unrest and popular revolts. This gave an opportunity to the Manchu army to move from the north and invade the Chinese empire. Chongzhen committed suicide in 1644 ending the Ming dynasty. A new Manchu Emperor Shun Chih occupied the imperial throne, beginning the Qing (Ch'ing) dynasty. The new Manchu government soon became aware of the advantages of western science and technology and maintained good relations with the Jesuits. A new prediction of an eclipse on 1 September 1644, in the timing of which the Chinese astronomers erred by an hour, but which Schall correctly predicted, convinced the new government of the superiority of western astronomy. Schall was appointed director of the Imperial Observatory with the dignity of a Mandarin of the Fifth Class, and his calendar for 1645 was finally approved and officially proclaimed.

Jesuits directors of the Imperial Observatory

The Board of Astronomy was an official institution created during the Earlier Han dynasty (206 BC to AD 9) and was dependent on the Tribunal of Rites. It was formed of four departments, of which the most important was in charge of the calendar. This department was divided into three sections, the first of which was dedicated to the astronomical part of the calendar and the other two to other aspects, namely, fixing the days which were propitious for services, feasts, rites and ceremonies and to the prediction of the weather, earthquakes and other phenomena. Schall was the first Jesuit to be appointed director of the Board of Astronomy by the Manchu Emperor Shun Chih in 1644. The appointment was confirmed by the Tribunal of Rites the following year. Jesuits occupied this post from 1644 to 1805, except for the period 1665-1668. The Society of Jesus was suppressed in 1773, but four ex-Jesuits continued in this post and as members of the Board until 1805.

Schall reorganized the Board of Astronomy and reduced the five calendars which existed during the Ming dynasty to two. The first was for the imperial administration and contained the ephemerides of the sun, moon and planets and other astronomical observations. The second was for the general public and gave more simple astronomical information, established the propitiousness of various days and included predictions of the weather and other phenomena. The astronomers of the Board were divided into two schools, Chinese and Muslim. The first followed the traditional Chinese astronomy and the second the Arabic astronomy introduced during the Mongol domination in the thirteenth and fourteenth centuries. Schall tried to create a third Western school, but soon abandoned the project owing to the strong opposition his initiative encountered. His influence in the court continued to grow and the young Manchu Emperor Shun Chih became so attached to him that he called Schall "honorable father." In 1658 he bestowed upon him the highest honour of Mandarin of the First Class. This honour was held only by the imperial ministers and princes. Sun Chih often came to visit Schall and received instruction from him about western culture and science. It can be said that no westerner in the whole history of China ever enjoyed as much influence as Schall did.

During this time, Schall worked tirelessly; besides his duty of publishing annual calendars, he taught western astronomy to Chinese astronomers, made many astronomical instruments and published as many as thirty books on astronomy in Chinese. These books dealt with the telescope, the theory of solar and lunar eclipses, trigonometric tables, catalogues of stars and a

Figure 6. The platform of the Imperial Observatory of Beijing, China, showing the six instruments built by Ferdinand Verbiest (about 1670).

summary of European astronomy, including the work of Copernicus, Tycho Brahe, Galileo and Kepler. One was a handbook on astronomy which included the work of Terrenz, Rho, and Longobardo and consisted of 150 volumes divided into three parts *Theoretical and Practical Astronomy, Astronomical Tables* and *Auxiliary Sciences.* Schall was not alone in this work. Among his Jesuit collaborators in the astronomical work were Lodovico Buglio (1606-1682), Gabriel Magalhaes (1611-1677), Nikolas Smogulecki (1610-1656) and Johann Grueber (1623-1665). One of the main advantages the Jesuit astronomers enjoyed over their Chinese colleagues was their use of better astronomical tables. Ricci brought the traditional *Alfonsine Tables*, completed in 1252, under King Alphonse X of Castille. Later, Jesuits brought to China more modern tables, such as the *Ephemerides Brandeburgicae,* published by David Origanus in 1609, the *Tabulae Frisicae* (Amsterdam, 1611), those included in *Astronomia Danica* (1622) by Christian S. Longomontanus (Longberg) and the *Tabulae Motuum Coelestium Perpetuae* (1632) by Phillipus Lansbergius. The latter two were both based on the system of Tycho Brahe. In 1646, Schall received the *Tabulae Rudolphinae*, published by Kepler in 1627, and he began to use them in his calculations.

The favour of the emperor Shun chih, the many honours and the real influence that Schall enjoyed in the court were the cause of strong animosity on the part of some Chinese astronomers, specially of Yang Guangxian (Yang kuan hsien) and Uming Huen of the Muslin school. In 1661, the emperor Sun chih died at the early age of twenty three and was succeeded by a council of regents. The enemies of Schall took advantage of this occasion to present three accusations against him, conspiracy against the state, preaching a religion dangerous to Chinese society and the promotion of western astronomy that was full of errors. The trial began on 26 September 1664 and lasted seven months. Schall and his collaborators, Jesuits and Chinese, were put in prison. On 16 January 1665 the relative value of western and Chinese astronomy was put to the test by the prediction of a solar eclipse. Schall, helped by his new Jesuit assistant Ferdinand Verbiest, determined the exact time. The Chinese astronomers failed by more than half an hour. Nevertheless, Schall was sentenced to death and the other Jesuits to exile. Only a large earthquake that destroyed part of Beijing convinced the judges of the Jesuit's innocence. Such a calamity was for them a sign of a serious injustice being perpetrated. Schall died a year later in 1666, but his innocence was not officially proclaimed until 1669 by an imperial edict of the new emperor.

Schall also had problems with his Jesuit companions. Not all of them agreed with his work on the calendar, his dignity as a mandarin and his influence at court. In 1649, Buglio and Magalhaes sent letters to Rome, denouncing the superstitious elements included in the calendar, especially the setting of propitious and unpropitious days and other astrological aspects. Schall defended himself, arguing that he was only responsible for the astronomical part of the calendar. In Rome a commission of professors of the Collegio Romano was formed to study the case. The commission gave a favorable decision in 1659, confirmed by a new commission in 1664. Pope Alexander VII confirmed the commission's ruling and gave his approval to Schall's position as Director of the observatory and to his dignity as mandarin. This was an important approval which affected future Jesuit directors.

Ferdinand Verbiest (1623-1688), born in the Flemish village of Pitthem, arrived at Beijing in 1660 and took the second position in the observatory under Schall's direction. Verbiest was a faithful collaborator of Schall; he was put in prison with him and defended him in the controversy over the calendar. In 1665, after the imprisonment of the Jesuits, Yang Guangxian, the principal accuser of Schall, was named Director of the observatory and given charge of the calendar. A year later Kangxi (Kang hsi), who was to be the greatest Manchu Emperor, assumed all power and suppressed the council of regents. Verbiest pointed out to the new Emperor the many errors contained in the calendar proposed by Yang Guangxian for the year of 1669. The Emperor took the problem into his own hands. Verbiest in his presence explained the errors of the calendar, which the Chinese astronomer could not answer, and challenged Yang Guangxian to predict the length of the shadow of a vertical rod on a determined day and at a determined hour. Yang Guangxian refused the challenge. Verbiest successfully repeated the test several times in the presence of the Emperor, thus showing the superiority of western astronomy. In 1669 Verbiest was finally appointed Director of the observatory and was officially asked to correct the calendar.

Verbiest, Director of the observatory for nineteen years, engaged in intense activity, preparing each year's calendar, teaching European astronomy to Chinese astronomers, and in making new astronomical instruments to replace the old ones. He wrote more than 20 books on astronomy in Chinese. Two of the most important, known by their Latin titles, are *Liber Organicus Astronomiae Europeae* (1668) and *Astronomia Perpetua Imperatoris Kam Hi* (1683). The latter contained the ephemerides of the sun, moon and planets, and tables of solar and lunar eclipses for 2000 years. Besides his works on astronomy, he also wrote two books on the thermometer and the barometer. Although Schall had built several astronomical instruments, it was Verbiest who, in 1669, began to replace the old instruments on the astronomical platform of the observatory. Some of the new ones can still be seen there, together with some others made by Verbiest's successors. The main instruments made under Verbiest's direction, all of them finely cast in bronze, were an ecliptic armillary sphere supported on four dragon heads, an equatorial armillary sphere supported on the arched back of a dragon, a large celestial globe encased in a horizontal framework with four pedestals, a horizon circle for azimuth measurements, a quadrant supported on a vertical shaft with upper and lower bearings, and a sextant on a single pedestal. All these instruments are described in his work published in 1673 in Chinese, *De Theoria,Uusu et Fabrica Instrumentorum Astronomicorum et Mechanicorum*. The purpose of these instruments, in addition to their use in the astronomical observations, was to serve as a demonstration of the superiority of western astronomy.

In 1678, Verbiest was appointed director of the Tribunal or Board of Astronomy and promoted to a higher mandarin class. From 1676, his influence in the court increased. He became a regular counselor of the Emperor and his interpreter in negotiations with European embassies. For example, Verbiest played an important role in the dealings with the Russian embassy headed by Nikolai G. M. Spathary in 1676. He was even given the task of constructing 150 pieces of light artillery. On this occasion he wrote a book on the construction of cannon in Chinese. In 1673 the Jesuit astronomer

Figure 7. Ferdinand Verbiest, second Jesuit Director of the Imperial Observatory of Beijing, China, 1678-1688.

Claudio Filippo Grimaldi (1639-1712) arrived in Beijing and, in 1684, Antoine Thomas (1644-1709), who became Verbiest's closest collaborator in the last years of his life. Verbiest's influence at court, and his ideas about the accommodation of Christian ways to Chinese customs created for him some problems with other Jesuits and missionaries. However, Pope Innocent XI approved his work in a papal brief of 1681, which praised Verbiest's "use of the profane sciences for the salvation of the Chinese people and for the increase of Christian faith." Verbiest died in 1688 and was buried near the graves of Ricci and Schall with solemn funeral rites decreed by the Emperor.

Verbiest was succeeded by Grimaldi who was Director from 1688 to 1707 and again from 1709 to 1712. During a long trip to Europe (from 1688 to 1694) Grimaldi was replaced by Thomas and Tomé Pereira (1645-1708). In 1702 Thomas made the first measurement of a degree of latitude in China, near Beijing. From this measurement he set the equivalence of the unit of length, the "li," taking one degree of latitude being equal to 200 li. In metres the li equals 556.5 metres and the degree measured by Thomas was 111,300 metres. Thomas was also the author of a new map of China. Around this time the famous controversy about the Chinese rites began. Basically this controversy refers to the permission for Christians to take part in the ceremonies of reverence to ancestors and to Confucius. The Jesuits defended the position of the civil character of these ceremonies, while other missionaries insisted on their religious character and the consequent need for their prohibition to Christians. The controversy did not go well for the Jesuits and the rites were condemned by Pope Clement XI in 1707 and in 1715, and again by Pope Benedict XIV in 1742. This condemnation angered Emperor Kangxi, and he withdrew his favour from the missionaries. The Jesuit astronomers participated in the controversy, but this did not affect their scientific work. Even in the most difficult times, when the Emperor forbade the preaching of Christian doctrine, the Jesuits continued directing the astronomical observatory.

The successors of Grimaldi, until the suppression of the Society of Jesus in 1773, were all German Jesuits. Kasper Kastner (1665-1709) held the post for only two years from 1707 to 1709. He was on good terms with the Emperor who named him Preceptor of the Imperial Heir. Kastner took an active part in the rites controversy, defending, in a trip to Rome, the Jesuit position of accommodation to Chinese customs. His successor, after the short second term of Grimaldi's, was Kilian Stumpf (1655-1720) who was director from 1712 to 1720. His most important work was the construction of astronomical instruments, especially a new altazimuth quadrant installed on the observatory platform. These were actually obsolete instruments by European standards of the time, and served more as show pieces than for observations. He also took an active part in the rites controversy, especially during the visit to China of the Papal Delegate Charles Maillart de Tournon; a writing of Stumpf was later condemned in Rome. In 1720 Ignaz Kögler (1680-1746), born in Landsberg, Germany, who had been a professor of mathematics in the University of Ingolstadt, was appointed Director and he occupied the post until 1746. He was held in great regard by Emperor Kangxi, who died in 1722, and his successor, Yong-zheng (Yung chen), who promoted Kögler to Mandarin of the Second Class, a member of the Tribunal of Rites and Preceptor to the Imperial Heir. His main scientific work was the renewal of the astronomical instruments. For example, his innovations included an elaborate equatorial armillary sphere, constructed in 1744, and installed on the observatory platform. He was also responsible for the publication in Chinese of astronomical and mathematical works, such as tables of logarithms, catalogues of eclipses, and observations of Jupiter's satellites. In 1736, Ch'ien-lung ascended the imperial throne. He maintained the Jesuit astronomers at court, even though the fresh condemnation of the Chinese rites by Pope Benedict XIV in 1742 made life difficult for Chinese Christians.

The last Jesuit, strictly speaking, to be Director of the Board of Astronomy was also an eminent astronomer, Augustin von Hallerstein (1703-1774), born in Laybach, Germany. Hallerstein had arrived in China in 1739 and had formed part of the observatory's staff from 1744. During his long presidency, from

1746 to 1774, he was very active, especially in astronomical observations, in particular, those of Mercury in 1746 and 1747. He introduced tables for the determination of ephemerides and eclipses using Newton's calculations and he maintained correspondence with the Royal Society of London and the Imperial Academy of St. Petersburg. He collected the astronomical observations made between 1717 and 1752 by Jesuit astronomers; these observations were published by Hell in Vienna in 1768. Using part of the work of previous Jesuit astronomers, Hallerstein published a collection of astronomical tables and observations in 35 volumes in Chinese. Hallerstein died of a stroke after receiving the news of the suppression of the Jesuits in 1773.

Although the Society of Jesus was suppressed in 1773, four Portuguese ex-Jesuits continued working in China after Hallerstein's death and were Directors of the Board of Astronomy till 1805. The first was Hallerstein's collaborator Felix da Rocha (1713-1781), who succeeded him from 1774 to 1781. Da Rocha had done cartographic work in the regions west of China and Tibet for which he was promoted to Mandarin of the Second Class in 1755. From 1781 to 1783 he was succeeded by Jose da Espinha (1722-1788), a collaborator of Hallerstein and da Rocha in astronomical and cartographic work. The last two Presidents were Andre Rodrigues (1729-1796) between 1783 and 1796, and José Bernardo de Almeida (1728-1805) from 1796 to 1805. Rodrigues and Almeida had established an academy of astronomy in Beijing in 1792, which trained many students for the observatory. These four ex-Jesuits closed this long tradition of Jesuit astronomers which had begun with the arrival of Ricci in Beijing in 1600.

The French mission

Some of the members of the French mission constituted another group of Jesuit astronomers and mathematicians established in Beijing. Although they never held any official position, their work also was very significant in the transfer of western science to China. The origin of this mission was a letter written by Verbiest, in 1678, to his superiors in Rome asking for material and personnel for the Chinese mission. In France the Jesuit royal confessor, François de la Chaize (1629-1709), convinced Louis XIV of the importance of financing a mission of French Jesuit mathematicians and astronomers to China. The leader of the expedition was to be de Fontaney, a professor of mathematics at the College Louis-le-Grand in Paris. De Fontaney contacted Giovanni Domenico Cassini, Director of the Paris Observatory and the leading astronomer of the Académie des Sciences in order to organise the expedition. Cassini, a former student and friend of the Italian Jesuits Riccioli and Grimaldi, was interested in testing his method of determining longitude using the observations of the satellites of Jupiter and this expedition offered him a good opportunity to test it. The expedition was therefore undertaken as part of the Academy's astronomical program. Besides de Fontaney, the other Jesuit members of the expedition were Guy Tachard (1648-1712), Joachim Bouvet (1656-1730), Claude de Visdelou (1656-1737), Louis Daniel le Comte (1655-1728) and Jean François Gerbillon (1654-1707). They were admitted by a particular privilege into the Academy under whose commission they made the journey and they received the title of Mathematicians of the King of France. They left Brest in March 1685 in the royal ship Oiseau and the frigate Maligne carrying an assorted number of scientific instruments, such as telescopes, quadrants, seconds-pedulums, thermometers and barometers, supplied at the king's expense. During the journey they made astronomical observations, determinations of longitude and of the magnetic declination. At the Cape of Good Hope they determined the longitude, by observing the satellites of Jupiter using Cassini's method and observed that one of the stars of the Southern Cross (Alpha Crucis) was a double star. This was one of the first observations of a binary star. In September 1685 they arrived in Siam (Thailand) where they were received by the king. Tachard remained in Siam and the rest, after many difficulties arrived in China in 1687. The Tribunal of Rites, always suspicious of foreigners, forbade their entrance, which was only permitted after Verbiest's intercession with the Emperor. The group entered Beijing in February 1688, shortly after Verbiest's death.

In 1700 Bouvet and Gerbillon founded a residence and a separate French mission in Beijing. The French Residence was built near the imperial palaces on grounds given by the emperor Kangxi. The residence soon became an important scientific centre. It possessed a good library, and instruments for astronomy, physics and chemistry. De Fontaney made observations of the sun and of Mercury and of a comet in 1699.

Among the French Jesuit astronomers Antoine Gaubil (1689-1759) and Michel Benoist (1715-1774) excelled. They established an observatory in 1745. Gaubil, who had studied astronomy in Paris Observatory with Cassini and Jacobo Maraldi, was the author of the first European history of Chinese astronomy, *Histoire de l'Astronomie Chinoise avec des Dissertations*, published in France in 1732. Needham (1959) refers to him as "the interpreter general and father superior of the history of Chinese astronomy" and Alexander von Humboldt praised him as the wisest of the Jesuit missionaries. Benoist, is reported to have introduced the Copernican system to Chinese astronomers in 1761. French Jesuits in China maintained close ties with France and beginning in 1688 published some of their observations in scientific journals, especially in the Jesuit *Mémoires de Trévoux* and the *Mémoires de l'Académie Royale des Sciences*. The last Jesuit of the French mission was Jean Joseph Marie Amiot (1718-1793), who had been a professor at Toulon. He developed a method for a comparison of weights with established standards, and published meteorological observations made in Beijing from 1757 to 1763. Amiot wrote a number of works about the language, culture and history of China. He remained in Beijing after the suppression of the Society and succeeded in arranging continuing financial support from the King of France for the residence. In 1785 the direction of the residence was given to the Lazarists. One of them, M. Roux, an astronomer, continued working at the observatory and became later Director of the Imperial Observatory. In 1755 there were two other small observatories in Beijing at the Portuguese College and at St. Joseph Residence; there was another one at the Portuguese mission of Shanghai. No information about them has been found.

The transfer of western science

In his work on science in China, J. Needham (1959) opens his chapter on the Jesuits saying: "In the history of intercourse between civilizations there seems no parallel to the arrival in China in the 17th century of a group of Europeans so inspired by religious fervour as were the Jesuits, and, at the same time, so expert in most of those sciences which had developed with the Renaissance and the rise of capitalism." There is no doubt that the Jesuits brought to China many new European developments in mathematical and astronomical knowledge not known to Chinese astronomers. In mathematics, among the most important, one finds the Euclidean spherical geometry necessary to understand the motion of heavenly bodies, the stereographic projection and the new developments of algebra and computing methods. In astronomy they brought the geometrical analysis of planetary motions, the idea of a spherical earth and its divisions by meridians and parallels, and the methods for predicting eclipses. Jesuits in China also did important and extensive work in astronomical observation which has in part already been mentioned in the short comments about the different directors of the Imperial Observatory. This work included, in particular, catalogues of the positions of fixed stars, observations of the planets, of the satellites of Jupiter and of comets, and of solar and lunar eclipses. These observations were published in Chinese in voluminous works under the authorship of the different Jesuit directors. Some of them were also published in Europe, as has already been mentioned.

However an important problem in the Jesuits' transfer of European astronomical science to China was their failure to introduce the heliocentric system. As we have seen, Jesuit astronomers in Europe did not accept Copernicus' system as representing the physical reality, but only as a hypothesis to calculate the positions of the planets. After the ecclesiastical condemnation in 1616, this position was firmly held. The three systems of

Ptolemy, Tycho Brahe and Copernicus were used as hypotheses. They excluded from the real situation the motion of the earth proposed by Copernicus. However, according to Needham, before the condemnation some Jesuits in China, for example, Wenzel Kirwitzer (1588-1626), who arrived with Terrenz and Schall, but died young, and Smogulecki, who was a Pole, were Copernicans. We have seen that from about the period between 1620 and 1650, the system of Tycho Brahe was generally accepted by Jesuit astronomers in Europe, and thus brought also to China. Although some Jesuits may privately have held to the Copernican system, it was not until much later, around 1750, as we have seen, that Jesuits, such as Boscovič, accepted it publicly. This explains how Jesuit astronomers in China in the seventeenth and the first part of the eighteenth centuries used the astronomical works of Copernicus, Kepler, and later of Newton, for the calculation of the relative positions of celestial bodies, but retained geocentric cosmology. Kögler and Hallerstein used Newton's method to calculate lunar positions. It must be noted that the Copernican system was not wholly accepted by European astronomers during the seventeenth and early eighteenth centuries, as is often presumed. For example, Cassini, first director of the Paris Observatory still held to the geocentric system in 1659. However, it must be admitted, ecclesiastical obedience in this case had negative consequences that affected the transfer of European astronomical knowledge to China by the Jesuits.

Needham also adds other points such as the use of ecliptical coordinates instead of the polar-equatorial ones, which were already used by the Chinese; and the failure to use Chinese records. He concludes his discussion on this subject saying that on the whole the Jesuit contribution was not an "unmixed blessing". He states that in order to accomplish their religious mission, science was a means to an end for the Jesuits. Their aim was naturally to support and commend the "Western" religion by the prestige of the science from the West which accompanied it. In spite of these shortcomings, important as they may be, there is no doubt in regard to the positive value of the presence of Jesuit astronomers in China. They arrived at a time when science in general, and mathematics and astronomy in particular, were at a very low level there, contrasting with the birth of modern science in Europe. They made an enormous effort to translate western mathematical and astronomical works into Chinese and aroused the interest of Chinese scholars in these sciences. They made very extensive astronomical observation and carried out the first modern cartographic work in China. They also learned to appreciate the scientific achievements of this ancient culture and made them known in Europe. Through their correspondence European scientists first learned about the Chinese science and culture. There is no doubt that the scientific transfer between West and East carried out by the Jesuits over two hundred years benefited both sides. Needham himself recognizes that "all in all, the contribution of the Jesuits, chequered though it was, had qualities of a noble adventure."

The French expedition to Siam

We have seen how on their way to China the Jesuits of the French expedition stopped in Siam in 1685. There they were received by the king Phra Narai and his first minister Constantio Phaulkon, who was a Christian. They went to Louvo (Lop Buri) where the king had a palace; there the Jesuits installed their astronomical instruments and the king was able to observe a lunar eclipse through the telescope. They remained in Siam for eight months. Tachard stayed while the other five continued their journey to China. The king wanted to found a scientific academy and an observatory in Siam similar to those of Paris and Beijing. For this purpose Tachard, accompanied by Siamese ambassadors, returned to France with the task of recruiting a group of Jesuit mathematicians and astronomers.

Tachard arrived at Brest in June 1686. The Siamese were well received by Louis XIV and Tachard gathered 14 Jesuits ready to travel to Siam. They were given the title of Royal Mathematicians and the support of the Academy which provided them with diverse instruments, such as quadrants, telescopes of 1.8 and 3.6 meters and pendulums. The names of the 14 Jesuits were: Claude de Bèze (?-1695), Jean Venant Bouchet (1655-1732), Charles de la Breuille (1653-1693), Jean Colusson (?-1722), Patrice Comilh (1658-1721), Charles Dolu (1655-1740), Jacques Duchatz (1652-1693), Pierre d'Espagnac (1650-1689), Marcel Leblanc (1653-1693), Jean Richaud (1633-1693), Louis Rochette (1646-1687), Abraham le Royer (1646-1715), Pierre de Saint-Martin (?- 1689), and François Thionville (1650-1691). The expedition, led by Tachart, left Brest in March 1687 in a fleet of six warships with the returning Siamese delegation and a French ambassador extraordinaire to Siam. During the journey the Jesuits made astronomical observations and corrections to their star maps of the southern hemisphere. On 11 May 1687 they observed a solar eclipse to the great interest of the Siamese on board. At the end of September the expedition arrived in Bangkok, except for Rochette, who had died during the journey. The Jesuits were received enthusiastically by the king, and it seemed that the project was going to be very successful. The king had promised the construction of an observatory, a house and a church.

Tachard returned in January 1688 to France, accompanied by twelve young Siamese of noble families sent to study at the Collège Louis-le-Grand in Paris. In April the Jesuits organized in the palace the observation of a solar eclipse, in which the king participated. The construction of the observatory began. A month later everything suddenly went wrong. A palace revolt led by the Mandarin Pitracha succeeded, and King Phra Narai was murdered. The first minister Phaulkon tried to defend himself with the aid of French officers and soldiers, but they were overcome and Phaulkon was beheaded. The Jesuits were not harmed, but they left Lop Buri and went to Bangkok, where they boarded a ship to India. San Martin and D'Espagnac died during the journey. Only Breuille remained in Siam, with the Portuguese Jesuits in Ayuthia. Breuille, Duchatz, Leblanc and Richaud died in 1693, probably in a shipwreck after having spent four years in India.

Jesuit astronomers in India

The work of Jesuit astronomers in India is less well known than that carried out in China, but not of less interest. Led by Francis Xavier, Jesuits established their first mission in Goa in 1542. There is little information about astronomical work being done by Jesuits in India before the arrival of French Jesuits about 1700. In 1613 Antonio Rubino (1578-1643), though he was more interested in geography and cartography than in astronomy, wrote to Grienberger asking for some books on astronomy and information about the newly discovered telescope. He is credited with being the first to introduce a telescope into India. Late in the sixteenth century French Jesuits established themselves in Pondicherry in southern India. They brought astronomical instruments with them and carried out observations between 1688 and 1735. Two survivors from the Siam expedition, Bouchet and Richaud reached Pondicherry in 1689. Bouchet covered the Coromandel coast on foot, made astronomical observations and prepared maps and sketches. Richaud made some of the earliest observations with a telescope in India and discovered in 1689 that the star Alpha Centauri was a double star. Pierre François Xavier Duchamp (1693-1740) observed a lunar eclipse on 2 December 1732. He became interested in Hindu astronomy, studied some of its texts, such as the *Grihacandrika*, and published a small treatise on the subject. He compared the predictions of the 1732 lunar eclipse using Hindu and western methods and found some discrepancies between them; he attributed these to errors in the Hindu tables he used. He sent his book to Gaubil in China, who, as we have seen, was studying Chinese astronomy.

From about 1730 a group of Jesuits began making astronomical observations in Chandernagore, near Calcutta, in West Bengal, under the direction of Claude Stanislaus Boudier (1686-1757). They had several telescopes of lengths ranging from 2 to 5 meters, a quadrant of 60 cm radius, and other instruments and clocks brought from France. Boudier used the new astronomical tables of de la Hire for his calculations (Philippe de la Hire had published in Paris his astronomical tables in 1682 and 1702). Boudier made observations of the sun for several years beginning in 1731.

Figure 8. Manuel Figueredo showing the European astronomical tables to the Maharajah Sawai Jai Singh at the Astronomical Observatory of Jaipur, India.

He also made observations of several lunar and solar eclipses, including the lunar eclipse of 2 December 1732, which was observed also by Duchamp in Pondicherry. Boudier was a very diligent observer, but sometimes his observations did not achieve the accuracy of his contemporaries in Europe. He complained that he lacked the appropriate instrumentation, and often requested that better equipment be sent from France.

In 1720 the Maharajah of Amber and Jaipur, Sawai Jai Singh, initiated a program in astronomy and built five observatories in northern India, the most important in Jaipur, where he had established his new capital. The Jaipur observatory, like the other four, consisted of large stone and mortar structures, in the form of quadrants and sextants and of other types for carrying out observations of the sun, stars and planets. Work on the observatory was finished in 1728. The large structures are still preserved and make an impressive sight. Jai Singh was concerned about the accuracy of the Indian astronomical tables and asked for the collaboration of the Jesuits. In 1728 he sent a scientific delegation to Europe to obtain new astronomical tables and instrumentation and named Manuel Figueredo (1689-?), the Rector of the Jesuit

College of Adra, as leader. The delegation returned in 1730 with the new tables of de la Hire, other books and some instruments. The Maharajah advised his astronomers to use the new tables, but in order to solve some discrepancies he wrote to Boudier and invited him to come to Jaipur. In 1734 Boudier, accompanied by Jean François Pons (1698-1752), made the long journey from Chandernagore to Jaipur carrying a 6 m long telescope, a 60 cm quadrant and other equipment. During the journey, the Jesuits determined the latitude and longitude of the towns along the way, using observations of the altitudes of stars and of the occultations of the satellites of Jupiter. They also made a survey of their route from Agra to Allahabad. They made 50 determinations from Chandernagore to Delhi and 10 from Jaipur to Agra. Longitudes were measured with respect to the Paris meridian. They visited the observatory of Delhi (similar to that of Jaipur), where they observed the solar eclipse of 3 May 1734. Boudier compared the observed time of this eclipse with those of de la Hire's tables and found some discrepancies. Boudier remained at Jaipur for two years and helped the Maharajah and his astronomers with their observations. He corresponded with Gaubil in China, to whom he sent his observations.

Sawai Jai was very interested in mathematical astronomy and made the necessary arrangements to bring two Jesuit mathematicians Andre Strobl (1703-1770) and Anton Gabelsberger (1701-1741) from Europe, who reached Jaipur in 1740. Both Jesuits helped the Maharajah's Hindu astronomers with their observations and with the computations of the lunar positions. Sawai Jai died in 1743 and there is no information about the work of the Jesuits in Jaipur subsequently. The observatory was abandoned and fell into ruin; it was restored in 1901. In 1743 Joseph Tieffenthaler (1710-1770), a Tyrolean Jesuit, arrived in India with the idea of working at the Jaipur observatory, but the Maharajah had died a few months earlier and activity at the observatory had ceased. That year Tieffenthaler traveled to Delhi, where he met Stroble, and they made joint astronomical observations. Afterwards Tieffenthaler visited the observatory of Ujjain, another of the observatories founded by Sawai Jai, and traveled extensively throughout India determining latitude and longitude and doing cartographic work. Tieffenthaler was a tireless explorer. He wrote that "next to the salvation of souls and their conquest for God, nothing has afforded me greater pleasure than the study of the geographical position of places, the variations of winds, the nature of soils, and the character and manners of the regions through which I am travelling." He stayed in India after the suppression of the Jesuits and wrote several books on the geography of India, the most important of these was *Descriptio Regionis Indiae.* These books were used by European geographers as the basis for their maps of India and for regional geographical work there. From 1785 to 1788 Johann Bernoulli translated into German and published several of Tieffenthaler's works

We have seen how Jesuits had a profound influence on Chinese astronomy through their direction of the Beijing Observatory and the Board of Mathematics. According to Sharma (1982) this was not the case in India. For example, Indian astronomers failed to appreciate the cross-hair fitted telescopic instruments brought by the Jesuits and did not use the telescope for their observations. Sharma also points out the failure of Jesuit astronomers to introduce heliocentric astronomy into India. We have discussed this problem already in the case of China. Since Jesuit astronomers were mainly concerned with observational astronomy, this was not a major drawback. But it has to be admitted that it was an obstacle to the spread in India, as in China, of modern cosmological theories.

An observatory among the Guaranies

The last observatory founded by Jesuits in mission countries was in the famous Jesuit "reducciones", or enclosed missions, of Paraguay among the indigenous Guarani people. This was a little-known observatory, lost in the territory only inhabited by these primitive people. Thanks to Guillermo Furlong (1929), we have a detailed description of the observatory and of the life of its founder and only Director, Buenaventura Suárez (1679-1750). Suárez, born in Santa Fe, Argentina, went in 1701 to the mission of San Cosme y Damián. This mission, founded in 1634, was located first among the Tape mountains in Ybytymini, in the present territory of Rio Grande do Sul, Brazil. In 1638 the mission was moved to the other side of the river Paraná, and located between the Aguatepey river and the Candelaria mission in what is now Argentina. In 1718 the location was moved again, by Suárez himself, about five kilometers to the east. In 1740 it was moved again to the north of the Río Paraná near the village of Itapúa, and, in 1760, to the present location of the village of San Cosme in Paraguay. All these locations are less than one hundred kilometers apart. Notwithstanding that he was in charge of the mission, Suárez found time among his many other occupations for his great interest in astronomy.

Suárez established the first observatory about 1706 at the second site of the mission. Though seldom acknowledged, this was the first astronomical observatory of the western hemisphere, almost a century and a half before the first installed in North America. In the first period of the observatory, from 1706 to 1745, Suárez used instruments he himself made. He had several telescopes with two convex lenses of lengths of from two meters to six meters, a pendulum clock and a quadrant. He even polished the lenses himself from natural quartz crystal. The observatory was probably installed on a wooden tower, which also served as the church tower. With these primitive instruments Suárez made the observations for his most famous work, *Lunario de un siglo,* first published in 1743 and reedited in 1748, 1751 and 1856. The book contains the ephemerides of the sun and the moon, the times of the solar and lunar eclipses and other information for the period from 1740 to 1840. It also gives the method for extending the calculations from 1842 to 1903.

The second period of Suárez's activity from 1745 to his death was marked by the acquisition of professional astronomical instruments. These were two telescopes of lengths of 2.52 meters and 5 meters, and two clocks by Martiron, all made in England and bought in Lisbon. They arrived in Buenos Aires in July 1745 and then transported to San Cosme. The fact that his superiors agreed to spend a considerable amount of money on these instruments indicates that Suárez's astronomical work was highly valued after the publication of his *Lunario*. Suarez continued his observations, especially those of the satellites of Jupiter, lunar eclipses and comets. These observations were praised and used by Peter Wilhem Wargetin, a professor of the University of Uppsala, Sweden, in his work about the satellites of Jupiter. During this time Suárez also made meteorological observations and weather forecasts. Though Suárez was in remote Paraguay, he kept contact with astronomers in Europe, in particular with the Jesuits Grammatici in Ingolstadt and Kögler in Beijing and his library included the works of Copernicus, de l'Isle and de la Hire whose tables he used. After his death his work was discontinued. The Jesuits were forced to leave the missions in South America in 1767 when the Society of Jesus was suppressed in Spain.

Chapter 4

EUROPE: JESUIT ASTRONOMY AND GEOPHYSICS (1814-2000)

Observatories in Europe were started soon after the restoration of the Society of Jesus in 1814, most of them before 1900 (18 out of 24). The majority were connected with schools or faculties of philosophy for Jesuit students or with seminaries for secular priests, while other were attached to universities, colleges or secondary schools. In both cases their main purpose was to serve as an aid for teaching science courses and specifically courses on astronomy. However, some of them began with the intention of their being real research facilities. These later acquired an identity independent of the schools to which they were at first attached, as in the case of the observatories of the Collegio Romano (Italy), Stonyhurst (England), Kalocsa (Hungary), Cartuja and Ebro (Spain). The main interest of these observatories was initially in astronomy and meteorology, responding to the practice at the late part of the XIX century, when these two types of observations were usually carried out in the same observatory. These observations were followed later by those of geomagnetism and seismology. The main astronomical observatories were those of the Collegio Romano (Italy), Stonyhurst (England), Kalocsa (Hungary), Gianicolo (Italy), Valkenburg (Holland), and Cartuja (Spain). The Vatican Observatory, the direction of which is entrusted to the Jesuits, has a different character in that it is directly dependent on the Holy See. The observatories usually had one main telescope with equatorial mounting installed within a rotating dome, with other smaller telescopes and complementary instruments. Solar and spectral observations became soon also very common in Jesuit observatories, and for this purpose spectroscopes were added to the telescopes. Meteorological instrumentation was, generally, very complete with observations of atmospheric temperature, pressure and humidity, of rainfall and of wind direction and velocity done several times a day. The observatories of Mondragone (Italy), La Guardia (Spain), Gozo (Malta), Jersey (England) and Comillas (Spain) were dedicated primarily to meteorology. Geomagnetic observations started early in the Collegio Romano (Italy), Stonyhurst (England), Kalocsa (Hungary), and were the main interest in Ebro (Spain). In most cases magnetometers for absolute measurements, and variometers giving continuous recordings of the variations of the three components of the Earth's magnetic field were installed. Seismological observations were started around the beginning of the 20th century in Mondragone (Italy), Stonyhurst (England), Cartuja (Spain) and Ebro (Spain). Later in Cartuja seismology became the main interest. The observatories of Travnik (Bosnia) and Rathfarnham Castle (Ireland) were set up exclusively as seismological stations.

Collegio Romano

The first observatory of the modern Society of Jesus was that of the Collegio Romano, one of the oldest Jesuit colleges founded in 1551 and a true university since 1553. The origin of the observatory, as has been mentioned in chapter 2, can be traced back to the astronomical observations carried out since the time of Clavius, Scheiner, Grienberger and other professors of mathematics and astronomy at the beginning of the 17th century. Astronomical observations were made during the 17th and 18th centuries, but no formal observatory existed. As we have seen there were plans by Borgondio and Boscovič to established one, but they were never carried out. When the Society of Jesus was suppressed in 1773, the Collegio Romano was transferred to the secular clergy. The following year, Pope Clement XIV ordered the establishment of an observatory with the title of the Osservatorio Pontificio del Collegio Romano and named Giuseppe Calandrelli, a diocesan priest, as Director. For more than ten years this project existed only on paper, although meteorological observations were started in 1782. In 1786, the then aging Boscovič was asked to participate in the project and tried to give an impulse to the realization of the old plans. In 1787 a tower was built for the observatory, not on the roof of the church, but on the eastern corner of the front of the Collegio, where in 1804 some astronomical instruments were finally installed. The astronomical observations begun by Calandrelli, A. Conti and G. Riechenbach, were published in eight volumes of the Opusculi Astronomici.

In 1824 the college was returned to the Society, which had been restored in 1814, and the observatory became the first Jesuit observatory in the restored Society. Therefore this observatory constitutes a link between the observatories of the old and the new Society. Calandrelli resigned as Director and a new Director was named: the Jesuit Dominique Dumouchel (1773-1839), who had studied in the Ecole Politechnique of Paris. A few of the instruments of the old observatory, for example, a Reichenbach transit, a small Dollond equatorial telescope and two chronometers remained, and a new Cauchoix equatorial telescope was installed, which was a gift of the Jesuit Father General Luigi Fortis. In 1835 Dumouchel and his assistant Francesco de Vico (1805-1848) predicted and later observed the return of Halley's comet. In 1840 de Vico was named Director and he initiated a great impetus in the observatory. His collaborators were Benedetto Sestini, (1816-1890), Paolo Rosa de Conti (1825-1874) and Victor della Rovere. In 1842 a new Ertel meridian circle was installed, a gift of Father General Jan Roothaan. Soon de Vicos's work became to be known in astronomical circles. Among his work we can mention the observation of the satellites of Saturn, Mimas and Enceladus, the discovery of eight comets between 1844 and 1847, which earned him a prize from the King of Denmark, and the compilation of a star catalogue down to magnitude 11, which he could not complete. Bernardino Gambara (1814-1884) was in charge of the meteorological observations, which he carried out carefully seven times a day, continuing the series of measurements begun in 1782. The observatory was then well established as one of the ten existing in Italy; the other nine were those of Turin, Milan, Florence, Bologna, Padua, Naples, Palermo and Capitolio (Rome). However the political situation in Italy became disturbed when, in 1845, Vittorio Emanuele II, King of Piedmont began a movement towards the unity of Italy. A revolution broke out in 1848 in Rome which installed a short-lived Roman Republic, and which resulted in the expulsion of the Jesuits from the city. De Vico, Rosa and Sestini traveled first to Paris, then to London, and finally to Washington where De Vico was named director of the recently created Astronomical Observatory of Georgetown University. The same year De Vico made another journey to London with the aim of acquiring some instruments. He died there after a short illness.

In 1849 the Jesuits returned to the Collegio Romano and Pietro Angelo Secchi (1818-1878) was named Director of the observatory. He is one of the most outstanding Jesuit scientists of the modern Society and made the observatory a first class astronomical institution. His interest in science began during his studies of theology, in 1848, in Stonyhurst College, England, where he started to collaborate with Alfred Weld, in the observation of sunspots. The same year he traveled to Washington to work at Georgetown Observatory, but finally returned to Rome the following year. Secchi revived Boscovič's project of building the observatory on the roof of the Church of Saint Ignatius, but on a more modest scale. On the roof, the engineer A. Vescovali built two small circular and elliptical rooms to house the telescopes as well as other rooms for library and work space. These rooms were positioned on top of the four large columns, designed to support a dome that was never built. Completed in 1853, the new observatory was solemnly inaugurated by Pius IX. In 1854 a new equatorial telescope, one of the best instruments then available, was acquired. Made by Merz it was equipped with a lens of 24 cm aperture and a focal length of 4.35 m. The telescope was installed in the circular room fitted with a rotating dome. Two other rooms housed the Ertel and Cauchoix telescopes. From the beginning the observatory kept a meteorological station where measurements were continued. In 1854 Pius IX granted the observatory the title of "Pontifical", and gave it the mission of keeping the official time for the Papal States. Each day at midday a ball was dropped down a mast on the roof of Saint Ignatius Church to signal the exact time. Magnetic observations began in 1858 using a set of magnetometers, an inclinometer and a declinometer donated by Pius IX. They were installed in a room on the roof of the church, separated from the other rooms and specially built with non-magnetic materials.

Figure 9. Angelo Sechi, Director of the Astronomical Observatory of the Collegio Romano, Rome, Italy, 1850-1878.

Secchi decided to dedicate the observatory to the growing field of what was then known as physical astronomy, or new astronomy, that is, the present discipline of astrophysics. In 1800 a new line of research had been initiated by William Herschel with the study of the spectrum of the sun; and in 1814 Joseph von Fraunhofer observed the continuous spectrum of the sun interrupted by dark lines. Soon spectroscopy became an important tool in physics and astronomy. Until this time astronomy had been limited to the observation of the location and size of heavenly bodies. Now spectroscopy provided a wealth of information on the chemical composition and physical conditions of the sun, stars, nebulae, comets and planets, making possible the new science of astrophysics. In 1860, Gustav Robert Kirchhoff and Robert Wilhem Bunsen made the first chemical analysis of the sun's atmosphere, introducing spectrochemical analysis into astronomy. Secchi saw the potential of this new line of research and dedicated his efforts to the study of the constitution of stars, star groups, double stars and nebulae. In 1862 he installed a Hoffmann spectroscope attached to the Cauchoix refractor. With this arrangement he made more than 4000 observations of the spectra of stars down to magnitude 7. He is considered a true pioneer in this work which constitutes his main scientific contribution. His scientific recognition was shown by his election a fellow of the Royal Society of London in 1856.

In 1867 he established a classification of the spectra of stars in four classes, namely, class 1, with strong hydrogen lines which includes blue and white stars; class 2, with numerous lines which includes yellow stars; class 3, with bands rather than lines, sharp toward the red and fuzzy toward the violet including orange and red lines; class 4, with bands sharp toward the violet and fuzzy toward the red, including red lines. This classification extended and modified by E. Pickering and A. Cannon, has become one of the basic tools of astrophysicists. In 1872 Secchi was one of the founders of the Società degli Spettroscopisti Italiani created to promote these studies in Italy. Another field to which he dedicated many of his efforts was the study of the sun, measuring the changes in temperature across the solar disk, its spectrum, sunspots, prominences and other characteristics. These he discussed in *Le Soleil,* a book published in French in 1870, and later translated into German and Spanish. In 1866 Secchi made detailed investigations and confirmed the Wilson effect, that the central part or nucleus of a sunspot is beneath the level of the sun's spherical surface. He measured the depth of the depressions and found that they were relatively shallow. He also showed that sunspots radiate less heat than the surrounding photosphere. In this way Secchi continued, Jesuit studies of the sun which had begun with Scheiner's work in the XVII century. We will see that this work became a favorite subject among Jesuit astronomers. Secchi's ingenuity in instrumentation can be seen in the design of his "Universal Meteorograph", an instrument for the simultaneous recording of several meteorological measurements: pressure, temperature, humidity, and wind velocity and direction. This instrument was awarded a gold medal in the Paris World Fair of 1868. In 1854 Secchi was commissioned to execute the measurement of a geodetic base-line, extending over an arc of two degrees, between Rome and Rimini along the Via Appia, repeating with greater accuracy the measurements made by Boscovič in 1751. This work was left incomplete. Secchi was a prolific writer with many articles in scientific journals and besides his book on the sun he published two others: *Le Stelle,* on the stars, and *L'Unitá delle Forze Fisiche*, on the unity of the physical forces. Secchi considered always his religious commitment as the true motive that spurred him on in his astronomical studies. Thus he wrote: "To whisper to oneself how magnificent it is to reveal the works of the Creator: this is a stimulus which lasts when all other motives fail".

This flourishing scientific work did not last long. On September 20, 1870 the troops of Vittorio Emanuele entered Rome and the Jesuits were driven into exile. The Collegio Romano was expropriated and declared state property. The worldwide reputation of the observatory saved it from this fate and Secchi was allowed by a special Act of the Italian Parliament to remain in his post. He and

Figure 10. Interior of the dome of the Observatory of the Collegio Romano with a Merz equatorial telescope.

his assistants were exempt from the decree of banishment. This situation continued for eight years until Secchi died in 1878, after some years of illness. Pope Pius IX said of him "Father Secchi has always known how to unite science with religious virtue, but the two virtues which shone most in him were humility and obedience". His assistant for many years, Gaspare S. Ferrari succeeded him only for a short time, since in the following year, he was expelled from the observatory which was declared state property and given the new name of Regio Osservatorio al Collegio Romano. Its first lay director was Pietro Tachini, a former student and friend of Secchi, who carried on the work on solar physics with great distinction until his death. We will see him in connection with the founding of the Observatory of Calcutta (Chapter 7). The observatory then languished on for some years until it was closed in 1923. The expropriation of this first modern Jesuit observatory by a government decision, motivated by antireligious and anti-Jesuit prejudices was to be repeated in other cases, as we will see, in the history of Jesuit observatories. It must be added, that in the case of the observatory of the Collegio Romano, this happened when it was at its best, directed by as eminent scientist as Secchi was, and it could have still continued producing important work under Jesuit direction for many years.

Ferrari was not satisfied with the end of Jesuit work in astronomy in Rome and decided in 1880 to establish a small observatory in the Villa Cecchina, on the slope of the Gianicolo Hill, which he named Osservatorio Astronomico Privato sul Gianicolo. In 1880, he installed a Merz equatorial telescope of 23 cm aperture and 3.82 m focal length in a rotating dome, and later added a smaller one of 10.8 cm aperture and 150 cm focal distance. The observatory was associated with the Gregorian University, the new name taken by the Jesuit university in Rome, which opened in 1873 in the palazzo Borromeo, after the confiscation of the Collegio Romano. Ferrari was Professor of Astronomy and Director of the observatory from 1879 to 1894, when he left the Society of Jesus. He was succeeded by Adolph Müller (1853-1939) who was also a professor of astronomy in the University from 1894 to 1913. The observatory tried to continue Secchi's work on solar observations, but never reached the same importance as its predecessor and it was finally dismantled; the foundation of the dome serves today as the base of a statue of the Sacred Heart of Jesus in the gardens of the Jesuit General Curia (Borgo S. Spirito).

Stonyhurst

A more peaceful story was that of the second observatory, founded in 1838 in Stonyhurst College, Lancashire, England, a boarding school for boys with a long tradition, located near the town of Preston. The history of Stonyhurst College goes back to 1593 when the English Jesuit Robert Person (1546-1610), with the financial support of King Philip II of Spain, established in St. Omer, France, a school for English boys who owing to the persecution suffered by Catholics under Queen Elizabeth I, could not get a Catholic education in England. After some changes of location in France and Belgium, the school settled in England in 1794 in Stonyhurst, on the estate of the Shireburn family, donated to the college by Sir Thomas Weld, who had been a former student. The college was run by former Jesuits during the time of the suppression from 1774 to 1814 as a Missionary College for England. Later it recovered its whole Jesuit character. The original buildings were enlarged between 1876 and 1889 to achieve the impressive appearance one can admire today. An avenue half mile long leads in a straight line to the main entrance. The imposing west-facing frontage is dominated by the Gatehouse, with its twin towers, originally built in 1592 and completed in 1712. For some time the school of philosophy and theology for Jesuit students was in the adjoining St. Mary's Hall built in 1843. The founding of the observatory is not altogether clear, but it can be dated with certainty from 1838, when a small pavilion was built in the garden for meteorological observations. In that year the first instruments were installed, acquired through James Glaisher, chief meteorologists at Greenwich Observatory, who calibrated the instruments and afterwards maintained an active collaboration with the observatory. Charles Irvine (1801-1843) is often quoted as the founder, but actually he had left the College in 1836. Although the first meteorological observations started late in 1838, the first Director who held this title in 1842 was Henry McCann (1801-1888), Professor of Natural Philosophy, Mathematics and Chemistry. In 1844 new meteorological equipment was installed, and in 1845 the first telescope, a meridian circle of 8.2 cm aperture. Later, in 1848, a telescope with equatorial mounting of 12 cm aperture was acquired. Both telescopes were made by Jones in London. These were the years when de Vico was carrying out his astronomical work in Rome.

In 1846 Alfred Weld (1823-1890), grandson of the founder of Stonyhurst and a teacher of science, was named Director and started the truly scientific work in the observatory with systematic meteorological observations and observations of sunspots. He got Secchi, then a student of theology, exiled from Rome in 1848, interested in astronomy. In 1851 Weld went to study theology and returned to occupy the post of director of the observatory from 1856 to 1860. He initiated scientific research in the observatory with work on meteorology, terrestrial magnetism, sunspots, as well as star positions and occultations, and made the observatory known in the scientific circles of Great Britain. Weld was elected fellow of the Royal Astronomical Society in 1849, the second Jesuit to be elected after de Vico (in 1841), and one of the first members of the British Meteorological Society. In 1857, two new reflecting telescopes of Cassegrain and Newtonian type were acquired. In 1834 Edward Sabine, President of the Royal Society and famous for his Arctic expeditions and work in geodesy, astronomy and terrestrial magnetism, commenced the first systematic magnetic survey of the British Islands. In 1858 Sabine installed one of his temporary magnetic stations in Stonyhurst, at Weld's request, and suggested that a permanent station be installed there as part of the world network of stations he, Sabine, was promoting. The same year Weld acquired from Jones and Borrows the magnetic equipment of the Kew type for making absolute measurements of the horizontal component of the intensity of the magnetic field (H), magnetic declination (D) and magnetic dip or inclination (I). A similar set of instruments was acquired also for the Collegio Romano at Secchi's request and paid for by Pius IX. Thus Stonyhurst, together with the Collegio Romano became one of the earliest geomagnetic observatories. Through the influence of Sabine, who maintained a cordial relationship with Stonyhurst, the observatory became nationally and later internationally known. Weld's scientific work was cut short when in 1860 he was named Novice Master and, four years later, in 1864, Provincial of the English Jesuit Province. In this position he was able to help the observatory by giving his successor all his sympathy, encouragement and active support.

Figure 11. Stephen J. Perry, Director of Stonyhurst Observatory, Lancashire, England, 1868-1889, an influential figure in the early development of Jesuit observatories (SJBPA).

In 1860 Stephen J. Perry (1833-1889), still a Jesuit student, was named Director for the first time, remaining as such until 1862. Perry is without doubt the most outstanding figure of the Stonyhurst Observatory and of the English Jesuit scientists. He was specifically trained for this post, attending lectures in mathematics, in 1858, in London from De Morgan and, the following year, in Paris from Bertrand, Liouville, Cauchy and Serret. In 1862 he went to complete his theological studies and from that year to 1868 Walter Sidgreaves (1837-1919) who was to be Perry's close collaborator and successor occupied the post of Director. In 1866 the observatory, on Sabine's recommendation, was elected as one of the seven first class meteorological stations of the British Meteorological Office. Sidgreaves installed new equipment: barographs and thermographs with photographic recording, a Robinson anemometer and a pluviometer. In 1863 Sidgreaves began monthly systematic absolute magnetic measurements of the magnetic components H, D and I, which, in 1865, were reported on, for the first time, by Sabine in a meeting of the Royal Society. These observations were continued until the closing of the observatory in 1974 and constitute one of the longest series of magnetic observations ever done at the same site. In 1867, with a grant from the Royal Society, a special underground room was built at the back of the observatory building for the magnetic observations. There Sidgreaves installed a set of variometers with photographic recording, to record continuous variations of the horizontal and vertical components of the geomagnetic field and magnetic declination. Thus Stonyhurst became one of the six magnetic stations with regular observations in Britain. The same year a new building for astronomical observations was added, at some distance from the first. It was fitted with a metallic dome to house a new telescope with equatorial mounting of 20 cm aperture made by Napier and Carey with a lens of Troughton and Simms.

The time of Perry's directorship from 1868 to 1889 was without doubt the most brilliant period of the observatory. Perry's personality as well as his scientific dedication and excellence greatly contributed to this. Besides his work at the observatory and many scientific lectures to all kinds of audiences which were very popular, he was a professor of mathematics of the Jesuit students at St. Mary's Hall and of the boys at the College. The observatory became the training ground for visiting Jesuits who afterwards became directors of observatories around the world. Among them were Dechevrens (Zikawei), Faura and Cirera (Manila), Viñes and Gangoiti (Havana), Berloty (Lebanon) and Goetz (Bulawayo). Perry also helped them to buy astronomical and magnetic equipment for their observatories. The national scientific prestige of the observatory is reflected in the fact that the British Association for the Advancement of Science held its meetings of 1883 and 1887 at Stonyhurst. One of Perry's main scientific activities was in the field of geomagnetism. He carried out three magnetic surveys: two in France, in 1868 and 1869, in the west and the east, assisted by Sidgreaves with 15 and 21 stations respectively; and a third in 1871 in Belgium helped by William Carlisle (1842-1910), a Jesuit brother, with 20 stations. In each of these three surveys, at each station, careful measurements of the horizontal component of the magnetic field (H), the magnetic declination (D) and inclination or dip (I) were made. In Belgium Perry found large magnetic anomalies related to coal mines. George B. Airy, then president of the Royal Society, considered this last survey one of the best for a limited region. In order to study the relation between solar and terrestrial magnetic activity, in 1881 Perry began a series of observations of sunspots by means of drawings of the solar disk on a large scale (27 cm diameter) of which 3800 in all were made. In his solar studies he used also photography and spectral analysis. For this purpose he installed two direct-vision spectroscopes and an automatic Browning spectroscope with six prisms as well as a photographic grating spectrometer by Hilger. The solar observations of Stonyhurst were ranked in quality with those of Zurich, Greenwich and Mount Wilson. With these instruments Perry carried out very complete studies of sunspots, faculae and prominences and tried to show their relation with terrestrial magnetic activity. This was still a controversial subject. In 1852 Sabine had noted that minima and maxima in magnetic disturbances

Figure 12. View of Stonyhurst College, Lancashire, England. The astronomical observatory with its dome, the observatory main building and the entrance to the underground vault are on the front of the picture (SJBPA).

corresponded to minima and maxima of the sunspot periodicity of 11 years discovered in 1843 by Heinrich Schwabe. Studies of this relation were carried out in 1859 by Richard C. Carrington and in 1861 by Balfour Stewart. The direct influence of the sun on the magnetic field of the earth was still denied by Lord Kelvin as late as 1892. It was not until 1904 that Walter Maunder finally proved that magnetic disturbances have their origin in the sun. Perry's last great project of collecting and comparing all his magnetic and solar data together was never completed due to his untimely death. Other astronomical work of his involved observations of the satellites of Jupiter; occultations of stars by the moon; observations of meteorites; and determination of the position of comets.

Perry is best known for his participation, in many instances as the leader, in scientific expeditions to carry out astronomical and geophysical observations. This activity made the observatory known in and outside the United Kingdom. His first expedition as chief scientist in 1870, accompanied by Sidgreaves, was to observe a solar eclipse in Cadiz, Spain. His most important expedition was to Kerguelen Island in the southern Indian Ocean in 1874 to observe the transit of Venus across the sun. As we have seen, the previous ones in 1761 and 1769 had been observed in Lyon by Beraud and in northern Norway by Hell, Director of the Observatory of Vienna. Perry, commissioned by the Royal Society, and again accompanied by Sidgreaves, was put in command of the British expedition by Airy, at that time Astronomer Royal. During the long sea voyage he took many magnetic observations with the ship at anchor. After observing the transit of Venus, he remained five months on the island, taking many astronomical, magnetic and meteorological observations. In 1882 he observed the second transit of Venus in Madagascar. In 1886 and 1887 he observed two solar eclipses in Carriacou, the Lesser Antilles and in Pogost, Russia. His last expedition took place in 1889 to the isles of Salut, off French Guyana near Cayenne were, after completing his observation of the solar eclipse, he died of severe fever, caused by the unhealthy climate of the island. He had insisted on staying on land to minister to the convicts in the prison and to the sick in the hospital against the opinion of the ship commander who considered it a risk for his health. In all these journeys Perry's affable character and sense of humor endeared him to the men of the Royal Navy, and he was always warm in his praise for the enthusiasm and help he received from the officers and men of the ships assigned to transport his expeditions. Perry was member of and held offices in several scientific societies. In 1874 he was elected a Fellow of the Royal Society for his work on terrestrial magnetism, an honor shared only by three other Jesuits: Boscovič, Poczobut and Secchi. After his death the journal *Observatory* stated that Perry "was always eager to gird on his armor in the sacred name of science: the discomforts and anxieties, nay, the real dangers of the crusade never daunted him for a moment; and we can claim for him all the laurels due to the soldier who pays for victory with his life and dies bravely, cheerfully, nobly, at the moment of success".

Perry's successor till 1919 was Sidgreaves, his collaborator and companion in several expeditions. Sidgreaves worked mainly on stellar spectrography and continued Perry's work on solar physics and geomagnetism. During some years his assistant was William McKeon (1851-1920) a Jesuit Brother, who became a real expert in photography. In 1890 it was proposed to honor the memory of Fr. Perry with the installation of a new telescope. For this purpose it was necessary to raise 2700 pounds to buy a telescope of 38 cm aperture and to install a new metallic dome. Only 607 pound were raised, and, as a compromise, a new lens of 38 cm made by Howard Grubb of Dublin was fitted to the old equatorial telescope. The thus upgraded telescope was called the "Perry Memorial Telescope." It was inaugurated in 1893. In 1892, a young Jesuit, Aloysius Cortie (1859-1925) who had been a student at Stonyhurst College, joined the observatory and he was to succeed Sidgreaves as Director. He accompanied Sidgreaves to Vinaroz, Spain in 1905 to observe a solar eclipse, together with other Jesuits. As we will see later, a number of Jesuits observed this eclipse in different parts of Spain. In 1909 the work of the observatory was augmented by the installation of a Milne seismograph, replaced in 1924 by a Milne-Shaw instrument, on loan from the Royal Society. Cortie who dedicated himself mainly to astronomy, continued Sidgreaves' work on stellar spectrography and also the work begun by Perry on the relationship between solar activity and terrestrial magnetism. He participated in scientific expeditions to observe the solar eclipses in Tonga (Pacific) in 1911 and in Sweden in 1914. In 1925 Cortie was succeeded as director by Edward D. O'Connor (1874-1954) who had collaborated in the observatory since 1902 and continued the work on stellar and solar spectrography. Failure to renew the instrumentation and diminished scientific interest began a decline in the work of the observatory. O'Connor was named rector of the College and was succeeded, in 1932, by James P. Rowland (1875-1948) who had been working at the observatory from 1919. His main interests were seismology and meteorology, so that the astronomical work was practically abandoned. In 1919 Rowland had complained: "my impressions of the state of the observatory are that it is suffering badly from years of neglect and is almost in the last estate of decay, both in instruments and fabric". Rowland turned to meteorology, publishing weather forecasts in the local newspaper, under the name of "The Wizard of the North". He fixed the seismograph that had been out of service for years and kept the seismological station running, being an active member of the Seismological Committee of the British Association. The condition of the observatory, especially after the war, seriously deteriorated and it was officially closed in 1947. The meteorological observations, however, were continued through the painstaking work of James Lawrence (1898-1960), who, until his death, kept sending them to the Meteorological Office.

In 1957, on the occasion of the International Geophysical Year, there was an attempt to reopen Stonyhurst Observatory under Lawrence's direction, with the purpose of continuing the magnetic observation, as these had been done for almost a century. The Jesuit astronomers, Patrick J. Treanor, then working in the Observatory of Oxford University, and Daniel O'Connell, Director of the Vatican Observatory, insisted on the importance of continuing the work in the observatory, at least on terrestrial magnetism. New equipment was acquired and installed in 1960. To calibrate the new instruments José O. Cardús, from the Ebro Observatory, Spain, spent a month at Stonyhurst. Lawrence died the same year and James Worthy was named director and he continued the meteorological and magnetic observations. Most of the rest of the equipment was lost and in 1966, Perry's Memorial Telescope was sold. Finally the observatory, or what was left of it, was definitely closed in 1974, when Worthy left the College and the Jesuits. Some years later, a group of local amateur astronomers, the East Lancashire Astronomical Society, began to get interested in the facilities of the observatory for their own use. In 1979 its President Phil Horrocks made an agreement with the college to renovate the astronomical building that had been abandoned for many years and was very dilapidated.

The building and the dome were fixed and put to use the following year. A small telescope from Alvan Clark of 18 cm aperture, which had been donated to the observatory in 1932, and whose lens had miraculously survived through the years of abandonment was installed. In 1976 the students of the college had formed "The Astronomy and Space Research Society". Under the direction of the lay science professor, Fintan O'Reilly, they joined with the amateur astronomers in the renewal of the astronomical premises and used the telescope for their demonstrations. In 1989 on the occasion of the centenary of the death of Fr. Perry a search was started to recover the old Perry Memorial Telescope from its purchaser. The telescope was finally found and bought back and after its renovation it was installed again in the original dome. The telescope is still used by the students and the local amateur astronomers

Kalocsa

Far from England, another Jesuit observatory was founded at Kalocsa, Hungary, a town about one hundred kilometers south of Budapest and an important agricultural center. In the founding of this observatory the initiative did not come from the Jesuits themselves, but from the archbishop of Kalocsa, Cardinal Lajos Haynald (1816-1891). He was interested in astronomy and had acquired a Merz telescope of 6 cm aperture. In 1877 he decided to donate the instrument to the Jesuit College (Stephaneum Gymnasium) with the idea of helping the science teachers create a small observatory. At that time in Hungary there were only two astronomical observatories, since the old University observatory near Budapest on Gellert Hill had been destroyed in the war in 1849. Miklós Konkoly Thege had created an observatory in 1871 at Ogyalla, and another was established in 1879 at Hereny by Jeno Gothard. The State Minister Trefort, a friend of Haynald, encouraged him to create a full scientific observatory in Kalocsa, not just a small one for teaching purposes. Haynald made this decision and a tower was built on the roof of the college for the observatory, which was finished in 1879. The astronomer Konkoly Thege took part in the design and construction of the different facilities of the observatory. The observatory was named after its founder Haynald Observatory and Carl Braun (1831-1907), a German, professor in the Seminary of Fulda, was named Director. He had studied mathematics and astronomy in Paris and had spent some time with Secchi in the Observatory of the Collegio Romano. Braun completed the installation of a fine Merz telescope of 18 cm aperture with equatorial mounting by Browning and installed a Hilger spectroscope for the study of the sun following Secchi's footsteps. The observatory also had a complete meteorological station. Braun began what would be Kalocsa's most important contribution through the study of sunspots, prominences and faculae, and designed for this purpose a spectro-heliograph that he never completed.

Figure 13. Gyula Fényi, director of Haynald Observatory, Kalocsa, Hungary, 1885-1913. Memorial on the front of his birthplace in Sopron, Hungary. The inscription says: "In this house P. Fényi Gyula in January 8, 1845 saw for the first time the light of the sun of which he became a famous researcher." The head is the work of Elisabeth Haics. (Courtesy of J. Verö, Geodetical and Geophysical Institute of the Hungarian Academy of Science).

The first working spectro-heliograph was actually made by George E. Hale in 1892. Due to poor health Braun left the direction of the observatory in 1884 and retired to the college of Mariasheim in Bohemia.

In 1885 Gyula Fényi (1845-1927) was named director. Fényi had already worked in the observatory between 1871 and 1874, while he was teaching mathematics and physics at the college. Fenyi was the key figure at Kalocsa, as Perry was at Stonyhurst, and during the time he was Director, from 1885 to 1913, the observatory experienced its best days. Fényi was a very dedicated person and very exact in his observations. Between 1885 and 1917, every day, when the weather allowed for it, he made solar observations completing about 40000, of which 6000 were published. The study of the sun had, thus, become a Jesuit specialty as we have seen with Secchi and Perry. Fényi concentrated all his efforts on the study of sunspots and prominences. He not only made very detailed observations, but also proposed some ideas about their nature. He concluded that both have the same origin due to the motion of the material on the surface of the sun, but with different characteristics. He found that the chromosphere was generally depressed over sunspots, and proposed that prominences are formed by currents of incandescent gas projected into the vacuum. These fall again owing to the gravitational attraction of the sun forming the faculae by their collision with the surface. He observed that the distribution of prominences at the time of a sunspot maximum was greater in the polar regions and he presented an explanation for the shift toward the red and blue in their spectra as being due to currents in opposite directions. He also studied the relationship between large prominences and the occurrence of magnetic storms, concluding that the latter were created by the presence of metallic gases in the currents produced on the sun's surface. As we have seen, these were still at that time, controversial questions. Fényi published his observations and interpretations between 1886 and 1925 in 145 articles in French, German, Italian and Hungarian scientific journals. Sometimes the same article was published in the four languages, as well as in the bulletins of the observatory.

During this time the meteorological station was upgraded with better equipment, and from 1904 it received government funding. Affiliated with Kalocsa were the meteorological stations installed in Mozambique, in the localities of Boroma (Tete) and Zumbo (chapter 6). The observations of these stations between 1896 and 1905, among the first in southeast Africa, were published at Kalocsa by Fenyi. Johann Schreiber (1843-1903) worked as assistant to Fényi between 1891 and 1903, and collaborated in the solar observations and studies of the atmospheric electricity. With Fényi he developed an instrument to record the electrical discharges in the storms. They named the instrument "Gewitterregistrator". Schreiber also took care of the meteorological station and worked about the history of Jesuit astronomers.

Fényi developed an illness that affected seriously his eyes and in 1913 was succeeded as Director by Theodor Angehrn (1872-1952), a Swiss Jesuit who had studied at the University of Vienna and had been Fényi's assistant between 1906 and 1909. Fényi was able to continue his observations until 1918, when in that year the condition threatened to destroy his sight completely he retired to the villa of the College. There he continued to be intellectually active, preparing the publication of his work until his death in 1927. For a while Anghern continued Fenyi's work on solar prominences, but owing to his feeble sight he dedicated himself more to meteorological observations. Fényi had plans for moving the observatory outside the town, but these were never carried out. After the First World War in 1918, the foundation that supported the observatory economically lost most of its funds and its continuity was seriously

Figure 14. Haynald Observatory, Kalocsa, Hungary, with two rooms with conical roof for the telescopes.

threatened. Nevertheless, the observatory continued, though in a precarious state, even during and after the Second World War. In 1946 Mátyás Tibor (1902-1995), who would be the last Director, came to the observatory and was named Director in 1948. He had worked in the Vatican Observatory from 1937 to 1940, mainly on the observation of the Milky Way, and was later Professor of Cosmology in Collegium Martyrum Cassioviensis. Recognizing that most of the instruments were already obsolete, Tibor tried to acquire a new telescope to continue his astronomical work. The renewal of the observatory, however, was not possible because two years later, in 1950, it was confiscated by the Communist government and the college was transformed into a state school. The instruments and library of the observatory were taken to Budapest and distributed to various state institutions. As we will see, the same was to happen to the Jesuit observatories of Zikawei (China) and Belen (Cuba)

Vatican

The expropriation in 1870 by the Italian government of the Observatory of the Collegio Romano, which had been given the title of Pontifical Observatory, left the Holy See without its own astronomical institution since another pontifical observatory, created in 1827 by Pope Leo XII in Campidoglio was also expropriated. The interest of the Popes in astronomy played an important role in demonstrating that the Church is not uninterested in science and still less opposes it, as at that time it was accused of being from within rationalist ranks. In 1890 Pope Leo XIII founded a new pontifical observatory with the name of Specola Vaticana, directly dependent on the Holy See. At its inauguration in March 1891 the Pope made it explicit that " this plan is simply that everyone might see clearly that the Church and her Pastors are not opposed to true and solid science, whether human or divine, but they embrace it, encourage it, and promote it with the fullest possible dedication". The first director of the new Observatory was Francesco Denza, a Barnabite, former director of the Meteorological Observatory of Moncalieri. Ferrari participated in the Governing Board of the Observatory. The observatory was installed inside the Vatican in the Tower of Winds, built by Gregory XIII, on top of which a small dome was added. Among the first instruments were two Merz telescopes, one of 10.2 cm aperture and equatorial mounting, and the other of 10.6 cm aperture and altazimuth mounting together with a meridian circle by Stark. The observatory also had meteorological and magnetic sections. The first project in which the observatory participated was the international program known as "Carte du Ciel". This was a large project, proposed in 1852 by Admiral E. B. Mouchel, Director of the Paris Observatory. Its primary aim was to produce a photographic atlas of the sky, down to the stars of 14th photographic magnitude, and its second aim was to produce a catalogue of all stars down to 11th magnitude, known as the Astrographic Catalogue. Fifty six scientists from nineteen nations attended the Project's first meeting in Paris in April 1887, sponsored by the French Académie des Sciences. The Specola was one of the eighteen observatories participating in the project. To carry out the observational work, a photographic double equatorial telescope of 33 cm aperture, made in Paris by the brothers Paul and Prosper Henry, was installed in 1891 in a rotating dome of 8 m diameter on the Leonine Tower. By participation in this program, the observatory took the first step towards becoming an important institution on the international astronomical scene. Another important instrumental contribution was the installation in 1893 of a heliograph of 14 cm aperture, which produced images of the sun of 28 cm diameter on photographic plates, the first of its kind in Italy. After Denza´s death in 1894, Giuseppe Lais, an Oratorian, took charge of the observatory as Vice-Director for four years. Both Denza and Lais had worked with Secchi. After Lais the post was occupied for eight years by Angel Rodriguez de Prada, a Spanish Augustinian, who was more interested in meteorology and who partly neglected the astronomical work.

In 1906, Pope Pius X, at the suggestion of Cardinal Pietro Maffi, President of the Governing Board of the Observatory since 1904, designated Johann Georg Hagen (1847-1930), Director of the Observatory of Georgetown

Figure 15. Johann G. Hagen, Director of the observatories of Campion College, 1881-1888, and Georgetown University, 1888-1906, in the United States, and first Jesuit Director of the Vatican Observatory, 1906-1930.

University, as Director of the Specola Vaticana, the first Jesuit to occupy this post. Hagen had been working in astronomy, from 1880 in Campion College, Wisconsin, and then from 1888 in Georgetown, mainly in the observation of variable stars (Chapter 5). The appointment of a Jesuit as director of the Specola did not fail to bring out the usual reaction from the Italian anticlericals as evidenced by the letter published in a newspaper of Bologna with the title "The Jesuits invade Rome, in the astronomical observatory". From his first years in the Specola Hagen gave a great impetus again to astronomy and suppressed the meteorological and magnetic sections of the observatory. In 1909 he acquired a large telescope from Merz of 40 cm aperture and 6 m focal length with equatorial mounting. This telescope was installed in a new dome of 8.8 m diameter built on one of the towers of the old Leonine fortifications. In fact the whole length of about five hundred meters of the wall with all of its three towers was made available to the observatory. The other instruments and facilities were also moved to the new location, so that the old rooms in the Tower of Winds were definitely abandoned. Hagen took great care to create a good astronomical library and his first project was to complete the part of the

Astrographic Catalogue assigned to the Specola. This was finished in 1928 and published in ten volumes. It listed the brightness and positions of 481215 stars. For this purpose Hagen got help from the Sisters of the Istituto di Maria Bambina who, between 1910 and 1921, made most of the measurements with great care. The Specola was the fifth among the eighteen observatories which had taken responsibility for the mapping to finish its section. The same success was not repeated with the maps for the zone assigned to the Specola in the Carte du Ciel. When Hagen died only 107 of the 540 maps were published. Hagen also continued with his personal project of observation of variable stars mapping 484 variables with 579 stars charts and 24000 reference stars. His results were published in the six volumes of the *Atlas Stellarum Variabilium*, a work that brought him wide recognition. Rapidly, thanks in great part to Hagen, the Specola Vaticana had acquired a high reputation in international astronomical circles. At that time relations between the Vatican and the Italian government were still delicate. In 1922, on the occasion of the meeting of the International Astronomical Union in Rome, Hagen and Cardinal Maffi, representing the Specola, were present at the opening ceremony. Thus, thanks to astronomy, a Cardinal and the king of Italy, Vittorio Emanuele III, appeared side by side in public for the first time since the entry to Rome of the Italian troops in 1870.

In 1930 Johan Stein (1871-1951), a professor of mathematics and physics in Amsterdam and former collaborator of Hagen, was named Director. The increase of night lighting in Rome required the transfer of the observatory out of the city and in 1932 the decision was made to move it to Castelgandolfo. The new observatory was installed on the roof of the Papal palace in the Villa Barberini. In 1935 the work was finished with two domes, one of 8.5 m diameter where a new Zeiss equatorial telescope of 40 cm aperture and 6 m focal length equipped with a Graff photometer and a Danjon interferometer was installed and a second dome of 8 m diameter where a new Zeiss double astrographic telescope consisting in a refractor of 40 cm aperture and 2.4 m focal length and a reflector with a parabolic mirror of 60 cm diameter and focal length equivalent to 8.2 m was installed. Other complementary instruments were also installed, so that the new observatory was completely modernized. In the installation of the new equipment Paul Guthnick, director of the Observatory of Berlin, took a large part. Offices, work rooms and the library were installed on the fourth and fifth floors of the Papal Palace. Alois Gatterer (1886-1953) installed a very complete laboratory for stellar spectrography, with a GH Steinheil instrument of three prisms together with others by Zeiss and Halle. The inauguration of the new observatory at Castelgandolfo by Pope Pius XI took place on the 29th of September 1935. In his discourse the Pope praised astronomy as a science, which more than others, bring people to religious considerations and he gave the Specola the motto "Deum creatorem venite adoremus", inscribed in a marble plate on the wall of the large dome. On this occasion Pius XI formally entrusted the Specola to the Society of Jesus and this was accepted by the Father General Wladimir Ledochowski. Thus, the observatory became an official work of Jesuits. An international community was formed of Jesuits active in the Observatory that continues to this day.

Among the first tasks of the new observatory were the conclusion of the work of the *Atlas Stellarum Variabilium* started by Hagen and the conclusion of the part assigned to the Specola in the international project of the Carte du Ciel. In 1941 the astrographic telescope was transferred to Castelgandolfo and installed in a dome built in the gardens of the Villa Barberini. Interrupted during the war, the work on the Carte du Ciel was completed by the end of 1955, though the project itself was not finished until 1962, when the last zone was published. Another project was the study of the structure of our galaxy, the Milky Way, in which, as we have seen, Mátyás Tibor, who was to be the last director of Kalocsa, participated together with Josef Junkes (1900-1984), Walter Miller (1904-1973) and Jules de Kort (1910-1994). In 1935 the widow of Adrien Charles, Marquis of Mauroy donated her husband's rich collection of meteorites to the Specola. Little by little the collection has been enriched by new gifts and exchanges and it is still considered by specialists to be one of the richest such collection because of the quality and variety of its specimens. Today the collection consists of 1022 pieces of 449 different meteorites, with a total weight of 130 kilograms. It is arranged in a display case in the Museum on the ground floor of the Specola. In 1972 a small specimen of lunar rock, brought back by the Apollo XVII mission and presented to Pope Paul VI by President Richard Nixon, was added to the collection.

After Stein's death in 1951, he was succeeded by Daniel O'Connell (1896-1982) who was then director of the Riverview Observatory, Australia. Instrumentation was improved in 1957 with a new Schmidt telescope by Cox, Hargreaves and Thomson of London with a mirror of 65 cm diameter and focal length of 2.4 m; it was installed in a new dome of 8.5 m diameter in the gardens of the Villa Barberini. The telescope was adapted with three prisms for use in stellar spectroscopy and it was used from 1962 to 1982 to study the evolution of cumulus of stars by means of stellar spectroscopy and polarimetry measurements. O'Connell continued in the Specola the research he had began at Riverview on eclipsing binary stars. In this field he became a real expert and was twice elected president of the corresponding Commission of the International Astronomical Union. O'Connell was popularly known for his study of the effect at sunset and sunrise known as the "green flash". In 1958, he published a book with hundreds of color photographs of the "green flash", the first publication of its kind. In 1987 the Russian astronaut Georgy Grecko attracted by this study came to the Specola to obtain a copy of this book. Patrick Treanor (1920-1982) who was incorporated to the Specola in 1961

Figure 16. The Vatican Observatory at Castelgandolfo (top) and at Arizona (bottom).

became director in 1970. Treanor had obtained his doctoral degree in Oxford University and had worked in the observatories of Lick and Yerkes in the United States. He was a very fine astronomer and worked mainly on the spectra of stars, their polarized light and the interstellar medium. For the research in stellar spectrography new instruments were acquired from Jarrell-Ash of Boston and Bausch and Lomb of Rochester. In 1960 the first computer center was installed with an IBM1620 computer that was modernized in following years with an IBM5100 in 1978 and a Vax3100 in 1990 with all the necessary auxiliary equipment

Treanor died relatively young and was succeeded in 1978 by the present director George V. Coyne (1933-), who had for some years been a professor in the University of Arizona, Tucson and a member of the staff of the Specola from 1969. In 1980, owing to the worsening of conditions for astronomical observations in Castelgandolfo, Coyne established an agreement with the University of Arizona, whereby the astronomers of the Specola were guaranteed office space, services and access to the Steward Observatory telescopes. They also got access to other telescopes installed by the same

Figure 17. Jesuits attending the 3rd meeting of the International Astronomical Union at Leiden in 1928. First row: O'Connor (Stonyhurst), Stein (Vatican), Philips (Georgetown). Second row: Angehrn (Kalocsa), Rodés (Ebro), Lejay (Zikawei), Rowland (Stonyhurst) (EO).

University, the National Optical Astronomy Observatory and the Smithsonian Institute, in the desert around Tucson, where quality of the atmosphere and climate for astronomical observations was exceptional. The astronomers of the Specola in Tucson became known as the Vatican Observatory Research Group (VORG). Thus, after two periods in which the Specola carried out its work in the Vatican and then in Castelgandolfo, a third period began in which the observational and research work was to be carried out in Arizona. In order not to depend entirely on equipment from other institutions the Specola decided to install its own new telescope of special design, known as VATT (Vatican Advanced Technology Telescope) on Mount Graham. During its installation an unforeseen difficulty occurred due to a protest of a group of Native Americans who consider this mountain as a holy place and who objected that the telescopes would disturb the spirits. The Jesuits finally convinced them that there was nothing to fear in this respect. The spirits will have no reason to worry because of the presence of the telescope and the astronomers. Finally, the new telescope was installed in a new building with a rotating dome and was inaugurated in 1993. Observations began in 1995 with several research projects on the galactic and intergalactic medium. At present, all the observational work of the Specola is done in Arizona.

A recent new activity of the Specola concerns the philosophy and history of science and its theological implications. In 1987 the observatory, together with the Center for Theology and the Natural Sciences (CTNS) of Berkeley, California, initiated a series of research conferences in the interdisciplinary areas of science, philosophy and theology. This has proved to be an important activity due to the growing interest in the relations between science and religion. Up to the present time, four conferences have been held and their papers published in four volumes. In the educational field, since 1986, the Specola has organized the biannual Vatican Observatory Summer School in astronomy and astrophysics at Castelgandolfo, a month long school for promising young scientists, especially from developing countries, with lectures by leading astronomers. The observatory also publishes the *Studi Galileiani* a series of current research on Galileo and the Copernican controversy, a subject of continuing and great interest. At present about 13 Jesuits are actively working in the observatory on different programs, namely, theoretical studies in astrophysics and cosmology, extra-galactic research, the galaxy and galactic objects, and planetary sciences.

Cartuja

The end of the nineteenth century was a moment of great splendor in the history of Jesuit observatories. The Spaniard Luis Martín, elected General of the Society of Jesus in 1892, had a great appreciation for scientific studies among Jesuits and favored the creation of observatories. In Europe the observatories of Stonyhurst and Kalocsa and in the mission lands those of Manila, Belen, Tananarive and Zikawei were at their peak. In Spain, from about 1860, several small observatories, dedicated to meteorology, were created at colleges and faculties of philosophy for Jesuit students; but it was at the beginning of the twentieth century that two observatories, Cartuja and Ebro, were founded which became leading institutions in seismology and solar terrestrial relations. At that time in Spain there were only two astronomical observatories, the Naval Observatory of Cadiz, created in 1753 and moved to San Fernando in 1793, and that of Madrid founded in 1790. By 1900 the Naval Observatory of San Fernando also had sections of meteorology, magnetism and seismology. There were a number of meteorological stations, but no other geophysical observatories. The idea of creating an observatory in the Faculties of Theology and Philosophy in the neighborhood of Cartuja, Granada came first to Juan de la Cruz Granero (1849-1917), Rector and Professor of Cosmology, on the occasion of the sun's eclipse of 28 May 1900. The principal

Figure 18. Cartuja Observatory, Granada.

aim of the observatory was to contribute to the scientific formation of Jesuit students, but it was also intended to carry out relevant observational work. A donation by the Osborne family, who were owners of an important winery in Jerez de la Frontera and had a Jesuit son, made possible, in 1902, the construction of the building in the grounds of the Faculty of Philosophy. Designed by the architect Enrique Fort, the building, of attractive appearance, is in the form of cross with a dome of 8 m diameter resting on a circular tower in the center, and a Doric classical front with the inscription "Coeli enarrant gloriam Dei." From the beginning the observatory had three sections, namely, astronomical, meteorological and geodynamical or seismological. The astronomical section had an equatorial telescope of 32 cm aperture and 5.35 m focal length by Mailhat, Paris, and was installed in a rotating dome. In addition, it had two other telescopes of 16 and 10 cm aperture, a meridian circle and a spectroscope of 12 prisms from Grubb, Dublin. A second dome of 3 m diameter housed a photoheliograph of 9.4 cm aperture and 1.5 m focal length. G. Bigourdin, an astronomer of the Paris Observatory participated in the installation. The meteorological section had very complete instrumentation including Richard thermographs, anemometers and Ducretet barometers. The geodynamical section had two horizontal Stiatessi seismographs with Zöllner suspension and 208 kg mass, a Vicentini microseismograph and a vertical pendulum. The installation of the seismological instruments was supervised by Raffaello Stiattesi, director of the Observatory of Quarto Castello with the collaboration of Guido Alfani of the Osservatorio Ximeniano of Florence and José Algué of the Manila Observatory. The location of Granada in one of the most seismically active regions of Spain made this section, with time, the most important part of the observatory. This was the second seismographic station installed in Spain, after that of the Naval Observatory of San Fernando and its importance was recognized in 1904 by Professor Georg Gerland of the University of Strasbourg. The state seismographic stations did not begin operation until about 1910. Although the founder of the observatory was Granero, the first Director was Ramón Martínez (1847-1911), a professor of physics and chemistry. The following year the direction of the observatory was divided between Ramón Martínez, in charge of meteorology and seismology and José Mier y Terán (1878-1942), in charge of astronomy. Mier y Terán worked on the observation of solar activity, which was, as we have seen, a popular subject among Jesuit astronomers. He organized the observations by Jesuit scientists of the solar eclipse of 30 August 1905 in Carrion de los Condes, Palencia, with the participation of Fenyi and Angehrn from the Observatory of Kalocsa. As we will see later, this was part of a large program of Jesuit observers along the zone of totality that crossed Spain, coordinated by Cartuja and Ebro observatories.

In 1907 Manuel M. Sánchez Navarro-Neumann (1867-1941) took charge of the seismological section and introduced a fresh impetus. As we have seen in other cases, Navarro-Neumann soon became the key figure of the observatory and gave it great relevance in national and international seismological circles. In 1908 Ricardo Garrido (1878-1959) became director of the astronomical section and continued the solar observations, but this section soon began to decline. In 1915 Navarro-Neumann became director of both the seismological and astronomical sections and several Jesuits took charge of the meteorological section. In 1921, the Rector of the Faculties, Manuel Grund assumed the direction of the observatory with Navarro-Neumann as subdirector of seismology and Garrido as subdirector of meteorology and astronomy. This anomalous situation only lasted two years and in 1923 the directions were again separated until 1931, Navarro-Neumann remaining director of the section of seismology and at times of the whole observatory. While the astronomical section continued to decline, the seismographic station progressed. Under the direction of Navarro-Neumann, owing to its location and to his careful observations, Cartuja became one of the most important seismographic stations in Spain.

Navarro-Neumann is recognized as one of the most important Spanish seismologists of his time. His most important contributions were the first modern studies of the seismicity of Spain, with the first modern catalogue of earthquakes published in 1917 and 1920 and with many articles about the development of instrumentation and other seismological subjects, for example, one of the first determinations of earthquake energy release. From 1908 he kept modernizing the seismological instrumentation, designing and building new seismographs by himself. The first were mechanical instruments with smoked paper recordings, similar to those developed by Mainka in Germany, with vertical and horizontal components. In 1920 Navarro-Neumann built a large inverted pendulum seismograph similar to a Wiechert instrument with a mass of 3000 kg and a magnification of 1000, one the largest magnifications ever obtained by a mechanical instrument. Lack of funds did not allow him to acquired the new electromagnetic seismographs developed in 1910 by Prince Galitzin in Russia, so in 1924 he built by himself similar instruments, with many modifications and improvements for the vertical and horizontal components with 7.5 kg masses, galvanometers and photographic recording. With these instruments he obtained a performance of equal or even superior quality to that of the original instruments. In 1920 the seismographic station was declared of public service by the government of Spain. The expulsion of Jesuits from Spain by the first Republican Government in 1931 resulted in the exile of Navarro-Neumann to Portugal. The observatory was confiscated by the state and passed until 1938 under the control of the Instituto Geográfico Catastral, the Spanish institution which operates the network of state observatories. Once again a Jesuit observatory was confiscated, in this case only for a few years, by an unjust government decision.

After the Spanish Civil War the observatory was returned to the Jesuits and in 1940, Antonio Due Rojo (1898-1975) was appointed Director. In 1942 the observatory was associated with the newly created Consejo Superior de Investigaciones Científicas (CSIC; High Council of Scientific Research), the Spanish state research institution. During this time Due continued the seismological studies of Navarro-Neumann who, failing in health, died in 1941. However need of instrumental renewal and lack of funds produced a decline in the importance of the observatory. The astronomical instruments had not been renewed from its foundation and were already obsolete, and the seismographs built by Navarro-Neumann were no longer operational. In 1953 the observatory changed its name to "Observatorio Astronómico y Geofísico", but its importance continued to diminish as the state seismological stations increased in number and with more modern instrumentation. In 1965 Teodoro Vives, who had been trained in astronomical institutions in England, Germany and Italy, came to Cartuja as Subdirector and was appointed Director in 1967. Vives tried to give a new impetus to the astronomical section. He recognized the need to move the astronomical part of the observatory outside the city and built a small observatory in the Sierra Nevada, near Granada, at a height of 2506 m. This was the first observatory built in Spain at high altitude and its example would later be followed by other observatory built in the same Sierra

Nevada, in Calar Alto (Almeria) as well as the observatory built in the top of the Teide Peak (Tenerife, Canary Islands). In the Sierra Nevada Vives installed a Cassegrain reflecting telescope of 32 cm aperture, donated by the Observatory of Georgetown University, Washington. Jean Rosch astronomer of the Observatory of Pic du Midi, France helped in the installation of this telescope. Vives established cooperation agreements with the Royal Observatory, Greenwich, and the Max Plank Institut für Aeronomie at Lindau, Germany; the astronomers at these observatories were interested in the excellent conditions of the new observatory. This was the first time European astronomers appreciated the good climatic conditions in the high mountains of southern Spain for astronomical observations; this later led to the joint observatories of Almería, Tenerife and La Palma. These developments seemed to indicate a new bright future for Cartuja Observatory, but in 1971 Vives left the observatory and later also the Jesuit Order. Radical changes in the interests of the Jesuits in Spain, which placed a high priority on pastoral and social work, led to the final decision to close the observatory. In 1971 it was transferred to the University of Granada. The observatory adopted the new name of Observatorio Universitario de Cartuja and in 1987 the section of seismology became the Instituto Geofísico de Andalucía dependent on the University of Granada and the Regional Government of Andalusia; it continues today active seismological research. After nearly 70 years, the Jesuits abandoned this observatory, a pioneer in some aspects of the seismological and astronomical research in Spain.

Ebro

We have seen that the founding of the Cartuja Observatory had its origin in the great interest in Spain, at that time, in the scientific formation of Jesuit students. In the Jesuit Province of Aragón, which included Aragón, Cataluña, Valencia and the Balearic Islands, the Provincial Luís Adroer, conceived the idea of establishing, three laboratories, of physics, chemistry and biology, at the Faculty of Philosophy and Theology in Roquetas, Tarragona, near the Ebro River. Those of chemistry and biology, founded in 1904, were moved in 1914 to Barcelona and became scientific institutes for lay students. The organization of the institute of physics was entrusted to Ricardo Cirera (1864-1932), who from 1890 to 1894 had been in charge of the magnetic section of the Observatory of Manila (Chapter 7). Cirera proposed that, instead of an institute of physics, it would be better to create an observatory, dedicated mainly to the study of the relationship between solar activity and terrestrial magnetic and electric phenomena, a subject then of great scientific interest. The new observatory was to be divided into three sections: solar physics, geophysics and meteorology. The first of these was to study solar activity; the second was to concentrate on terrestrial magnetism, telluric electric currents and seismic movements, and the third was to focus on weather observations, with emphasis on the electric phenomena in the atmosphere. Cirera made intense preparations for his future work in the observatory, studying from 1900 to 1903 in Paris, Stonyhurst and Louvain and visiting the most important European observatories.

The site for the observatory was a hill behind the faculty building. There in 1903 ten small buildings were built for seismology, magnetic variations, magnetic absolute measurements, meteorology, telluric currents, atmospheric electricity, solar spectrography, offices and a workshop. Construction was supervised by Mariano Balcells (1864-1911) who had been an engineer before joining the Society of Jesus and who would take charge of the solar section until 1911. Construction was possible, thanks to the financial help from the Gil Moreno de Mora family. The geologist and astronomer, José J. Landerer was an active supporter of the observatory and he always kept closely in touch with it. In his will he left his library and archives to the observatory. The observatory was inaugurated in 1904 being given the name of Observatorio de Física Cósmica del Ebro (Observatory of Cosmic Physics of Ebro), though the public inauguration took place the following year, to make it coincide with the solar eclipse of 30 August 1905 with the zone of totality crossing Spain from north-west to south-east. On this occasion Ebro together with Cartuja organized a large program of observations with the participation of more than 70 scientists, most of them Jesuits from Spain, France, Germany and Italy, among them, Cortie from Stonyhurst, Esch and Baur from Valkenburg, Algué from Manila, and Fenyi and Angehrn from Kalocsa. It can be said that this was truly a Jesuit solar eclipse. Ten sites were occupied for observation from Gijon to Palma. Sixteen observers were located at Ebro Observatory and three other sites. Besides solar observations, magnetic and meteorological measurements were made. Observations were coordinated and later published by Ebro and Cartuja. The establishing of Ebro Observatory was welcomed by the international geophysical community, who admired its modern approach to the study of solar terrestrial phenomena. In 1910 the publication of the Bulletin of the Observatory began with separate sections for solar physics, geophysics and meteorology. From the time of its foundation the observatory has been recognized by Spanish public authorities as an institution of public interest, and in 1918 was declared to be a private learning institution (Entidad de Beneficiencia Particular Docente). In 1913 it acquired an independent public identity with its own foundation and separated from the Jesuit Faculty which moved to Barcelona in 1915.

Figure 19. Some of the buildings of Ebro Observatory. The dome houses the solar telescope (EO).

At the time of its foundation the observatory was very well equipped. The magnetic section had two pavilions: one for Mascart variometers with photographic recording and direct measurements of the Z and H components of the earth's magnetic field and the magnetic declination D, and the other for absolute measurements with Dover-Kew magnetometers for H and D, and an induction inclinometer by Schulze, Potsdam. The solar physics section was installed in a building in the form of a cross, with a rotating dome at its center. It had a telescope of equatorial mounting of 16.2 cm aperture by Mailhat adapted to take photos of the sun for the study of sun spots and prominences; a meridian circle, an Evershed heliospectrograph, the first in Spain, and a Rowland spectro-goniometer with photographic camera. The electric section was installed in a separate building with two Mascart instruments to measure the electric potential of the atmosphere, two Deprez-d'Arsonval galvanometers to measure telluric currents with two perpendicular lines in N-S and E-W directions with lengths of 1280 m and 1415 m respectively; an instrument to measure electric discharges in the atmosphere developed in Kalocsa and another to measure ionization of the atmosphere by Gerdien and Elster-Geitel. The seismological section had Vicentini seismographs of three components and two horizontal Grablovitz seismographs. The meteorological section had very complete equipment with Tonnelot and Fortin barometers, Richard barographs, an Angström piroheliometer to measure solar radiation, a Cornu photopolarimeter, anemometers, thermometers, etc. The observatory began as a collective effort of Jesuit scientists from different countries and some of them participated in the installation and calibration of the equipment. Among them, were Balcells in the solar physics section, Comellas who would later go to the Observatory of Manila in the astronomical section; the French Jesuits A. Willot and Etienne Merveille in the magnetic section; the German Jesuit Dressel from the observatory of Valkenburg in the electric section; Berloty, founder of the Observatory of Ksara, Lebanon, in the section of time determination; and Stein, from the Vatican Observatory who made the determination of the observatory coordinates. During the direction of Cirera until 1920, several Jesuits collaborated in the observatory for different periods of time. Balcells who was sent to the Massachusetts Institute of Technology, Boston to complete his scientific studies, and died there in 1911. This was a great loss for the observatory.

In 1914 Luís Rodés (1881-1939) arrived at the observatory and later, (1916 to 1919) received a sound scientific formation in astronomy and astrophysics in the United States at the universities of Harvard and of Chicago as well as at the observatories of Yerkes and Mount Wilson. In 1920 he was appointed director, succeeding Cirera. Rodés expended his tireless energy on the development of the observatory and its complex and many sided program of helio and geophysical observations for nearly twenty years, of all of which he was the soul and for most of which he was personally responsible. His main work was on the various forms of solar activity, especially on sunspots and prominences and their influence on the terrestrial magnetic and electrical fields. He tried to extend this influence together with that of the moon to the occurrence of earthquakes, a controversial subject. In 1927 he published his book on astronomy *El Firmamento* which was very popular and a second edition of which was published in 1939. The observatory maintained a policy of updating its equipment. In 1914 substituted new Mainka seismographs, a vertical of 300 kg mass and horizontals of 150 kg masses each, for the old seismographs. The personnel of the observatory changed with new persons who took charge of the different sections. In 1925 Ignacio Puig (1887-1961) was named Vice-Director until 1934 when he went to Argentina where he founded the Observatorio de Física Cósmica de San Miguel, an observatory similar to Ebro (Chapter 6). In 1930 the 25th anniversary of the observatory was celebrated with the inauguration of the new Lánderer building for the library, museum and archives. In 1931 when the Jesuits were expelled from Spain, the observatory, according to the statutes of its foundation, passed to the control of the Bishop of Tortosa, Esteban Bilbao, who confirmed Rodés as Director. In this way the observatory continued its work under Rodés' direction during the difficult years from 1931 to 1939. From 1931 to 1936 only three other Jesuits with Rodes, kept the observatory functioning, though with great difficulty. After the beginning of the Spanish Civil War in 1936, Rodés remained the only Jesuit in the observatory. In 1938 in the last days before Franco's advancing army entered Roquetas, the observatory was occupied by the retreating republican soldiers and most of its instruments were packed and taken to Gerona, by order of the Republican Government. After the end of the Civil War, in 1939, Rodes retired to Majorca for a rest but died there a few days later.

In 1939 Antonio Romañá (1900-1981) was appointed Director. He had been at the observatory, since 1934, in charge of the sections of magnetism and electricity. Romañá had obtained his doctoral degree in exact sciences at the University of Madrid and worked for some time in the observatory of the University of Vienna with Professors Graff, Berheimer and Krumpholz. His first task was to find the instruments which had been taken from the observatory and reinstall them. In 1939 the sections of seismology, meteorology and solar physics began to function, while the functioning of those of magnetism and telluric currents was delayed until 1943. Publication of the Bulletin was resumed again in 1942. This required an enormous effort given the precarious situation of Spain after the war. Romañá was very influential in the setting up of the CSIC, the research council created after the war by Franco's government; he held an important position in the CSIC. He established an association of the observatory with the CSIC which assured its financial support and made possible the renewal of the instrumentation of the different sections. Thus, in 1950 new QHM and BMZ magnetometers and in 1954 La Cours variometers with photographic recording were installed. The same year a new line of ionospheric research was established, with the installation of an ionosonde made in France, the first to arrive in Spain. The observatory significantly participated in the programs of the International Geophysical Year, in 1957, especially in geomagnetism and aeronomy. As a result of this participation a new Lyott spectroheliograph was installed to replace the old Evershed instrument. Romañá enjoyed a great prestige in national and international scientific circles. He held office on several commissions of the IAGA (International Association of Geomagnetism and Aeronomy) and was President of the Spanish committee of the International Geodynamic Project. He was very keen that the observatory maintained good international relations with other observatories and that it participated in international projects. The observatory was always represented at meetings of the International Union of Geodesy and Geophysics (IUGG) and of the International Astronomical Union (IAU). About this time an enthusiast group of young Jesuits joined the observatory, among them José O. Cardús (1914-) and Eduardo Galdón (1915-) some of whom eventually went to work at the Vatican Observatory. Ebro Observatory participation in international programs led to its designation in 1958 as the base of the IAGA Commission for rapid magnetic variations. Upgrading of magnetic equipment continued: in 1960, new Ruska and Askania magnetometers replaced the old equipment and, in

1966, a new generation instrument, an Elsec proton magnetometer, and a new Magnetic AB ionosonde were acquired. Seismological instruments were also changed with the acquisition of Benioff short period and Sprengnether long period seismographs. A new program of measurement of the electron content in the ionosphere began in 1964 in collaboration with American institutions, the National Aeronautic and Space Administration (NASA) and the Air Force Office of Scientific Research (AFORS). Between 1968 and 1970 a new group of Jesuits arrived at the observatory, Antonio Echevarría, who had been Director of the Observatory of Sucre, Bolivia, Luis F. Alberca (1930-) and Ernesto Sanclement (1933-), who took charge of the sections of seismology, the ionosphere and geomagnetism, respectively. These were to be the last Jesuits to be incorporated into the Observatory.

In 1970 Cardús was named Director and continued the policies begun by Romañá, who continued in the observatory for ten more years. The observatory was divided now into five sections, namely, solar physics, the ionosphere, geomagnetism, meteorology and seismology and three services, namely, the library, computation, and time. In 1972 a new solar radiotelescope with three channels of 6 cm, 11 cm, and 21 cm donated by NASA and, in 1973, a polarimeter to measure the total content of electrons using artificial satellite signals were installed. In 1967 the first computer, a PDP-8, was acquired; it was replaced in 1980 by an Interdata and more recently by several workstations. Seismological instrumentation was also upgraded, by means of a new short period station in 1987, in collaboration with the Instituto Geográfico Nacional, and more recently by a digital broad-band seismographic station; these stations were both installed outside the grounds of the observatory. During this time, the observatory continued its participation in several international projects and Cardús held office at the IAGA for many years. In 1985 Alberca was appointed Director and non-Jesuit scientific personnel began to be employed at the observatory. Since 1996 the Observatory has been associated to the newly formed private Universidad Raimund Lull of Barcelona. Recently, a shortage of Jesuits dedicated to the work in the observatory has led to the establishment of a new organizational status in order to ensure the future of the observatory, even in the absence of Jesuits. A foundation was created in 2000 with the participation of several Spanish state and private institutions, which will contribute the necessary financial support and in 2001 a layman was named director. Today only three Jesuits, Cardús, Alberca and Sanclement are actively working at the Observatory. Their replacement is not very likely and in a few years there will be no Jesuits present at the observatory. A token presence of the Society of Jesus in the foundation will be all that will be left of a glorious past of nearly a century of Jesuit scientific work in Ebro Observatory.

Valkenburg

As we have seen, the origin of the observatories of the Collegio Romano, Cartuja and Ebro was in part motivated by the teaching of science courses in the faculties of philosophy for Jesuit students. Other smaller astronomical observatories were also established in Europe for this purpose. The most important of these observatories was that of the Faculty of Philosophy and Theology in Valkenburg, Holland, founded by the German Jesuits in 1894, after they were exiled from Germany by Bismarck's government in 1872. The observatory was included in the original blueprints of the College and was built in the northwest corner of the part of the building allocated to the Faculty of Philosophy, with a tower 30 m high with a rotating dome of 5 m diameter. The main instrument was a telescope by Saegmueller and Clacey of 23 cm aperture and 2.8 m focal length with equatorial mounting. The telescope was bought in Chicago by Hagen who was then Director of Georgetown Observatory in Washington and who came to Valkenburg to supervise its installation. Other instruments were a telescope by Floess, Vienna, of 15.3 cm aperture and altazimuth mounting, a Breithaupt theodolite, chronographs, and spectrographs. The observatory was inaugurated in 1896 and was used both for teaching and for astronomical observations.

The most active directors of the Valkenburg Observatory were Joseph Hisgen (1868-1910) from 1897 to 1902, Alfred Baur (1872-1912) for only two years from 1902 to 1904 owing to his poor health, and Michael Esch (1869-1938) director from 1906 until 1938, with an interruption from 1908 to 1918 when he studied mathematics and astronomy in Vienna and taught astronomy in Innsbruck, Austria. The principal research line of the observatory was the observation of variable stars in collaboration with Hagen, who from 1904 was director of the Vatican Observatory. Hisgen, Esch and Baur participated in this work, between 1896 and 1919. Hisgen collaborated in two volumes and Esch in four volumes of Hagen's *Atlas Stellarum Variabilium.* Esch continued the observations alone from 1919 until 1937. In 1901 a meteorological station was established; its observations were sent to the Royal Meteorological Institute of Holland in Utrecht.

Jesuits of the Faculty of Philosophy of Valkenburg had a great interest in the natural sciences besides astronomy, especially in physics and biology. Francis X. Kugler (1862-1929) was famous for his studies of the Summerian and Babilonian astronomy. Among the professors of physics Ludwig Dressel (1840-1917), and Theodor Wulf (1868-1946) were outstanding. Dressel was the author of a very popular textbook of physics *Elementare Lehrbuch der*

Figure 20. Jesuits attending the 7th meeting of the International Astronomical Union in 1948. First row: Junkes (Vatican), Bertiau (Vatican), Treanor (Oxford, later Vatican), O'Connell (Riverview later Vatican), Miller (Vatican). Second row: de Kort (Vatican), Stein (Vatican), Romañá (Ebro), Plassard (Ksara) (EO).

Physik nach den neusten Anschuauungen, which went to four editions between 1895 and 1913. Wulf published several books about physics, among them the textbook *Lehrbuch der Physik* in 1926 and 1929, and an early book on Einstein's theory of relativity in 1921. Wulf also developed a very sensitive electrometer to measure the radiation in the atmosphere which was used with some modifications by Werner Kohlhörster, a professor in Potsdam, in his pioneer work on cosmic rays. In 1870 Dressel and Wulf went to Quito, Ecuador to take charge of the Polytechnic School founded by the President García Moreno, returning to Germany in 1876 after the death of the president. Dressel and Baur spent some time at Ebro Observatory in 1905 and 1908 respectively and Wulf and Esch, as we have seen, took part in the observation of the solar eclipse of 30 August 1905 in Spain.

The observatory was closed in 1941 when German SS troops took over the Faculty, after the invasion of Holland. The building was used first as an SS Reichschule, and after the war as a hospital by the U. S. Army, subsequently as a prison and later for some time by a religious congregation. Nothing is known of the fate of the astronomical instruments. The Jesuits never returned to Valkenburg and the Faculty of Theology was installed in Frankfurt in 1950. After the war there was never any attempt by the German Jesuits to establish an observatory, they apparently having lost their interest for the natural sciences.

Other astronomical observatories

Other minor astronomical observatories were established in Europe in several faculties of philosophy, seminaries and colleges. The first was in the College Notre-Dame de la Paix in Namur, Belgium, where in 1838 Antoine Maas (1795-1879) installed a physics laboratory with two telescopes, one by Dolland and the other by Merz of 9 cm aperture, which were used by Jesuit students of philosophy. In 1857 a meteorological station was added. When the faculty of philosophy was moved to Louvain in 1867 the telescopes were also moved there. The observatory, however, continued and in 1880 Victor Van Tricht (1842-1897) carried out some meteorological work which he published in the *Annales de la Societé Scientifique de Bruxelles*. In 1883 a tower with a dome was added to the college building in which a telescope by Frères Henry, Paris, of 20 cm aperture and 1.20 m focal length was installed which was used by Van Tricht until 1886. In 1912 the observatory organized the observation of the solar eclipse of 17 April, in which several Jesuits participated, among them Wulf from Valkenburg. The meteorological observations continued until 1940, when the observatory was closed. Another astronomical observatory was founded in Belgium in 1884 in the Jesuit Faculty of Philosophy and Theology in Louvain. The observatory was located in a tower with a rotating dome built in the new building in which an equatorial telescope of 15 cm aperture and a meridian circle of 7.5 cm aperture were installed. The observatory had auxiliary equipment and was used mainly for student demonstrations. In 1929 the faculty was moved to Eegenhoven and the observatory was also installed there. There were directors assigned to the observatory from 1887 to 1940. In 1958 the faculty was moved again to a new building in Heverlee, outside Louvain, where a new observatory was built. The observatory served primarily for teaching purposes and no directors were appointed. The professors of astronomy in the faculty of philosophy and Jesuit students used the observatory until about 1970 for observations and demonstrations. Since that time the observatory has been used sporadically by students of the University of Louvain.

The two observatories established in Austria and Holland in a college and a seminary are a different story. The College Stella Matutina in Feldkirch, Austria, was founded in 1856; a Jesuit college had existed there from 1645 to 1773. The college program emphasized the humanities and classical studies, but it also had science classes. The observatory began with meteorological and astronomical observations in 1889. The observatory had a Merz telescope with equatorial mounting of 15 cm aperture and 2 m focal length and a spectroscope which were used for demonstrations for the students. It was unusual for a secondary school to have such instruments. There were no directors appointed to the observatory, but some Jesuits were assigned to the college as a teacher of astronomy. Joseph Paffrath (1854-1935), a teacher of mathematics, studied the climate of the region using the meteorological observations. The college was seized in 1936 by the Nazi government but returned to the Jesuits after the war. The college was closed in 1979 and the telescope sold to an amateur astronomer. The diocesan seminary in Oudensbosch, Holland was put under the care of the Society of Jesus in 1878. In 1891 a new building was added to the seminary having a tower and a dome where a telescope of 8 cm aperture was installed. The observatory also had a meteorological station where measurements were made three time a day and were sent to the Meteorological Institute in Utrecht. Among the professors of mathematics, astronomy and cosmology, those who deserve mention are Victor Becker (1863-1898), professor from 1878 to 1898, and Henri Bolsius (1852-1924) the last one from 1901 to 1915. The Jesuits left Oudenbosch in 1929.

Meteorological stations

We have seen that meteorological observations were carried out in most of the observatories we have already described. Besides these, a number of small observatories were established that were dedicated exclusively or mainly to meteorology. It has to be born in mind that, in the nineteenth century, state meteorological services with networks of stations were not yet common in Europe and meteorological observations were made at observatories run by different institutions. During that century meteorology moved from purely observational work to the modern approach of trying to understand the physics of the atmosphere. The largest group of Jesuit meteorological stations was installed in Spain in faculties of philosophy and secondary schools. The first was installed in 1860 in the Jesuit Faculty of Philosophy and Theology of San Marcos in Leon which had been founded one year before. The station was at that time one of 28 which existed in Spain. The observations were sent to the Spanish central office and published in the *Anuario Estadístico* from 1865 to 1867. The Director of the station from 1863 to 1868 was Antonio Cabré (1829-1883), a professor of physics and astronomy who had been the first director of the Belén Observatory in Havana, from 1857 to 1860. The station was closed together with the faculty in 1868 when Jesuits were expelled from Spain by the Provisional Government installed after the exile of Queen Isabel II. After the return of the Jesuits to Spain with King Alphonse XII, another meteorological station was installed in the Faculty of Philosophy and Theology of the Province of Aragón in Tortosa, Tarragona, in 1880. The same year it began observations of temperature, atmospheric pressure and rainfall. Observations were made by Jesuit students under the direction of the professors of science. In 1904 the faculty was moved to Roquetas where the Ebro Observatory was created and where the records of these observations are kept. The meteorological observatory established in 1880 in the Colegio Apóstol Santiago, a secondary school, in La Guardia, Pontevedra near the River Miño was more important. The observatory was installed in a hexagonal pavilion in the gardens of the school. It had a very comprehensive instrumentation, and observations were made twice a day and sent to the Instituto Central Meteorológico in Madrid as well as being published in its annual bulletin from 1883. From 1888 to 1906 Baltasar Merino (1845-1917) was director; he published several studies about the climate of the region and about the storms on the western coast of Galicia, based on the observations made at the observatory from its foundation. Merino is better known for his pioneer studies of the flora of Galicia. The observatory was closed in 1916 when the school was moved to Vigo. A similar meteorological station was installed in 1883 in the Colegio de Nuestra Señora de la Antigua in Orduña, Vizcaya, founded in 1870. Observations were sent to the Instituto Central Meteorológico and published in its annual bulletin until 1931, when the station was closed with the exile of the Jesuits.

Another observatory established in a Jesuit faculty of philosophy and theology was that founded in Oña, Burgos, in 1882, in the Faculty of Philosophy and Theology of the Jesuit Province of Castilla, established in an old Benedictine monastery in 1880. Meteorological observations began in 1882 and were sent to the Instituto Central Meteorológico in Madrid and published in its annual bulletin until 1931. In 1885 this was one of the 64 meteorological stations in Spain. The directors of the observatory were the professors of science in the Faculty of Philosophy. One of them, Bonifacio Fernández Valladares (1851-1928) published, in 1896, a climatological study of the region from observations from 1882 to 1895. In 1905 Fernández Valladares and other Jesuits from Oña participated in the observation of the solar eclipse of 30 August together, with the team of the Ebro Observatory. As a result from this interest in astronomy, in 1906 a telescope of 20.3 cm aperture and 3 m focal length with altazimuth mounting was installed in a rotating dome. Although the observatory received then the name of Observatorio Astronómico y Meteorológico interest in astronomy did not last. Meteorological observations continued until 1931 when Jesuits had to leave Spain. After the Civil War they returned to Oña and the meteorological observations continued between 1942 and 1967, although no directors were assigned. In 1967 Oña was closed and the Faculty was moved to Bilbao. A parallel case was that of the meteorological station installed in 1900 in the Seminario and afterwards the Universidad Pontificia de Comillas, Santander, on the northern coast of Spain. Closed in 1931 the station was opened again after the Civil War from 1942 until 1967 when the university was moved to Madrid. The professors of science in the

Faculty of Philosophy were in charge of the station. Observations were sent to the Servicio Meteorológico Nacional.

An interesting development of Jesuits involvement in meteorology in Spain was the installation in several secondary schools of a small network of stations in 1902. The origin seems to have been a visit of José Algué, Director of Manila Observatory to Father General Luís Martín in 1901, in which Algué proposed the creation of a network of Jesuit meteorological stations in Spain. As we have already mentioned, Father Martín was very interested in the scientific work of Jesuits and fostered the creation of observatories. The central station was established in 1902 in the Colegio de Nuestra. Señora del Recuerdo, in Chamartín, Madrid. Its first Director was José A. Pérez del Pulgar (1875-1939) who later was the founder and Director of the Instituto Católico de Artes e Industrias (ICAI) in Madrid, an engineering school of great prestige which is now part of the Universidad Pontificia Comillas. The network was formed by eight stations in the Jesuit schools of Puerto de Santa María, Málaga, Orihuela, La Guardia, Oña, Orduña, Comillas, and Gijón and in 1905 one more in Villafranca de los Barros. Between 1903 and 1906, the central station published a joint monthly bulletin with the observations of all stations. The network did not last long and stopped functioning in 1906, though some of the stations continued as individual stations. The meteorological station in Chamartin continued until 1931 when the Jesuits were expelled from Spain. Besides the normal observations, the observatory also made measurements of the ozone content in the atmosphere.

In Italy the Jesuits operated two meteorological observatories. The first was the Osservatorio Tusculano, installed in the Collegio de Mondragone, founded in the old Villa Borghese, in Frascati, in 1865, and originated in a proposal by Secchi. He considered this observatory necessary in order to study the weather in the Lacio region. Observations were started in 1868 but the observatory was formally established in 1879 by Felice Ciampi (1826-1889) who had been Director of Belen Observatory in Cuba and later Professor of Physics in the Collegio Romano where he collaborated with Secchi. From its foundation the observatory had very comprehensive meteorological instrumentation. From 1891 to 1894 Giovanni Egidi (1835-1897), a member of the Accademia de Nuovo Lincei was director. He was also interested in seismology; he designed a seismometer which was installed in 1888 and collaborated from 1872 with Michele S. de Rossi, one of the pioneers of Italian seismology, Director of the seismological observatory of Rocca di Papa and founder of the Bolletino del Vulcanismo Italiano. In 1908 two seismographs were installed. The directors of the observatory were the Jesuit teachers of science in the college and they changed every two or three years. The observatory was closed in 1944. Another meteorological observatory was established in 1915 in the Collegio

Pennisi in Acireale, Catania, Sicily which lasted until the school was closed in 1978.

The Jesuits operated two meteorological observatories in Malta. The first was established in 1882, in the Sacred Heart Seminary, a diocesan seminary entrusted to the Society of Jesus in 1866 on the island of Gozo. Francesco Denza who had, as we have already seen, founded a meteorological observatory in Moncalieri, south of Turin and who in 1890 became the first Director of the Specola Vaticana was involved in the installation. The observatory was installed on the roof of the seminary with a very comprehensive set of meteorological instruments and it was associated with the Rete Meteorologica Italiana. Directors of the observatory were the professors of science in the seminary. In 1909 Jesuits left the seminary. The observatory continued for a short time and was later closed. The second observatory was established in Saint Ignatius College in Malta. The college had been founded in 1877 by the Jesuits of the English Province. The observatory was installed at Perry's requests and their observations were published in the Stonyhurst bulletins. The observatory was closed together with the college in 1907.

In 1893 on another small island, the island of Jersey in the English Channel, near the coast of France, Marc Dechevrens (1845-1923) after quitting as director of Zikawei Observatory in 1887 for health reasons, founded a meteorological observatory. There, in 1880, French Jesuits, after their expulsion from France by a decree of the anticlerical government, had established the Maison Saint Louis, a college for the study of philosophy. Pierre Teilhard de Chardin (1881-1955) studied philosophy there from 1902 to 1905, but there is no evidence of his having contact with the observatory or with Dechevrens. The most prominent part of the observatory was an iron tower 50 m high. Construction of this tower was difficult because of the strong winds in the Channel. A number of engineering firms were consulted, but finally the project went to T. Seyrig of the Société Anonime de Construction et des Ateliers de Willebroeck, Belgium at a cost of 31500 francs. In the tower a special instrument designed by Dechevrens to measure the horizontal and vertical components of the wind, called a universal anemometer, was installed. The observatory also had a variety of other meteorological instruments. During Queen Victoria's Diamond Jubilee in 1897 the tower was decked with lights over its whole length and a searchlight installed at the top which could be seen for miles around. The one-story observatory building contained the library, recorders, offices and a meridian circle. In 1897 instruments to measure magnetic inclination and declination were added and, in 1936, a Mainka seismograph, donated by the Institute de Physique du Globe of Strasbourg. From 1894 the observatory published an annual bulletin. In this observatory Dechevrens continued his studies about the movement of air masses, the vertical component of the wind, and the distribution of temperature and pressure with height. After Dechevrens' death in 1923, some instruments were sent to Zikawei and in 1929 the tower was dismantled, but observations continued. In 1934 Charles Louis Rey (1897-1981) was appointed director; he had worked with Dechevrens between 1917 and 1921 and was at Tananarive Observatory, Madagascar, during two periods, 1921-1924 and 1929-1932. He continued the work of the observatory during the Second World War despite the German occupation of the island. After the war the observatory passed to the control of the Meteorological Service of the United Kingdom with Rey as director. In 1954 Maison Saint Louis was closed and moved to France and the building and land were sold in 1974, but Rey continued at the observatory until 1979. At that time the observatory passed into the control of the Jersey Meteorological Department and is still in operation. The observatory continues today an important and uninterrupted series of meteorological observations, from its foundation to the present, a period of more than one hundred years.

Seismographic stations

We have seen how in several Jesuit European observatories seismological instruments were installed. The first one was the seismometer built by Egidi and installed in 1888 in the Osservatorio Tusculano. Seismographs were installed in Cartuja (1902), Ebro (1904), Stonyhurst (1909), Kalocsa (1920) and Jersey (1936). About this time the earliest seismological instruments were being developed by J. Milne, F. Omori and E. Wiechert. Cartuja, located in the seismically active area of south Spain, became the most important Jesuit seismographic station in Europe. Curiously, two seismographic stations were installed in Ireland, a non-seismic area. The first was in Mungret College, a secondary school near Limerick founded in 1892. In 1908 a small building was erected in the college grounds where two horizontal Bosch-Omori seismographs were installed. In 1911 William J. O'Leary (1869-1939), a teacher of physics since 1905, built a seismograph of his own design, with an inverted pendulum supported by three cables and with a mass of 600 kg, the movement of which was recorded on two horizontal components on smoked paper. In 1915 O'Leary was named Professor of Physics and Mathematics in the Jesuit Faculty of Philosophy established in Rathfarnham Castle in 1913. Thither he brought his seismograph from Mungret College, and built another of the same type with a larger mass of 1500 kg. Both instruments were installed in a small building. In 1929 O'Leary went to Australia and became director of Riverview Observatory.

The seismographic station of Rathfarnham Castle continued active and in 1932 a Milne-Shaw seismograph with photographic recording was installed. In 1935 Richard E. Ingram (1916-1967) took charge of the observatory. He had studied mathematics in Dublin and Baltimore and had specialized in seismology at the California Institute of Technology in Pasadena. In 1949 he returned to the observatory and installed a short period seismograph in 1950. He took care of the observatory and was Associate Professor of Mathematics in the University College of Dublin. He published several articles on theoretical aspects of seismology and collaborated in the edition of William R. Hamilton's mathematical work. After his death in 1967 the observatory was closed. Another seismographic station was installed during the short period from 1924 to 1928 by Peter Gärtler (1889-1969), a teacher of mathematics and sciences in the College Seminary of Travnik, in Bosnia Herzegovina.

Chapter 5

NORTH AMERICA: THE JESUIT SEISMOLOGICAL ASSOCIATION (1814-2000)

Jesuit observatories in the United States and Canada had the common characteristic that most of them were installed in colleges and universities and only a few, as in Europe, were in faculties of philosophy for Jesuit students. Those dedicated to astronomy were created in the second part of the nineteenth century and, with the exception of the Observatory of Georgetown University, were small and used more for demonstrations for students than for scientific work. Among theses were the observatories of Creighton College, Marquette University and the University of Santa Clara. Meteorological observatories also began very early, but never became very important. Those established in John Carroll University, University of Santa Clara, and Saint Louis University will be mentioned. The most important contribution of Jesuit observatories in North America was in the field of seismology and is associated with the Jesuit Seismological Service (JSS) and, later, with the Jesuit Seismological Association (JSA). This was a cooperative effort of a number of Jesuit colleges and universities which formed a network of seismological stations. The first central station was in Cleveland and was later in Saint Louis. As we shall see, a good number of American Jesuit seismologists working at the JSA stations made important contributions to this science in the first half of the twentieth century.

Georgetown

The first and most important Jesuit astronomical observatory in North America was the Observatory of Georgetown University in Washington founded in 1844. The origin of Georgetown University had an interesting link with Jesuit work in the observatories of the Society of Jesus in the seventeenth and eighteenth centuries which we have seen in chapters 2 and 3. This link was Giovanni Antonio Grassi (1775-1849), an Italian who had entered the Society of Jesus in 1799 in Italy, before its official restoration in 1814, and afterwards joined other Jesuits in Russia where the Society had never been suppressed. Former Jesuits in China wrote to the Jesuits in Russia, asking them to send some persons to continue the work of Jesuit astronomers in the Imperial Observatory of Beijing. Grassi, who was rector at the College of Polocz (Polotsk, Belarus), was selected with two other Jesuits. Together they began a long journey which took them first to Saint Petersburg and, after travel through Russia to China proved to be impossible, to Finland, Sweden, Denmark, England and Lisbon. In Lisbon they spent two years studying astronomy and mathematics, still hoping to reach China. Afterwards they went to Stonyhurst, where they continued their studies. In 1810 they realized that their hope of reaching China had to be abandoned. Grassi was sent to America as a professor of mathematics at Georgetown University, founded in Washington in 1789; he was Rector there from 1812 to 1817.

The idea of establishing an astronomical observatory in Georgetown University originated with James Curley (1796-1889), a professor of science from 1831. In 1841 the university authorities accepted Curley's proposition and took the decision to establish an observatory. In 1843, when consulted about the matter, the Jesuit Father General Jan Roothaan wrote indicating that he was not in favor of the observatory because of the great expense involved. Curley did not interpret the letter as negative, maybe because it was written in Latin, and wrote back: "It was a pleasure to me, when after recommending it to our Lord, we received your approbation and permission to build it (the observatory)". He went on to ask for some contribution, "some gift of your kindness for our observatory". Roothaan's answer was to emphasize his displeasure about the construction of the observatory and the fact that he never had given his permission ("Equidem significavi, displicere mihi constructionem speculae, sed propterea minime revocavi permissionem quam scilicet nunquam dedi"; "Indeed, I indicated that I was not pleased with the construction of the observatory; hence, I in no way revoked my permission because I never gave it"). He added that, naturally, he was not ready to contribute financially to the observatory with funds which could better be used in other works. Despite of all this, the view in the United States was different from the view in Rome and the project went forward. The building with three stories and a large dome as well as all the necessary installations were finished in 1844. The observatory was inaugurated the same year with Curley as the first Director. The final costs of building and equipment were about 27000 dollars, paid by a prominent Baltimore family, the Jenkins, whose son was a Jesuit. The first instruments were a transit telescope of 11 cm aperture by Ertel and a meridian circle of 10 cm aperture by Simms. In 1849 a telescope of 12 cm aperture and 2 m focal length with equatorial mounting by Troughton and Simms was acquired. Georgetown was the fourth astronomical observatory established in the United States, only eight years after the first one in Williamstown, Massachusetts, and was contemporary with the Naval Observatory, also in Washington.

In 1849, as we have already mentioned in dealing with the Osservatorio del Collegio Romano, Francesco de Vico and Benedicto Sestini arrived at Georgetown Observatory after being expelled from Italy (Chapter 4). De Vico died in London the same year while he was buying equipment for the observatory. Sestini remained permanently at Georgetown working at the

Figure 21. Georgetown Observatory, Washington, D.C. (GUA)

observatory and teaching physics at the university. The same year Secchi also spent some time studying physics at the university and collaborated at the observatory. Curley continued as Director until in 1888, when Johann Hagen arrived and succeeded him. Hagen had established a small observatory in Campion College, Prairie du Chien, Wisconsin, where he had begun his observations of variable stars, as we shall see. In 1891 Hagen installed a larger telescope of 30.5 cm aperture with equatorial mounting by Fauth in a new dome, and an additional building was added. He hesitated whether to dedicate himself to the study of variable or double stars until he was advised by Edward Pickering, Director of Harvard Observatory, to work on variable stars, because the resolution of his telescope was not good enough for observation of double stars. Some observatories in the United States had already installed refracting telescopes with apertures between 60 cm and 100 cm, suitable for studying double stars. Hagen, then, continued his observations of variable stars and in 1899 began the publication of the first volumes of his work *Atlas stellarum Variabilium.* In this work he engaged the help of Hisgen, Esch and Baur from Valkenburg (Chapter 4) and of Edmund Goetz from Bulawayo (Chapter 7). Hagen succeeded in attracting other Jesuit astronomers to work with him. In 1890 George A. Fargis (1854-1916), William Rigge and John T. Hedrick (1853-1923) arrived at the observatory and, in 1893, José Algué, the future Director of the Observatory in Manila (Chapter 7). Algué remained for two years and worked on the design of a new reflecting telescope, before returning to Manila. Fargis, Algué, Hedrick and Rigge worked with Hagen on the photographic observation of stars, which was then a new and rapidly developing astronomical technique. Fargis and Hedrick developed between 1889 and 1894 an instrument they named a "photochronograph," to determine the exact time photographically by means of a zenith telescope. In 1906 Hagen was named director of the Vatican Observatory, as we have already seen (Chapter 4), and was succeeded by Hedrick, who was director until 1916.

In 1927 Paul A. McNally (1890-1955) was assigned to the observatory and the following year was appointed Director of the Astronomical Section. McNally participated in astronomical expeditions to observe solar eclipses in 1932 together with other five Jesuits to Freiburg, Maine, in 1936 to Siberia, in 1937 to Canton Island, South Pacific, and in 1940 to Brazil. For some years, during the Second World War, astronomical work declined until in 1948 Francis J. Heyden (1907-1991) took charge of the astronomical section and gave it a new impulse, establishing a graduate program in astronomy in the university. He had previously been assigned to Manila Observatory, but this observatory was destroyed during World War II and after the war he remained at Georgetown. Helped by graduate students in astronomy, Heyden launched research projects in solar spectrography, radio astronomy and the study of planetary atmospheres. Heyden was considered for the directorship of the Vatican Observatory when Stein died (Chapter 4), but the President of Georgetown had the appointment blocked, commenting that Heyden had too many graduate students. This indicates the importance that astronomy had in the university at that time. Some of these students were Jesuits who later went to work in the Vatican Observatory. Heyden participated in scientific expeditions to observe solar eclipses in 1947 in Brazil and in 1948 in China. Late in the 1960s the university lost interest in astronomy and in 1968 decided to close the department of astronomy and the observatory. The first commission appointed for this purpose, on the contrary, recommended its continuation and further financial investment in the astronomy department. A second commission was then appointed, which recommended closure. This second decision was more in keeping with the wishes of the Board of Directors and was thus the one accepted. However, the decision was not carried out until 1972. That same year Heyden packed up and went to Manila Observatory, which he had always considered as his first responsibility. He continued his scientific work there for another twenty years (Chapter 7).

In 1909 Francis A. Tondorf (1870-1929), who had joined Georgetown Observatory in 1903 and had worked for some time with Hagen, installed the seismological station in an underground chamber with Wiechert, horizontal and

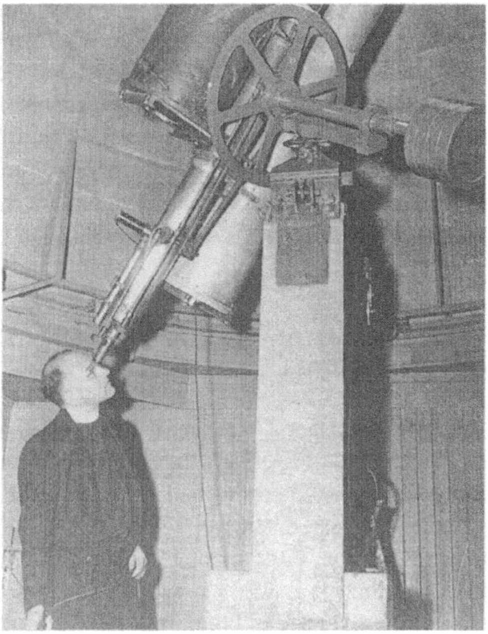

Figure 22. Francis J. Heyden, Director of Georgetown Observatory, 1948-1972 (GUA).

vertical seismographs of 80 kg mass, as part of the JSS program. Later we shall consider the full story of this Jesuit project in seismology. In 1912 the station was improved with new seismographs, horizontal and vertical Wiechert of 200 kg mass, horizontal Bosch-Omori and horizontal Mainka of 135 kg mass. The seismographic station functioned independently of the astronomical section with its own directors after 1910. In 1925 new three-component Galitzin electromagnetic seismographs were acquired, the most advanced instruments of the time and the first of their kind installed in North America. Tondorf dedicated himself totally to seismology and was widely esteemed by the seismological community. He and Odenbach may be considered as the fathers of seismology in the Jesuit colleges and universities of the United States. After his death in 1931 Frederick W. Sohon (1894-1972) became Director, having been on the staff of the observatory from 1927. He was an excellent professor of mathematics and the author of one of the earliest books on the theory of seismological instrumentation. The fine performance of the seismographic station prompted the installation in 1962 of a World Wide Standard Station Network (WWSSN) station. The WWSSN was a global network of 125 seismographic stations with the most modern instrumentation at the time, funded by the U.S. Government. The instruments were three component short and long period seismometers with galvanometers and photographic recording, built by Geotechnical Corporation. Stations were distributed throughout much of the world with the exception of Canada, the Soviet Union, China and communist countries. Records from all stations were sent to the Seismological Laboratory of the U.S. Coast and Geodetic Survey in Albuquerque, where they were copied on to easily readable 70 mm microfilm chips and made available to any researcher who requested them. This network represented a great advance in observational seismology. Ten of these stations were installed in Jesuit observatories, three of them in the United States. Georgetown Seismographic Station was closed together with the observatory in 1972.

Other astronomical observatories

As we have already mentioned, Hagen had installed in 1881 a small observatory in the Sacred Heart College in Prairie du Chien, Wisconsin, a boarding high school for boys where he had been a teacher of mathematics from 1880. He installed two small telescopes of 7.5 cm aperture and a good chronometer. Two years later he built a small observatory with a dome of 4 m diameter. He then began the observation of variable stars, collaborating with Prof. Holden of Harvard University in compiling his catalogue of stars and he also observed the occurrence of auroras. The observatory was abandoned after Hagen's departure in 1888. The observatory established in 1885 in Creighton College, Omaha, Nebraska was better equipped. Its founder was Joseph F. Rigge (1842-1913), a science teacher, who built a small one story building with a dome, where he installed a telescope of 12.5 cm aperture with equatorial mounting, a meridian circle of 7.5 cm and an altazimuth telescope, all made by Steward. He was succeeded in 1896 by his brother William F. Rigge (1857-1927) who had worked with Hagen in Georgetown and with whom he later collaborated. William Rigge worked mainly on observations of comets and planets and wrote many popular articles on scientific subjects. He used once his astronomical knowledge in court to help resolve a criminal case by fixing the hour at which a picture had been taken by the length of a shadow. He showed that the only witnesses could not have been at the scene of the crime and the case was solved. After his death interest in astronomy soon declined and no directors were appointed after 1932.

In 1899 a small astronomical observatory was established in Saint Mary's College, in the Jesuit Faculties of Theology and Philosophy. A telescope of 10 cm aperture with equatorial mounting was installed in a dome, as was a transit telescope. The observatory was used for student demonstrations in the classes of astronomy. In 1909 a Wiechert seismograph was installed as part of the JSS program but was destroyed in a flood the same year. The observatory was active until about 1910. In 1909 another small astronomical observatory was established at Marquette University with a dome on the roof of Johnston Hall, housing two telescopes of 15.2 cm and 13.3 cm aperture with equatorial mountings. The first director was James L. McGeary (1869-1945), a professor of physics and astronomy who had been an assistant of Algué in the Manila Observatory for two years from 1906. He did observations of sunspots, comets and planets. In 1937 a new telescope of 13.3 cm aperture was installed, but interest in astronomy was never high. Instruments were used mainly for demonstrations for students. In 1910 meteorological instruments were installed, but no research was done in meteorology. In 1909 a seismological station was installed in the basement of the same building as part of the JSS and JSA programs. Between 1932 and 1955, Joseph F. Carroll (1893-1955) took charge of the seismographic station. He was a professor of physics and had studied physics at the Universities of Munich and Bonn. In 1943 a local earthquake was recorded by the station; this brought public attention to the observatory. The station was closed after Carroll's death and the dome of the observatory was finally dismantled in 1983. A small astronomical observatory was established for a short time in Saint Louis University about 1880. It was used for a few years mainly for teaching purposes.

Santa Clara

Meteorological observatories were never very popular among the Jesuits of the United States. However that founded in 1895 by Jerome S. Ricard (1850-1930) in Santa Clara College (now Santa Clara University), Santa Clara, California can be attributed considerable importance. Besides all the necessary meteorological instruments there was a small astronomical observatory with a telescope of 20 cm aperture with equatorial mounting by Alvan Clark. Ricard, a Frenchman, who had come to California as a Jesuit novice in 1872, began teaching mathematics and philosophy in Santa Clara in 1891. He soon became interested in meteorology and founded the observatory where from, in 1907, he tried to investigate the relationship between weather and sunspots, a controversial subject at the time. His emphasis on this topic, which led him to believe that weather could be forecasted as much as a month ahead of time from observation of sunspots, provoked controversies with other meteorologists. Many noted scientists, among them the astrophysicist George E. Hale, Director of Mount Wilson Observatory, California , took issue with him, and rejected his theory. Ricard was never dismayed by adverse criticism and did not accept the opposing arguments. He made some of the earliest studies of the climate of California, especially its rainfall pattern, and published a local weather bulletin for which he was popularly known as the "Padre of the Rains". Although his theory of the influence of sunspots was not correct, his weather forecasts for the West Coast for periods of ten days in advance and even for periods of a month were remarkably accurate; sometimes better than those from the U.S. Weather Bureau.

As a mark of gratitude for his services to farmers by his weather forecasts, funds were raised in 1928 to build a new larger observatory. Ricard wanted to install in the new large dome which had been built a reflecting telescope of 150 cm aperture; this would have been one of the largest telescopes then in existence. The optician from Vancouver, Canada engaged to make the mirror defrauded Ricard of the money however, and after three years sent him a slab of concrete instead of the mirror. The decline in Ricard's health, which brought about his death, has been attributed to this unfortunate incident. It was later in 1941, that a new telescope, a refractor of 41 cm aperture by Alvan Clark, was installed in the dome. As early as 1907, after the major San Francisco earthquake of 1906, Ricard installed a seismometer of his own construction and Wiechert horizontal and vertical seismographs of 80 kg mass in 1909 as part of the JSS program. In 1930 a three-component set of electromagnetic Galitzin-Wilip seismographs were installed and the following year two horizontal Wood-Anderson torsion seismographs. All these instruments were moved to an underground vault in 1941.

After Ricard's death, Dr. Albert J. Newlin, his assistant from 1910, took charge of the observatory and kept up the meteorological and seismological observations, In 1941, a Jesuit, John A. Weber (1906-1982), was appointed Director; he carried out for some time astronomical observations of variable stars. After his departure in 1959 there was a decline in astronomical activities. The observatory building was for some years used as office and storage space by Bernard R. Hubbard (1888-1962), the famous "Glacier Priest", when not engaged in his exploration of glaciers in Alaska and the Aleutian Islands. In 1960, a group of students, under the supervision of Alexis I. Mei (1898-1983), again put the telescopes in operation for demonstrations for students. The seismographic station continued for some time until it was closed about 1968 after Mei's departure. In 1991, after years of neglect, the astronomical observatory reopened, following extensive renovation, sustained by financial backing from the local business community, for its use by students with instruction from the faculty.

John Carroll and the Jesuit Seismological Service

In 1895 a meteorological observatory was established in what was then Saint Ignatius College in Cleveland, Ohio, later to become John Carroll University. Its founder was Frederick L. Odenbach (1857-1933), who began to observe atmospheric electricity, clouds and electrical discharges in the atmosphere. For this purpose he designed an instrument he called a "ceraunograph". The meteorological station was improved in 1898, with the installation of a Secchi "Universal Meteorograph" donated by the Smithsonian Institution. In 1900 Odenbach began to be interested in seismology and installed two seismometers built by himself in the observatory; the first seismometers installed by a Jesuit in North America. Seismology proved to be his main contribution, for in 1908 he conceived the notion that the system of Jesuit colleges and universities distributed throughout the United States offered an excellent basis for a network of seismographic stations with uniform instrumentation. This was a novel idea, since at the time, stations had different kinds of seismographic instruments. In 1906 the International Seismological Center had been established in Strasbourg, France, where data were centrally reported and epicenters determined. Odenbach envisioned a network of Jesuit stations that could contribute significant data to this international enterprise.

Odenbach convinced the presidents of the colleges and universities and the Jesuit provincial superiors of the United States of the value of his proposal. As he put it in a letter addressed to the presidents of Jesuit colleges and universities, "It is this, to organize a system of observatories with the same kind of instrument handled in the same way and these stations to be fairly well scattered over the continent. ...This, dear Father, is the idea I proposed to Reverend Father Provincial a day or two ago and he was pleased with it and allowed me to write this circular letter to all colleges of our province." He ended his letter saying that "this movement is to the great benefit of your institution, the glory of our Society and the honor of that Church which has been proclaimed an enemy to all modern progress and enlightenment." In 1909 sixteen identical horizontal Wiechert seismographs of 80 kg mass were purchased from Spindler and Hoyer of Goettingen, Germany at a price of 112 dollars each. These instruments consisted of an inverted pendulum which recorded the two horizontal components of motion on smoked paper. They were distributed to fifteen colleges in the United States: Cleveland, Georgetown, Brooklyn, Fordham, Worcester, Buffalo, Spring Hill, New Orleans, St. Louis, Chicago, Milwaukee, St. Mary's (Kansas), Denver, Santa Clara and Spokane and one in Canada at St. Boniface (Manitoba). The station at Brooklyn College never actually worked.

In 1911 the organization was formalized and the sixteen stations formed a network which was given the name of the Jesuit Seismological Service (JSS). Individual stations were to process their own seismograms and send the readings to the central station in Cleveland. The data would then be collated and forwarded to the International Seismological Center in Strasbourg. The JSS, in effect, constituted the first seismological network with uniform instrumentation on a continental scale. As a matter of fact Jesuits succeeded in doing what the Government of the United States did not accomplish. A proposal in 1910 to create a Bureau of Seismology within the Smithsonian Institution was rejected in Congress. As a consequence there were only a few seismographic stations operated by the U.S. Weather Bureau until 1925, when the function was transferred to the Coast and Geodetic Survey. Prof. Willis

Moore, Chief of the Weather Bureau was one of those who encouraged the Jesuits to enter this field. In spite of the interest and collaboration of the Jesuits participating, many of the first stations, for a variety of reasons, floundered soon and the cooperative effort was never fully established. One element of tension among Jesuit seismologists was Moore´s offer to contribute funds if they agreed to their affiliation with the Weather Bureau. Odenbach was strongly against such an arrangement, but some, especially Tondorf at Georgetown University, were in favor. This created certain friction between Odenbach and Tondorf. In late 1911 Odenbach wrote to the different stations that he had lost interest in the original plans and advised them to act independently or to connect up with the International Bureau at Strasbourg. So the JSS came to a premature end. However, as we shall see, the program was reestablished fourteen years later.

The observatory continued after Odenbach's death, directed by Joseph S. Joliat (1882-1960) and from 1947 to 1962 by Henry Birkenhauer (1914-). In 1946 the seismographic station was renewed by the installation of three-component Sprengnether electromagnetic seismographs in an underground vault in the basement of Rodman Hall. The meteorological section had been declining and was closed in 1950. Between 1962 and 1983 Dr. Edward J. Walter, a lay professor, took charge of the seismographic station. From 1985, the station was again directed by a Jesuit, Richard Ott (1942-) who increased the number of instruments to eleven seismographs of short and long period. In 1992 Ott went to the Observatorio San Calixto, La Paz, and although he returned after a couple of years, the station was closed. Ott went to Weston Observatory where he became its Assistant Director.

Saint Louis and the Jesuit Seismological Association

Saint Louis University was founded in 1818 in St Louis, Missouri. A small astronomical observatory existed between 1855 and 1888, but little information about it remains. Earlier, in 1835, Judocus Van Sweevelt (1804-1841), a Belgian Jesuit, began meteorological observations, which he continued until his death. A complete meteorological observatory was established in 1860 directed by Francis H. Stuntebeck (1829-1898), a German Jesuit from the Oldenburg region, who continued observations until 1876, when the U.S. Government installed a meteorological station in the city. In 1909 meteorological instruments were installed and the first seismograph, a horizontal Wiechert of 80 kg mass which was set in the basement of DuBourg Hall, as part of the JSS program. The first director of the refurbished observatory was John P. Goesse (1869-1931), a native of St. Louis. Goesse's mother contributed to the observatory paying for the seismograph.

Figure 23. James B. Macelwane, Director of the Department of Geophysics and Meteorology at Saint Louis University, Saint Louis, Missouri, 1925-1956, and first President of the Jesuit Seismological Association (SLU).

Goesse was assisted by George E. Rueppel (1864-1949) who had worked with Odenbach in Cleveland from 1895 to 1900 and had installed a meteorological station between 1901 and 1908 in St John's College in Toledo, Ohio. Rueppel installed in St. Louis in 1920 the second commercial radio station in the U. S., the first west of the Mississippi River. He became director of the observatory, keeping meteorological and seismographic observations after Goesse's departure in 1920. About this time the University began offering courses in meteorology.

James B. Macelwane (1883-1956), a professor of physics between 1912 and 1919, was, together with other young Jesuits, interested in the work of the observatory by Goesse. Macelwane obtained his doctoral degree from the University of California, Berkeley, in 1923 with one of the first dissertations in seismology in the United States. He remained at Berkeley for four years teaching in the Department of Geology before returning in 1925 to St. Louis where he became Director of the Observatory. The same year he created the

Department of Geophysics (of which he was Director) with a program of graduate studies in meteorology and geophysics leading to doctoral degrees in both sciences, one of the earliest in the United States. The department was one of the founding members of the University Corporation for Atmospheric Research. The observatory was integrated into the new department and students participated in its observational work. In 1944 the department became part of the new Institute of Technology created in St. Louis University. Macelwane was the first dean of the institute. He gave a great encouragement to geophysical studies and was without doubt the most renowned of Jesuit seismologists.

One of Macelwane's early efforts was to revitalize the Jesuit seismographic network. The impetus for this renewal came not only from his own interest, but also from the urging of scientists at the National Research Council and of the Carnegie Institution in Washington. Navarro-Neumann of Cartuja wrote to Macelwane complaining of the bad state of the JSS and the need for a reorganization: "I believe it would be very much AMDG (Ad Maiorem Dei Gloriam; For the Greatest Glory of God) if our chain of seismographic stations in the United States would rise once and for all out of its present state, which in your great charity and in consideration of the intention which prompted me, you will pardon me if I dare to call wretched". In the Summer of 1925 Macelwane, with the backing of the four Jesuit Provincials of the United States, organized a meeting of all interested Jesuits at Loyola University in Chicago. Out of this meeting came forth the reorganization of the stations into the new Jesuit Seismological Association (JSA) with the seismographic station of Saint Louis University as the Central Station. Article II of the Constitution of the JSA, which was approved by Father General Wladimir Ledochowski, reads "The object of this Association is to foster cooperation among Jesuit institutions of the American Assistancy (the Jesuit administrative unit that groups all American Provinces) toward an organized Jesuit contribution to the science of seismology". In its Article III, it is said "In order to insure a united effort, all the Jesuit seismological stations in the American Assistancy are to maintain active membership in the Jesuit Seismological Association. This membership is in no wise to be construed as limiting the autonomy of the individual stations". In this way a cooperative effort was established allowing for the autonomy of each station.

The stations participating in the JSA program were: Buffalo, Chicago, Cincinnati, Cleveland, Denver, Milwaukee, New Orleans, Fordham, St. Louis, St Mary's, San Francisco, Santa Clara, Spokane, Spring Hill, Georgetown, Weston and Worcester. The Central Station at Saint Louis University assumed the responsibility on behalf of the JSA of collecting data from member stations and from other stations around the world, of locating earthquake epicenters and of publishing these for the worldwide seismological community. The Central

Station published a *Preliminary Bulletin,* which was circulated to a large number of institutions and seismologists in all parts of the world, and a summary of the data was published yearly in the *Transactions of the American Geophysical Union* under the title "Report of the Jesuit Seismological Association." This *Preliminary Bulletin* was highly esteemed. Macelwane wrote in 1950 that "it is consoling to read, in published articles and official bulletins, the epicenters accredited to the JSA, issued even by countries, such as Switzerland and Norway, which will not permit Jesuits within their borders. Jesuits, but not Jesuit science, are officially excluded from Norway by Article II of its constitution". The Central Station continued this service until the early 1960s when, with the advent of computer determination of epicenters, it was discontinued as an unnecessary duplication of the determinations made by the U. S. Coast and Geodetic Survey (later by the U.S. Geological Survey) in the United States, and by other institutions, for example the International Seismological Center in Great Britain. Although this service was ended, most of the JSA stations continued in regular operation until relatively recent times. The JSA also continued, and its presidents after Macelwane were J. J. Lynch of Fordham University, (1957-1970), D. Linehan of Weston Observatory, (1970-1986) and W. Stauder of Saint Louis University (from 1986 to the present). During Macelwane's presidency the JSA enjoyed a very active life as was attested by the celebration of its 25th anniversary in 1950. As we shall see, many Jesuits in the stations of the JSA worked actively in seismology until most of the stations began to close after 1970.

Under Macelwane, the seismographic station of Saint Louis continuously updated its instrumentation. In 1928, a new station was built, outside the city, in an underground vault, in the grounds of the Jesuit Novitiate in Florissant, where three-component Galitzin-Wilip seismographs and two horizontal Wood-Anderson were installed. The same year another vault was built in Saint Louis in the Gymnasium building, where Wood-Anderson seismographs were installed; and in 1938 seismographs by Sprengnether, which were updated in 1948. Outside St. Louis seismographic stations were installed in 1930 in Little Rock, Arkansas and in 1938 in Cape Girardeau, Missouri whose instruments were updated in 1948. In the Department of Geophysics Macelwane worked intensely teaching and doing research in geophysics and meteorology; he was always interested in promoting educational programs and encouraging young students. One of the foremost purposes he had in mind in establishing the academic program in geophysics was to train competent personnel for the exploration programs of the oil industry. Many Jesuit seismologists whom we shall encounter later in different observatories were his students, among them William C. Repetti, who went to Manila Observatory (Chapter 7); Victor J. Blum, who succeeded him in Saint Louis; Henry F. Birkenhauer of John Carroll University; Jesús Emilio Ramírez of Bogotá, Colombia (Chapter 6) and Patrick A. Heelan (1926-), an Irishman who later became a professor of philosophy at Fordham (1966-1970), the State University of New York, Stonybrook (1971-1990) and from 1993 at Georgetown University. During Macelwane's time, Saint Louis University became one of the centers of the best-known and most prestigious centers of seismology.

Macelwane worked on many aspects of seismology and it is difficult briefly to summarize his achievements. Earthquakes generate seismic waves which propagate through the earth and their analysis provides a powerful tool for investigating the structure of the earth's interior. These studies were in their infancy in Macelwane's time. Using the analysis of travel times of seismic waves, he and his students established the existence of layers of different composition in the interior of the earth, especially of a transition layer just outside the earth's core. Apart from the vibrations which arise from earthquakes, there are additional earth motions, called microseisms, which occur more or less continuously and these attracted Macelwane's attention. His studies, in collaboration with Ramírez, on the nature of microseisms and their relation to atmospheric storms attracted the attention of the United States Navy, which in 1943 established the Hurricane Microseismic Research Project, approved by the Joint Chiefs of Staff. Macelwane was an adviser to this project. The goal of the project was to obtain dependable advance warning of the path and intensity of hurricanes. The Navy, following Macelwane's ideas, established a number of tripartite stations in the West Indies and in the Pacific Ocean for the early detection and tracking of storms by means of microseismic observations. The method proved not to be a reliable detection system and the project was soon abandoned by the Navy when other more reliable methods were developed. Macelwane was also interested in geophysical methods of prospecting for oil and served as a consultant for several exploration companies.

In 1936 Macelwane published *Introduction to Theoretical Seismology,* the first textbook on seismology in North America and in 1947 *When the Earth Quakes,* a popular non-mathematical book on earthquakes. These and other works helped to extend his influence as a teacher and educator beyond the circle of his own students. Macelwane actively participated in scientific societies and regularly attended their meetings. He was President of the Seismological Society (1928-1929) and of the American Geophysical Union (1953-1956) and in 1944 was elected to the National Academy of Sciences. He belonged to the pioneer generation of American seismologists formed by Hugo Benioff, Beno Gutenberg, Charles F. Richter, Perry Byerly and Maurice Ewing. In 1962 the American Geophysical Union honored his memory with a medal awarded each year to a young geophysicist of outstanding ability. The American Meteorological Society also presents a similar award to three young meteorologists. As Prof. Perry Byerly of the University of California, Berkeley, expressed it, "Father Macelwane was remarkable for his enthusiasm, for his optimism and for his dogged persistence. He had a genial personality, which elicited spontaneously the cooperation which he habitually received and so generously gave."

Macelwane was succeeded in St. Louis by Victor J. Blum (1907-1979). In 1962 William Stauder (1922-2002) who had received his doctoral degree at the University of California, Berkeley, working under Prof. Byerly, who had been Macelwane's first student, became Head of Department. Under Stauder's direction the department updated its equipment; installed a local network of five seismographic stations around St. Louis, and established a WWSSN station at Florissant. Stauder began a research program on the focal mechanism of earthquakes, which had been the subject of his doctoral thesis. The graduate program in seismology attracted a number of students among them some Jesuits such as Agustín Udías (1935-) who has become a professor of geophysics at the Universidad Complutense, Madrid, Luis Fernández who worked for some years at Observatorio San Calixto, La Paz, Bolivia, and Gabriel Leblanc for some years at Montreal.

In 1971 the department was reorganized with the title of the Department of Earth and Atmospheric Sciences and it moved to a new location. Further development in seismological projects was the installation in 1973 of a network of 32 stations in the region of New Madrid, Missouri, where large earthquakes had occurred in 1811-1812. In 1988 Stauder was named Associate Academic Vice President of Saint Louis University. A non-Jesuit, Dr. Robert Herrmann, was appointed Director and the direct active presence of Jesuits was ended. The department continues today as a recognized institution with a graduate program in geophysics and meteorology and an active research program in seismology. The department became member of the Incorporated Research Institutes for Seismology (IRIS), an American organization to promote seismological research. Recent innovations have been the installation in 1991 of a broad-band digital seismographic station in Cathedral Cave, Missouri as a component of the Global Seismic Network (GSN). More recently a network of such stations have been installed in the central region of the United States. At present the seismological observatory consists of a complex network of seismological stations of different types, known as the Upper Mississippi Valley Seismic Network.

Fordham

In 1910, as part of the group of seismographic stations of the JSS, a Wiechert seismograph was installed in the basement of the administration building of Fordham University. In 1922 two new Milne-Shaw horizontal seismographs with photographic recording were installed. In 1927 the existing seismographs, along with new Galitzin electromagnetic seismographs, were moved to a small building with an underground vault. In 1910, the first Director was Edward P. Tivnan (1882-1927), but he was Director only for two years. Much the same happened to the seven directors from 1912 to 1928. Among them we find some familiar names as Repetti, who later went to Manila Observatory, and Sohon who moved to Georgetown.

In 1928 Joseph Lynch (1894-1987), was appointed Director of the observatory. He had previously occupied this post from 1920 to 1923, and remained Director for nearly fifty years until 1977. He was a careful observer, kept the seismographs correctly operating and carried out several studies in seismology. In 1931 Lynch moved the location of the seismographs to a new place with a larger underground vault housing new instruments: Wood-Anderson, Sprengnether, and Benioff seismographs of short and long period were also installed. The station had at that time a very complete set of up-to-date instruments. In 1946 Lynch went to the Dominican Republic, in the Caribbean, at the invitation of its government, to study the effects of the earthquake of 4 August that caused severe damage in the capital Santo Domingo. There he was the guest of its President, General Leonidas Trujillo, who had changed the name of the capital to Ciudad Trujillo (Trujillo City) and exercised absolute control over the country. On the occasion of a horse race, Trujillo suggested to Lynch the horse on which to bet, and naturally that horse won the race. During the visit Lynch drew up plans for a seismic observatory, modeled on the observatory at Fordham. A young engineer was sent to Fordham to be trained and a seismographic station was later established in the Dominican Republic. In 1956 Lynch collaborated with Linehan in archeological studies under Saint Peter's Basilica, in the Vatican, using seismic exploration methods. After his retirement in 1977 the seismographs were kept functioning for a short time by some students, but finally the observatory closed. The rich collection of seismograms was donated in 1983 to the Lamont-Doherty Seismological Observatory of Columbia University, New York, where they are kept today.

Weston

In 1928 began at Boston College, Weston, Massachusetts, what would become an important seismological research institute. The first instruments installed were two horizontal Bosch-Omori seismographs of 25 kg mass, donated by Georgetown Observatory. The same year an astronomical section was added to the observatory with the installation of a small refracting telescope of 23 cm aperture by Michel J. Ahern (1877-1951), who used it mainly for his classes in astronomy. In 1931 the astronomical section was

Figure 24. Participants in the joint meeting of the Eastern Section of the Seismological Society of America and the Jesuit Seismological Association held in St. Louis, 1937. Jesuits are: first row, Westland (Spring Hills), Macelwane (Saint Louis), Lynch (Fordham); second row, Ramirez (Saint Louis later Bogota); third row, Stechschulte (Cincinnati), Zegers (Saint Louis); fourth row, Linehan (Weston), Delaney (Weston) (SLU).

enlarged by the installation of a Clark telescope of 38 cm aperture with equatorial mounting in a rotating dome built on what had been the base of a water tank. Thomas D. Barry (1901-1959) was Director of the astronomical section from 1940 to 1959 which closed after his death. The seismographic station was improved in 1934 with the installation of a horizontal Wiechert seismograph of 80 kg mass from Holy Cross College, Worcester; and in 1936 further improved with a set of three-component electromagnetic Benioff seismographs of 100 kg mass with photographic recording.

In 1936 Daniel Linehan (1904-1987) began to work in the seismographic station at Weston although he did not officially assume the post of Director until 1950. He gave the station (which assumed the new name of Weston Seismological Observatory) its real importance. Linehan specialized in seismological methods for shallow exploration and established contact with oil exploration companies; these financed his work. In 1939, he received portable seismic reflection equipment with 12 channels from the Humble Oil Company, and in 1949 seismic equipment with 24 channels from the Century Geophysical Corporation of Tulsa, Oklahoma. With this equipment he carried out numerous shallow seismic explorations in order to study the shallow structure of the earth's crust. In the course of these studies he determined the thickness of the glacial deposits over the bedrock in New England; and made determinations of the thickness of the ice in Antarctica.

Linehan, accompanied by Lynch, also used this technique in their archeological studies in the Vatican, where their studies led to the discovery of tombs from early Christian times under the Basilica of Saint Peter. For this purpose they employed special equipment provided by the Century Geophysical Company. To generate vibrations they used small explosions when they were working outside and blows with a sledge hammer when they were indoors. They had an array of small geophones for recording the vibrations. Subsequent excavations found a simple tomb, which is thought to be probably that of Saint Peter. The Vatican Swiss Guards were amused to see the big American priest who swung a ten-pound sledgehammer to shake the ground inside the Basilica in order to record the seismic waves on his instruments. During their experiments in the Basilica, Linehan and Lynch had to shout out "Stand still!" in many different languages so that the crowd of tourists and worshippers would quiet down and allow them to record the waves generated by the hammer blows. Linehan participated in many scientific expeditions to different parts of the globe. Three of these were to Antarctica and one to the Arctic earning him the title of "explorer". He was the first to measure the thickness of the ice at the South Pole using seismic methods, and the first also to celebrate Mass at the Pole itself and the first to perform a Baptism in Antarctica. Several UNESCO seismological missions took him to different parts of Africa, Asia and South America. Linehan was a prolific contributor of papers and technical articles to scientific and engineering conferences and journals, both national and international. He was in great demand as a lecturer for both popular and scholarly talks on geology and seismology, as well as about his scientific explorations, especially, to Antarctica.

In 1948 Linehan founded the Geophysics Department at Boston College and was its Director until 1963. A number of young Jesuits studied there; they later worked in other observatories. Expansion of the Department led, in 1949, to a new building for laboratories and offices. In 1955, the observatory installed a seismographic network of six stations in the northeast of the United States and the east of Canada. The network was enlarged in such a way that in 1980 it consisted of about 30 stations. In 1957, on the occasion of the International Geophysical Year, magnetic equipment, in particular, Ruska variometers with photographic recording, were installed in the observatory to record the variation of the components of the earth's magnetic field. Also two magnetometers, of the type known as proton and rubidium magnetometers, were installed later to make absolute measurements. In 1962 research in geomagnetism led to the participation of the observatory with the U. S. Air Force in the program of measuring the earth's magnetic field by satellite-borne instruments. Linehan retired as Director in 1974, but continued as "Director Emeritus" until his death. As happened in other observatories, after Linehan's death, Jesuit participation in the observatory declined. For nearly twenty years, until 1993, the observatory was directed by James W. Skehan (1923-). On his retirement John Ebel, a lay professor, was appointed Director with Richard Ott as Assistant Director, who had been the last director of the seismological station at John Carroll University. The observatory and the department continued today as part of Boston College with a vigorous academic and research program.

Other seismographic stations

The first group of stations of the JSS had been established between 1909 and 1910 and all of them began, as it was already mentioned, with the standard horizontal Wiechert seismograph of 80 kg mass. Some of the stations renewed their equipment with modern instruments and continued in operation till the 1970s or 1980s. Only a few are operating today. The station at Canisius College, Buffalo, New York, at first had the seismograph installed in a corner of the old recreation building. The seismograph was moved in 1913 to an underground vault in a new building. In 1932 new electromagnetic seismographs by Galitzin-Wilip and in 1946 two horizontal long-period Sprengnether seismographs and a vertical short-period Sprengnether seismograph were installed. In 1962 a new system for the control of time was installed. In 1963 all records were sent to the U.S. Geological Survey (USGS) for their microfilming and are kept there. The directors were Jesuit science professors in the college, except from 1940 to 1960 when a lay professor was Director. From 1960 to the present a Jesuit, James J. Ruddick (1923-), a physics Ph. D. from Saint Louis University, has been in charge. In the north-west in the state of Washington, a seismographic station was established at Gonzaga University in Spokane; it was inaugurated in 1909. In 1930 the seismograph was moved to the Jesuit Philosophy Faculty at Mount Saint Michael's outside Spokane. In 1946 two horizontal Wood-Anderson seismographs were added. The station was supervised by science professors; it closed in 1970.

In 1909 a seismographic station was established at Regis College, Denver, Colorado, in the basement of the main building. In 1946 new Sprengnether seismographs were installed, two horizontal of long period and a vertical of short period. In 1962, three-component short period seismographs were added.

In 1966 the instruments were moved to a new underground vault and new instruments were added. Armand W. Forstall (1859-1948) was Director for nearly forty years from 1909 to 1948. Joseph V. Downey (1909-1990) was Director from 1948 to 1989. The station is still in operation, but not directed by a Jesuit. Also in 1909 a station was established at Loyola University, New Orleans, Louisiana. At first, besides the horizontal Wiechert seismograph, the station had also a vertical seismograph. These instruments were moved to an underground vault in Bobet Hall in 1924 and, in 1946, new Sprengnether instruments were installed. For forty five years, from 1938 to 1983, the station was directed by Karl A. Maring (1898-1983), a Jesuit born in Germany, who had been director of the station at Spring Hill College from 1926 to 1929. In 1985 the station was closed.

A station of the JSS program with a horizontal Wiechert seismograph was established in 1909 in Holy Cross College, Worcester, Massachusetts. The location of the instrument was changed several times until the station closed in 1934 and the instrument was donated to Weston Observatory. In 1910, a station was established in Spring Hill College, Mobile County, Alabama, with the standard Wiechert instrument. The station was closed between 1935 and 1939, but opened again in 1941 with the installation of McComb-Romberg horizontal seismographs. Between 1939 and 1947 a meteorological station was installed. From 1950 to his death Louis J. Eisele (1912-1988) was the Director. He performed his role with considerable innovation and devotion. In 1958 the recording system was changed to pen-and-ink with electronic amplification and a vertical instrument was added. In 1962 a WWSSN station was established. The station was closed in 1989. The last station of the JSS program was established in 1912 at Loyola University, Chicago, Illinois with the standard Wiechert seismograph in the basement of the science building. In 1957 a set of three-component short-period Sprengnether seismographs from Fordham were donated by Lynch. In 1983 the system of recording was changed and a short-period vertical Kinemetrics instrument was installed. There were plans to modernize the station, but these were never carried out and the station closed in 1990. The last Jesuit director was Donald J. Roll (1912-1986), who took care of the station until his death.

After the reorganization of the JSA by Macelwane, a new station was established in 1925 at Xavier University in Cincinnati. The Rector wanted it to be a first class station so an underground vault was especially built for this purpose in the new library building. The first instruments, two horizontal Wood-Anderson seismographs, were installed in 1927 and, the following year, a vertical Galitzin-Wilip seismograph. In 1932 Victor C. Stechschulte (1893-1955), who had obtained his Ph. D. in seismology from the University of California, Berkeley, was appointed director. Stechschulte was an active seismologist who worked on several topics in seismology, especially on deep focus earthquakes. He was president of the seismological section of the American Geophysical Union from 1944 to 1947. Xavier Observatory came into special prominence as a result of two earthquakes which shook West Central Ohio in 1937 and which were felt in Cincinnati. In 1949 a set of short-period instruments of local construction replaced the Wood-Anderson instruments. Edward A. Bradley (1923-1996) succeeded Stechschulte, but after he suffered a disabling accident the station declined and was closed in 1984. The last Jesuit seismographic station established in the United States was installed in 1950 at the University of San Francisco, California, in cooperation with the University of California. Two horizontal long-period and a vertical short-period Sprengnether seismographs were installed in a vault in Saint Ignatius Church of the University by Alexis Mei. The station was closed in 1962 after Meis' departure for Santa Clara.

Montreal

In Canada the first observatory was established in 1910 at Saint Boniface College, St. Boniface, Manitoba, in response to Odenbach's proposal of the JSS program. The station with the standard horizontal Wiechert seismograph of 80 kg mass had a short life since it was destroyed in 1922 in a fire that affected the College. Thirty years later the Canadian Jesuit Maurice Buist (1902-1986), a professor of science at Collège Jean de Brébeuf in Montreal, visited Weston Observatory. During this visit Linehan encouraged him to reestablish the Jesuit seismographic observatory in Canada. Buist installed the first instrument, a Sprengnether seismograph, in the basement of one of the buildings of the Montreal college and later in the same year a vertical short period Willmore seismograph, donated by the Dominion Observatory in Ottawa, whose Director, John Hodgson, was a supporter of the new observatory. In 1955 an underground vault was built to house the two instruments, where in 1975 a three-component set of Benioff short-period seismographs was also installed.

In 1955 Ernesto Gherzi, the last Director of the Zikawei Observatory in China (chapter 7), agreed to come to Montreal from St. Louis where he had been for a short time since his expulsion from China in 1950. Gherzi thought that his scientific work would be more useful in Canada, where few Jesuits were then active in science. Gherzi introduced studies of meteorology at the observatory and thus some meteorological instruments were soon installed. In 1957 the laboratories, offices and library of the observatory were installed in a new building of the College, Lalemant Hall, with a new underground vault. Gherzi gave a new impetus to the research aspect of the observatory. In 1958 the meteorological instrumentation was enlarged with an ionosonde, a meteorological radar, a recorder of solar radiation and other instruments. Gherzi continued his work on the ionosphere and tried to correlate changes in atmospheric electricity with weather conditions, especially, with the presence of masses of polar air. He also tried to correlate the outbreaks of poliomyelitis with certain meteorological conditions. In 1957, at Gherzi's initiative, the publication of the observatory bulletin, *Bulletin de Geophysique,* began. To this bulletin he was a frequent contributor with papers about his research on meteorology and atmospheric electricity. In 1961 the installation of a WWSSN station by the U.S. Coast and Geodetic Survey was offered to the observatory. Canada had not accepted any of these stations and the Dominion Observatory did not approve the idea. Instead the Dominion Observatory donated a set of three-component long-period Sprengnether seismographs to the observatory. These instruments, together with the short period Benioff seismographs, made the station similar to a WWSSN station.

In the 1960's three young Jesuits, Conrad East, Gabriel Leblanc and Florent Verreault (1924-1996) had been sent to study geophysics in preparation for their work in the observatory. When they returned in 1966, the absence of professors of geophysics in Canadian universities convinced Buist that this was a more important work than working at the observatory. The new geophysicists moved instead to the universities of Laval, Montreal and Quebec to teach, although they were also on the staff of the observatory. Between 1965 and 1967 there was a discussion about the future of the observatory and it was decided that it should continue as an independent institution. The situation proved not to be very favorable and with Gherzi's death in 1973, the plans for making Brébeuf Observatory a research center in geophysics practically disappeared. The observatory was practically reduced to observational work in seismology and meteorology. In 1975 Pierre Gouin, who had been Founder and Director of the Observatory of Addis Ababa, Ethiopia (chapter 7) arrived at the observatory. In 1978, Paul Emile Tremblay (1914-), who had been for some years a physics teacher in Senegal joined the observatory staff. In 1983 he became director, when Buist retired after nearly thirty years in the post. In 1984, M. Claude Degrandpré, a lay professor of physics, joined the observatory staff in turn becoming director in 1998, after Tremblay's retirement. The observatory continues today as part of Collége de Brébeuf, but with little Jesuit participation.

Chapter 6

CENTRAL AND SOUTH AMERICA: TROPICAL HURRICANES AND EARTHQUAKES (1814-2000)

Astronomical observatories

Jesuit observatories in Central and South America were not very numerous and some were relatively small. Except for the Observatory of San Miguel, Argentina, they were mainly dedicated to meteorology and seismology . The occurrence of tropical hurricanes in the Caribbean and of large earthquakes in the Andean region determined the orientation given to these observatories. Lack of scientific institutions in these countries at the end of the 19th century and at the beginning of the 20th century made the work of Jesuit Observatories in these two fields of special importance. We can say that in great part Jesuits helped to lay the foundations of meteorology and seismology in Central and South America. Recognition of Jesuit expertise in meteorology is shown by the fact that they were asked to organize the national meteorological services in Ecuador in 1870 and in Colombia in 1922. The most important of the meteorological observatories was that of Belén in Havana, Cuba. Seismological observatories were established later and those of greater importance were those in La Paz, Bolivia and in Bogotá, Colombia. Astronomy was never very strong in these observatories, owing, maybe, to the high costs of instrumentation. However, small telescopes were installed at Belén, Cienfuegos, Puebla, La Paz, and Bogotá. These instruments were used mostly for the accurate determination of time and for basic astronomical observations. In Bolivia Jesuits were in charge of the official time keeping service.

The only proper astronomical observatory directed by Jesuits in South America was really a government institution in Ecuador with an interesting history. Toward the end of the 19th century, President Gabriel García Moreno wanted to modernize the scientific institutions of his country and asked for the help of the Jesuits. He began in 1864 with the establishment of a meteorological observatory in the Colegio San Luis in Quito as the center of the National Meteorological Service. Its first Director was Federico Aguilar. García Moreno had visited the Ecole Polytechnique in Paris and wanted a similar institution for his country. Thus, in 1870 Jesuits were asked to take charge of the newly created Polytechnical School in Quito. A group of six young German Jesuits from Valkenburg came to Quito for this purpose as professors of different disciplines of sciences. In 1872, García Moreno decided to create also an astronomical observatory and asked Johann B. Menten (1838-1900), one of the German Jesuits teaching science at the Polytechnical School and Director of the Meteorological Observatory from 1870, to take care of this project. Buildings were finished in 1875, the same year that President García Moreno was murdered. This tragic event meant the end of all these projects. The new government stopped its support to Jesuits in Quito. The meteorological observatory was closed and the professors of the Polytechnical School returned to Germany. Menten remained in Quito, left the Jesuit Order in 1876 and continued in the Astronomical Observatory, which was inaugurated in 1877. Its main instrument was a refracting telescope with equatorial mounting by Merz. Menten made meteorological and astronomical observations and continued as Director of the observatory until 1882. In 1895 he abandoned Ecuador for Colombia, where he died.

Belén

The most important meteorological observatory was established in 1857 in the Real Colegio de Belén, as it was then known, a secondary school founded by the Spanish Jesuits in 1854 in Havana, Cuba. The idea came from the need to complement the science courses in the college and from the importance of observing weather conditions, especially tropical hurricanes that each year cross the Caribbean region causing very serious damages and many casualties. Meteorological instruments were installed on the roof of the school and the observatory occupied rooms on the second floor. The first director was Antonio Cabré (1829-1883), a teacher of physics and mathematics. The observatory had very complete instrumentation, such as four mercury barometers, thermometers, pluviometers and Robinson anemometers. Regular observations began in 1858. In 1860 Cabré returned to Spain and for the next ten years several teachers of science acted as directors. In 1862, Felice Ciampi (1826-1889), an Italian who later worked in the Observatory of the Collegio Romano, installed magnetic instruments, namely, a bifilar magnetometer, a declinometer and a magnetic balance. The installation of the magnetic station was in response to the program of Edward Sabine to establish a world network of magnetic stations. Belén was for many years the only magnetic station in the Caribbean region. We have seen the good relations between Sabine and Stonyhurst Observatory and this may have been the reason for this installation. The magnetic station was closed in 1905 owing to the interference by the electric tramways installed that year in Havana.

In 1870 Benito Viñes (1837-1893) became Director and continued during 23 years until his death. He gave the observatory its well deserved scientific fame for his studies on tropical hurricanes. These were afterwards continued by his successors. In his very first year as Director, Viñes made the first observations of two hurricanes and two years later published the first

Figure 25. Benito Viñes, Director of Belen Observatory, Havana, 1870-1893. Portrait painted by Gonzalo Carrasco, Director of Puebla Observatory, Mexico, 1901-1904.

collection of meteorological observations made in Belén between 1858 and 1871. Using all of these accumulated observations, Viñes began to study the characteristics of tropical hurricanes in the Caribbean. He analyzed every hurricane in detail for its organization, circulation and translation. Little was known at that time about the theory of formation of hurricanes, except for the early investigations of William Redfield and James Espy. Viñes' approach to the study of hurricanes was basically empirical; he paid special attention to the form and motion of clouds at different altitudes and formulated several empirical laws about the circulation and paths of hurricanes in the Caribbean region. He proposed two general circulation laws, the first on the counterclockwise rotation of the winds, and the second on the changes of air currents at different heights, convergent at low heights and divergent at high ones. He also formulated six laws regarding the cyclonic translation motion in the Caribbean. Among these was a law establishing the latitude of the point of rebound of the cyclones for different dates between July and October. His ideas and observations were published in 1877 in his first book, *Apuntes relativos a los huracanes de las Antillas en Septiembre y Octubre 1875 y 1876,* parts of which were translated into English and German, and in a second one, published posthumously in 1895. His first book had a wide distribution and made his name known among meteorologists working on the problem of cyclones. The English version published by the Signal Office of Washington had three editions and in 1885 R. H. Scott of the Royal Meteorological Society of London made a very laudatory mention of it.

Viñes was not only interested in the study of hurricanes, but he put this knowledge to the practice of their forecasting in order to prevent their disastrous consequences. The results were his timely and accurate forecasts of hurricanes which cross the Caribbean every year. His predictions were based mainly on observations of sea swells and the motion of the hurricane cirrus canopy that streams outwards from the storm center and often heralds the approach of tropical cyclones. One of his most important contributions was that he stressed the importance of upper-level winds for hurricane forecasting. He made his first forecast for the hurricane that crossed Havana on the 13th of September 1875, publishing a note in the newspapers two days before. This is, probably, the first modern successful prediction of a hurricane. From that year on, he made continuous forecasts of cyclones in the region for the authorities and local newspapers; their accuracy was soon appreciated, especially by maritime authorities and sailors. He was made aware of the importance of the forecasting of hurricanes, when, commissioned by the Academia de Ciencias of Cuba, he studied in the field the devastating effects of the 1877 hurricane in Cuba and Puerto Rico. There is no doubt that, during the 23 years when Viñes was in charge of the observatory, his timely and accurate forecasts saved innumerable lives and prevented large material losses. The *Boletín Mercantil* (5 October 1877) of Puerto Rico wrote after a successful forecast: "Father Viñes, whose voice for us has the authority of an oracle, calmed our souls with his timely notice. He well deserves the European reputation that he enjoys . Spain should be proud of him". Ship captains who navigated through the Caribbean were grateful for Viñes' forecasts and his selfless work was recognized in *The Times Democrat* of New Orleans in 1890: "Padre Viñes for all that long period, without a thought, or a hope or even a possibility of reward, but simply and solely for the love of humanity, has continued his labors and given the results of them gratuitously to the masters of the vessels and others whose business is affected by meteorological variations".

Viñes wanted the most modern equipment for the observatory and in 1873 he acquired a Secchi meteorograph, which was kept in operation until the closing of the observatory. The observatory was included in 1877 in the international network of meteorological information. In 1882 Viñes journeyed to Europe where he visited several observatories in Spain, France and Belgium and spent a few months at Stonyhurst Observatory with Perry. One of the reasons for the journey was to buy new equipment for the observatory. After his return, he installed a Cooke equatorial telescope of 10 cm aperture and new magnetometers for absolute measurements of the earth's magnetic field. The same year, 1882, he collaborated with Perry in the observation of the transit of Venus on the solar disk and used these observations to determine the distance from the earth to the sun. Although he kept carrying out astronomical and magnetic observations, Viñes' main interest was always the tropical hurricanes. However, his interest in other types of natural disasters led him to study in the field the damage produced by an earthquake in 1880 in the western part of the island. Large earthquakes are not common in Cuba and this is one of the first for which a detailed survey of its damage was made. Viñes was, thus, one of the earliest Jesuits to study earthquakes in the western hemisphere. In 1890 the Spanish government established a meteorological station in Havana; Belén continued without any official support. The new station made at times contradictory forecasts and was the source of many conflicts in Viñes' last years.

Viñes realized that one meteorological station was not enough to monitor impending hurricanes and in 1886 planned to organize a network of stations in the Caribbean region to improve the observations. The following year he succeeded in getting the support of Maritime Companies, the Telegraph Company and the Junta General de Comercio de la Habana to finance the telegrams from other stations to the observatory. At some times there were at many as 20 stations sending observations to Belén, but those reporting on a regular basis were seven, namely, Trinidad, Barbados, Martinica, Antigua, Mayaguez (Puerto Rico), Jamaica and Santiago de Cuba. To keep this network in operation was Viñes main work in the last years of his life. Cuba was still a Spanish colony and he found little support from its government. The British and French telegraph companies were more cooperative and from 1888 sent without charge to Belén the telegrams from twenty stations in the Caribbean region. Finally daily telegrams were reduced to those only of four stations, St. Thomas, Barbados, Jamaica and Santiago de Cuba, and the rest of the stations only sent data in case of hurricane occurrence. To detect the orientation of an approaching hurricane Viñes designed two instruments which he called "ciclonoscopio" and "ciclononefoscopio". The first was based on a compass card with indications of the types of clouds and orientation of the wind and the second added a compass and a reflection nephoscope. He never finished their construction, but the idea was used later by Faura and Algué in Manila for a very useful instrument, as we will see in the next chapter. After a visit to Havana in 1888, E. Hayden, editor of *Pilot Chart,* referred to Viñes as "the eminent Havana meteorologist whose scientific ability, indefatigable energy and unrivalled opportunities have linked his name so inseparably with all recent advances in our knowledge of these storms". His pioneer work on tropical hurricanes is recognized by G.E. Dunn and B. I. Miller in their work on Atlantic hurricanes with these words: " Viñes was a true scientist, making a significant contribution both to hurricanes theory and the art of forecasting hurricanes. He probably deserves the foremost position among all meteorologists in the western hemisphere who up to and through the nineteenth century contributed to our knowledge of tropical cyclones".

After Viñes' death in 1893, he was succeeded till 1928 by Lorenzo Gangoiti (1845-1933), who continued meteorological observations and forecasts of hurricanes. In 1897 the observatory was installed in two towers at both corners of the front of the new building of the college; one of the towers had a dome for the telescope. In 1902 Mariano Gutierrez-Lanza (1865-1943) came for the first time to the observatory; he had studied for two years astronomy and meteorology in Georgetown University. In 1905 Simón Sarasola (1871-1947) also came to the observatory. They became directors of Belén Observatory and directed other observatories in Cienfuegos, Cuba, and Bogotá, Colombia. After the American-Spanish war the U. S. Weather Bureau established an office in Havana, which soon showed a total disregard for the work of Belén. Its observer claimed that, before the arrival of the Americans, nothing of scientific value had been done in Cuba regarding the occurrence of hurricanes. In 1905 Walter M. Drum, a Jesuit from Georgetown University, published in the United States a short study refuting the charges and explaining the work of Belén Observatory. Drum concluded "We hope that what we have given will go far to save the fair name of the Observatory of Belén, and to show that for thirty years the Cubans had been thankfully receiving Jesuit forecasts before our Weather Bureau observer found it difficult to interest the people in his warning system". As a consequence the U. S. Weather Bureau acknowledged the work of the Observatory, ordered its office in Havana closed and asked Belén to send daily telegrams with meteorological observations to its main office in Washington during the cyclone season. In 1907 two Bosch-Omori horizontal seismographs were installed in Luyanó, outside Havana, on a property of the Colegio. These were the first seismographs installed in the Caribbean region and they were in operation until about 1930. In 1925 the observatory moved again to a new site on a tower of the large magnificent new building of the Colegio de Belén in the neighborhood of Marianao. The telescope dome, however, was not installed.

The Director from 1928 to 1943, Gutierrez Lanza, was a very active lecturer and published several works on the tropical hurricanes of the Caribbean and the climatology of the region. From 1935 weather news from the observatory were broadcast twice daily through the radio stations of Havana and of other Cuban cities, through the network R. H. C. Cadena Azul. In 1942 the observatory installed its own short wave radio station to broadcast meteorological information. In 1943, Sarasola returned from Colombia, where he had organized in 1922 the Servicio Meteorológico Nacional, and remained as Director till 1947. That year Rafael Goberna (1903-1985) , was named director; he had studied geophysics and meteorology at Saint Louis University from 1943 to 1945 and spent a summer at the University of Chicago. Goberna continued the work of the observatory, but after the Second World War its importance began to decline with the development of new methods of observation and tracking of hurricanes. However, the observatory did not slowly fade away, but was abruptly closed in 1961, when the college was confiscated by the communist government of Fidel Castro and most Jesuits were expelled from Cuba, among them Goberna. Paradoxically Castro had been a student of Belén's and was on good terms with some of his former teachers. Shortly after being expelled from Cuba, Jesuits from Belén began in Miami, Florida, Belén Jesuit Preparatory School, a high school, at first, for boys whose families had left Cuba for the United States and afterwards open to other students. In the new school, in the 1980's, Pedro Cartaya (1936-), a Cuban Jesuit, founded an observatory with meteorological and astronomical equipment, for the use of students, as a continuation and tribute to the old Belén Observatory. More than 50 students are at present enrolled in the Meteorology and Astronomy Clubs, which hold meetings every two weeks after school as well as night-watches for astronomical observations.

Two other observatories were established by the Jesuits in Cuba. The first in the Colegio de Nuestra Señora de Montserrat in Cienfuegos, where meteorological observations began in 1886 and which were sent to Belén. Gangoiti was in charge of the observations until 1893 when he moved to Belén. The observatory was formally begun in 1908 and inaugurated in 1910, with Sarasola as its first Director. Although mainly dedicated to meteorology with very complete instrumentation, it had also an astronomical section, with a Zeiss telescope of 13 cm aperture and 2.4 m focal length, a meridian circle by Sartorius and other equipment. In 1920 Sarasola traveled to Colombia; he was replaced by Gutierrez Lanza until 1925, when Lanza moved to Belén. The subsequent directors were science teachers in the school. The observatory worked together with Belén and collaborated with the U.S. Weather Bureau in Washington. It was closed together with the school in 1946. The second observatory was established in 1934 in the Colegio Dolores of Santiago de Cuba in the western part of the island by Santiago Viña (1876-1952) who was its Director till 1951. The observatory continued in operation until 1961, when the government of Fidel Castro confiscated the school.

Mexico

In Mexico Jesuits established two observatories, which were primarily meteorological, although they had also astronomical, geomagnetic and seismological instruments. The most important was established in the Colegio del Sagrado Corazón, founded in 1870, in Puebla, a town south of Mexico City. The first meteorological observations began in 1875, but regular observations with a formal observatory began in 1877 under the direction of the Italian

Pedro Spina (1839-1925). This was one of the first meteorological observatories in Mexico, before the establishment of the Mexican Central Observatory. The observatory followed the procedures established in Belen Observatory in its observations. The observatory had also magnetic instruments. The same year a seismograph was installed, consisting of a vertical pendulum which recorded on a smoked glass. This instrument was constructed by Gustavo Heredia (1869-1926), who collaborated in the observatory and was its Director from 1905 to 1910. This may have been the first seismograph installed in America, ten years before the first ones installed in California (Lick and Berkeley). Heredia published the seismological observations in Puebla for the large Mexican earthquakes between 1877 and 1906. In 1880 a small Dollond telescope was installed, which was in 1882 replaced by an equatorial telescope of 11.6 cm aperture and 1.8 m focal length in a circular room with a rotating dome. Astronomical work in Puebla began only two years after the foundation in 1878 of the Mexican National Astronomical Observatory of Chapultepec. Spina published astronomical observations from 1881 to 1886, especially those of the transit of Venus and of a comet in 1882. He made also observations of sunspots and faculae and published some works on the climatology of Mexico, one of then on the influence of Mount Malinzin on cloud condensation over the valley of Puebla. Spina proposed in 1883 a curious cosmological theory based on the, so-called, laws of universal rotation and impulsion.

In 1886 Spina was succeeded by the also Italian Enrique Cappelletti (1831-1899), who continued the meteorological observations until 1890 and published a book on cosmography in 1887. Heredia, director from 1905, continued the astronomical and meteorological observations till 1910. During this time, the observatory sent the observations to the Servicio Meteorológico of Mexico. The observatory of Puebla figured in the list of astronomical observatories by Lancaster published in 1887. It was closed when the Mexican revolution closed the school in 1914. The school was opened again in 1921, but the observatory was never reestablished. The second observatory was established in 1884 in the Colegio de San Juan Nepomuceno in Saltillo, a town of northern Mexico in the state of Coahuila and was only dedicated to meteorology. Spina, Heredia and Cappelletti, directors of Puebla, were also at alternate periods of time directors of Saltillo. The observatory of Saltillo, as that of Puebla, was closed by the Mexican revolution in 1914.

Central and South American Meteorological Stations

Other small meteorological observatories were established in Jesuit colleges in Central and South America. The first was founded in 1851 in the Colegio-Seminario in Guatemala. This was actually the first Jesuit meteorological station on the American continent. In 1856, Antonio Canudas (1814-1874), a science teacher, was in charge of the observatory and was officially named Director in 1862. The observatory published monthly bulletins and an annual summary of observations (temperature, atmospheric pressure and humidity) and variations in the magnetic declination with maxima, minima and mean values. The last Director was Justiniano Arrubla (1840-1901), who in 1881 went to the Colegio de Belén in Havana where he was a physics teacher. The school and observatory were closed in 1871, when Jesuits were expelled from Guatemala. The observatory founded in 1862 in the Colegio de S. Ignacio de Santiago de Chile by Cappelletti with meteorological and magnetic observations had shorter duration. In 1867 the magnetic observations were discontinued and in 1872 the observatory was closed. Cappelletti, as we have seen, went to Mexico in 1886 where he became Director of the observatories of Saltillo and Puebla. In 1904 an observatory was established in the Colegio de la Inmaculada in Santa Fe, Argentina. Directors changed often and were the science teachers in the Colegio. In 1916 it had very complete instrumentation with four Fortin barometers, a Richard anemometer, thermometers, etc. It was considered a second class meteorological station and sent its observations daily and monthly to the Oficina Meteorológica of Cordoba and to the Oficina Central of Buenos Aires. The Observatory was closed in 1967.

In Sucre, Bolivia, a meteorological and seismological observatory was founded in 1914 by Francisco Cerro (1857-1945) who was its director until his death. The seismographs were mechanical, a horizontal with a mass of 3000 kg and a vertical of 1500 kg with smoked paper recording made in the observatory. For some years the observatory published a meteorological bulletin. In 1945 Antonio Echevarría (1904-1976), who succeeded Cerro, reinstalled the seismographs which were out of order for some years. They were damaged in the earthquake of 1948 and never replaced. Echevarría left the observatory in 1955 and in 1967 was in the Ebro Observatory. The Observatory continued as a meteorological station with the name Observatorio P. Cerro until 1966 when it was closed. The establishment of a meteorological observatory in the Colegio de Cristo Rey in Asunción, Paraguay was an initiative of Luís Parola (1885-1978), the Jesuit provincial, supported by the President of the Republic Eusebio Ayala. José Ubach (1871-1935), a teacher of the Colegio del Salvador in Buenos Aires, who had been in charge of the magnetic section in the Observatorio del Ebro between 1911 and 1912, was in charge of the installation. The instruments were lent by the Servicio de Meteorología Nacional of Argentina. The observatory installed on the roof of the school began observations in 1935. Its Director was Cecilio Martín (1891-1971), who designed a synchronous graph to record together atmospheric pressure, temperature and humidity. There was a project to install four more stations that never materialized. Five years later in 1940 the observatory was closed, due to difficulties with the Meteorological Service of Argentina, which reclaimed the instruments and gave them to the newly created Meteorological Service of Paraguay.

The Observatorio Nacional in Bogotá, Colombia was an early institution, founded in 1803 by José Celestino Mutis, with sections of astronomy and meteorology. In 1920 at the death of its Director Julio Garavito Armero, there were difficulties in finding a suitable successor. The president of Colombia, Marco Fidel Suárez, following a suggestion of the Ambassador of Spain, J. Almeida, asked the Jesuits to organize the Servicio Meteorológico Nacional and to take the post of Director of the Observatorio Nacional. The man selected was Sarasola, who was then Director of the observatory in Cienfuegos. Sarasola traveled to Bogotá in 1921. There, owing to the poor conditions of the building of the Observatorio Nacional, he established the Central Observatory of the new Meteorological Service of Colombia in the Colegio de San Bartolomé, the Jesuit college in Bogotá. Meteorological instruments were installed and a Zeiss altazimuth telescope of 13 cm aperture and 2 m focal length. The observatory was inaugurated in 1922. A network of secondary stations was established with great difficulty. Of the proposed twenty stations, in 1924 there were only four in operation, in Bucaramanga, Ibagué, Neiva and Tunja. Despite difficulties, Sarasola kept the Meteorological Service in good working condition and began a fruitful collaboration with the U. S. Weather Bureau in Washington. In 1923 he installed two seismographs, the first seismographs in Colombia, a Wiechert horizontal of 200 kg mass and a horizontal instrument of 1000 kg mass made in Spain according to the design of those in Cartuja. The observatory published periodically Annals and Notices. In 1941 Sarasola quitted in his job and together with Ramirez founded the Instituto Geofísico de los Andes Colombianos.

La Paz

The second activity carried out by Jesuit observatories in Central and South America was seismology. This is natural given the intense earthquake activity in this region, which causes a large number of casualties and great material damage. We have already mentioned the installation of seismographs in some of the meteorological observatories, the first in Puebla in 1877 and the second in Belén in 1907. Two observatories were specially distinguished for their work in seismology one in La Paz, Bolivia, and the other in Bogotá, Colombia. The origin of the Observatory in La Paz was the installation of a meteorological station in 1892 in the Colegio San Calixto, which had been founded in 1882. Meteorological observations have been kept from 1892 until today. In 1911, in the second General Assembly of the International Seismological Association in Manchester, a resolution was approved in which the Jesuits were asked to

Figure 26. Pierre Descotes, first Director of San Calixto Observatorio, La Paz, Bolivia, 1912-1964 (OSC).

install a seismological station, somewhere in the central part of South America, since no station existed there. This recommendation recognized the work being done by Jesuits in seismology in many of their observatories. In response to this petition a seismograph was installed in the Jesuit high school of La Paz. This was a bifilar horizontal pendulum of 450 kg mass, with smoked paper recording of the type designed by Navarro-Neumann in the Observatory of Cartuja. The instrument was made following his directions by Esteban Tortosa, who had worked in Cartuja for some time and was installed in the crypt of the church.

In 1912 the French Jesuit, Pierre M. Descotes (1877-1964) arrived in La Paz, after spending a few months in Cartuja with Navarro-Neumann, and was named Director of the observatory, which was given the name of Observatorio San Calixto. In 1913 he installed two new horizontal seismographs of the same type as the ones used in Cartuja, made in the observatory, one of 1200 kg mass and 8 seconds period and the other of 500 kg mass and 15 seconds period, with amplifications of 100 and 300. A third instrument was also made in the observatory, with a pendulum of 1500 kg mass and 2.5 seconds period, which recorded the two horizontal components. Since there was no astronomical observatory in Bolivia, the correct time was determined by astronomical means using a Claude and Driencourt astrolabe and in 1922 a meridian circle telescope by Cooke. From that year the time determined by the observatory was the official time in Bolivia for many years. In 1922 Descotes in order to determine the exact coordinates of the observatory, determined very accurately the difference in longitude with the Observatory of Santiago de Chile. In 1926, Descotes, on the occasion of the International Program of Longitudes made a revision of the longitude of the observatory receiving time signals from the station in Annapolis, in the United States controlled by the observatories of Washington and Paris. From these measurements Descotes discovered that there was an error in the previous value and therefore that the longitude of the Observatory of Santiago de Chile had an error of 1.18 seconds. This was confirmed by observations made by the observatories of Santiago de Chile and of La Plata, Buenos Aires, Argentina. Descotes traced back this error to the longitudes calculated by the American Commission of Green and Davies in 1879. This commission had erroneously corrected in one second the previous determinations made by E. Mouchez between 1856 and 1863. Since the results of Green and Davies were adopted for all the longitudes of South America, Descotes discovered that all the geographical longitudes in the continent were displaced 400 meters to the west and had to be corrected.

A great improvement in the seismological instrumentation of the observatory was made in 1929 with the installation of three-component Galitzin-Wilip electromagnetic seismographs, the most advanced instruments at that time. On this occasion the underground vault for the instruments was enlarged. The careful maintenance of instruments and exact analysis of seismograms by Descotes made the observatory one of the most reliable seismographic stations in South America. In the famous book by B. Gutenberg and C.F. Richter, *Seismicity of the Earth*, published in 1949, the observatory was praised with the words: "La Paz (observatory) at once became and still remains the most important single seismological station of the world". In 1957 during the International Geophysical Year, the Department of Terrestrial Magnetism of the Carnegie Institution in Washington donated a new vertical Wilson-Lamison seismograph of one second period of high magnification, improving the detection capabilities of the observatory.

During these years several Jesuits collaborated in the observatory, but it was Descotes who carried most of the weight. In 1959 Ramón Cabré (1922-1997) arrived at the observatory. He had finished his licentiate in physics in the University of Barcelona, Spain, and spent some months in the Institute de Physique du Globe, in Strasbourg, with Jean P. Rothe and P. Peterschmidt. Cabré was named Subdirector, and Descotes continued nominally as Director until 1964, although he had moved to Cochabamba in 1958 due to health

Figure 27. Ramón Cabré, Director of San Calixto Observatorio, La Paz, 1965-1993, in the tunnel of the seismological station of Zongo (OSC).

problems. Cabré was named Director after Descotes' death in 1965. He carried out a very active program of scientific collaboration with American and French institutions and continued enlarging the observatory with new modern equipment. In 1960 the observatory participated in an important project of seismic refraction organized by the Carnegie Institution, which installed 11 portable seismic stations in Bolivia to record several explosions of 50 tons made in the mines of Chuquicamata. The experiment showed that the earth's crust in the altiplano produces very high attenuation of seismic waves. New improvements in the instrumentation were made in 1962 when a seismographic station of the global network WWSSN promoted by the United States government, with the most modern three component seismographs of short and long period was installed just south of La Paz. Further collaboration with American institutions resulted, the following year, in the installation of an array LRSM of 7 vertical seismographs by the AFOSR (Air Force Office of Scientific Research) in Peñas, located to the NW of La Paz. This array was operative until 1975. In the same location an array of very sensitive atmospheric microbarograph sensors has been installed in collaboration with French institutions to monitor nuclear tests.

In 1966 Luis Fernández came to the observatory after obtaining his doctoral degree in seismology at Saint Louis University. He gave a great impulse to the seismological research in the observatory with new projects and publications. Fernández left the observatory in 1970 to work in the Seismological Survey of South Africa and afterwards left also the Jesuit Order. The observatory continued under the Cabré directorship incorporating lay personnel, renewing and expanding its instrumentation and also carrying out research, mainly on the seismicity, tectonics and crustal structure of Bolivia and the Andean region. The observatory participated in several international programs and research projects with institutions such as the U.S. Geological Survey (USGS), the University of Paris and the CERESIS (Centro Regional de Sismología para America del Sur) of which Cabré was the first Director. In 1972 a tunnel was excavated in the rock in the valley of Zongo, on the side of the mountain Huayna Potosí, in order to install a new station of the most modern technology, known as HGLP (High Gain Long Period), by the USGS. Only a dozen of these stations were installed around the world. The station was modified in 1973 into a station of the ASRO, (Adapted Seismic Research Laboratory) type, which was given the name of ZOBO. The same year the offices of the observatory were moved from the college to a new location nearby in La Paz. With the support of the USGS several seismographic stations were installed in Bolivia outside La Paz with instruments of short period. In 1988 in collaboration with the University of Paris a tripartite station array was installed in Zongo. In 1993 in collaboration with the USGS, instruments of the new generation of digital broad-band seismographs were installed in Zongo. The station is part of the Global Telemetered Seismic Network (GTSN).

After 28 years as Director Cabré manifested his desire to resign and the Jesuit Provincial of Bolivia wrote a letter to other provincials asking for a substitute. Two Jesuits showed their willingness to come to La Paz , Lawrence A. Drake (1931-), who had been the last director at Riverview Observatory in Australia and William R. Ott, Director of the Seismographic Station of John Carroll University, in Cleveland, Ohio. Ott remained in La Paz for a couple of years, returning to the United States in 1992. Drake remained in La Paz and became Director of the Observatory in 1993. Drake continued the programs and collaborations began by Cabré, who died suddenly in 1997 in Cochabamba, where he had retired to do pastoral work. In the end of 2001 Drake returned to Australia owing to health problems. On this occasion Jesuits superiors in Bolivia decided to transfer the observatory to the Universidad Católica in La Paz. This was the end of 90 years of Jesuit seismological work in La Paz.

Bogotá

In order to be prepared for scientific work and to continue the work begun by Sarasola in the Observatorio de San Bartolomé in Bogotá, Jesús Emilio Ramírez (1904-1981), was sent to Saint Louis University, where he studied under Macelwane and obtained his M.S. degree in Geophysics in 1931 and later in 1939 his Ph. D. His doctoral thesis, directed by Macelwane, was on the topic of the detection of cyclones by microseismic observations with a tripartite array of seismographic stations. During his stay in Saint Louis, Ramírez conceived the idea of creating a new geophysical institute in Bogota dedicated to seismology and meteorology. After his return to Bogotá in 1939, he obtained the collaboration of Sarasola for this project, who quit his job as director of the National Meteorological Service of Colombia. The observatory was set up in the new buildings of the Colegio de San Bartolomé, La Merced, as the old ones became government property. The seismological station was installed in a vault excavated in the rock in the grounds of the school, the meteorological instruments on the roof of the school building and the offices in its upper floor. The new observatory was inaugurated in 1941, with the name of Instituto Geofísico de los Andes Colombianos. Ramírez and Sarasola were codirectors, in charge of the seismological and meteorological sections respectively. The meteorological instrumentation was very complete. The seismographic station began with a vertical Benioff seismograph of 100 kg mass and the Wiechert seismograph from the San Bartolomé Observatory, which had been closed. A set of three-component Galitzin-Wilip seismographs had been ordered from Estonia, but owing to the beginning of the war in Europe they were never delivered.

In 1943, Sarasola returned to the Observatory of Belén and Ramírez assumed the directorship during the following forty years. Actually the Instituto was always very strongly linked to his person. Ramirez acquired soon an important presence in the scientific circles of Colombia and South America. In 1949 he was named Director of the Servicio Meteorológico Nacional of Colombia, a post which he occupied for only one year, and, for 13 years from 1947 to 1960, he was Professor of Geophysics in the state Universidad Nacional of Bogota. For his classes he wrote, in 1957, in collaboration with Luis G. Durán, *Nociones de Prospección Geofísica*, a textbook in applied geophysics, one of the first written in South America. His main research work was on the study of seismicity and tectonics of the Colombian region, a topic on which he published many scientific articles. In particular he studied in detail the characteristics and effects of the most important earthquakes which took place in Colombia. He also worked hard to complete and renovate the instrumentation of the seismographic station. In 1946 two horizontal

Figure 28. Entrance to the seismological vault of the Instituto Geofísico de los Andes Colombianos, Bogota, Colombia.

Sprengnether seismographs and a Montana accelerograph were installed in place of the Galitzin-Wilip instruments which never arrived. The Instituto became soon the primary seismological institution of Colombia and was responsible for the technical direction of the Colombian Network of Seismological Stations, established in Chinchiná (1949), Galerazamba (1949), Fúquene (1957), Pasto (1971) and Bucaramanga (1973). From 1972 it was also responsible for the network of accelerographs with instruments in Cartagena, Barranquilla, Montería, Medellín, Manizales, Calí and Bogotá. In 1962 the Instituto was selected for the installation of a WWSSN station. This was, after La Paz, the second station of this type installed in a Jesuit observatory in South America. In 1975 a very advanced SRO (Seismic Research Observatory) seismographic station was installed by the USGS. This type of station had the seismographs installed in a borehole of 20 meters depth to minimize the seismic noise. Very few of these stations were installed in the world and this was the only one in South America. The meteorological section was also augmented with new instruments. In 1957 a sounder for the study of the ionosphere was lent by the U. S. National Bureau of Standards; it operated until 1967. Supervised by the Instituto, another ionospheric sounder was installed in Cartagena from 1961 to 1966.

The offices in the college were soon inadequate for the volume of work of the Instituto, and a special building was built in 1957 in the grounds of the Jesuit Universidad Javeriana, near the college, for the offices, workshop, library and a museum of minerals and fossils. The same year the meteorological station was moved to the building of the Engineering Department. The Instituto was definitively ascribed to the Universidad Javeriana in 1960 when Ramírez was named Rector of the University. From 1973 to 1979, Ramirez organized the Nariño Project for the study of seismicity and crustal structure in Colombia, with the participation of several national and foreign institutions. This was one of the largest field research seismological programs carried out in South America. The presence of Jesuits in the Instituto was never very large, except for Ramírez himself. From 1946 to 1949 Wladimiro Escobar (1922-1989) was Subdirector; he returned in 1957 to became Director from 1985 to 1989. In 1962 Rafael Goberna, who had been the last director of Belén and was expelled from Cuba by Castro's government, came as Subdirector. He became Director after Ramírez' death from 1981 to 1985. René van Hissenhoven (1936-) was prepared to substitute Ramirez with studies of geophysics at Boston College (Weston Observatory) and at the University of Wisconsin. He became Director from 1989 to 1993, but his interest for the observatory work did not last long and he soon left the observatory to do pastoral work. He was the last Jesuit director of the Instituto. In 1993 A. Amezquita, Professor of Civil Engineering of the Universidad Javeriana was named director until 1996. In 1995 the meteorological station was closed and in 1996 the institute was merged into the Engineering Faculty of the University and its name changed to Instituto Geofísico de la Universidad Javeriana. A chair in the Universidad Javeriana "Cátedra Jesús Emilio Ramírez" was created to honor Ramirez' name. Active Jesuit participation in the Instituto was finished in 1993.

Other seismographic stations

Two other seismographic stations were established by Jesuits in Jamaica and Chile. The Jamaica mission was founded by English Jesuits in 1837. In 1850 a school was founded in Kingston by a group Spanish Jesuits from Colombia who were later replaced by English Jesuits. The school was destroyed in the earthquake of 1907. In 1914 the school was reopened with the name of St. George's College in its present location in Winchester Park. In 1940 Linehan of Weston Observatory installed a seismographic station in the college with two Wood-Anderson seismographs. The reason was the absence of seismological stations in the Caribbean region; only two existed, in Puerto Rico and Martinique. Its director was John A. Blatchford (1897-1978), a native of Boston, Massachusetts, and a science teacher since 1931. He retired in 1961, but continued as Director of the seismological station until it was closed in

1975. Records of the station were sent to Weston. It seems that relations between Blatchford and Linehan were not very good and the station did not operate always correctly.

A seismographic station was installed in 1949 by Germán Saa (1909-1970) in the Colegio de San Luís, in Antofagasta , Chile. The first instruments were two Bosch-Omori horizontal seismographs of 120 kg mass, recording on smoked paper. In 1957 a Wilson-Lamison seismograph was installed by the Carnegie Institution, Washington, in collaboration with the Universidad del Norte. A small telescope was also installed and the station functioned as a seismological and an astronomical observatory until 1963. The importance of the station decreased when in 1962 a WWSSN station, run by the Universidad de Chile, was established outside Antofagasta. In 1965 Saa moved to Lima, where he worked with the Carnegie Institution. Later he planned to establish a seismological station in Arica, in the northern part of Chile, but he died in La Paz before he could install the instruments.

San Miguel

The origin of the Observatorio de Física Cósmica de San Miguel, in the town of San Miguel near Buenos Aires, Argentina, was the great interest for science, and specifically for astronomy of Monsignor Fortunato J. Devoto, Auxiliary Bishop of Buenos Aires. He had been Director of the Astronomical Observatory of La Plata and in 1933 was the founder and first President of the Consejo Nacional de Observatorios (National Council of Observatories). At that time there were two astronomical observatories in Argentina, La Plata and Cordoba. Mons. Devoto conceived the idea of complementing the work of these two observatories with a new one dedicated to astrophysics and solar-terrestrial relations, and thought of the Jesuits for its direction. The Jesuit Provincial of Argentina, Luís Parola, approved enthusiastically the idea and offered the grounds of the Jesuit Philosophy and Theology Faculty, Colegio Máximo de S. José, in San Miguel, near Buenos Aires, for the new Observatory. For its direction Ignacio Puig (1887-1961), subdirector of Ebro Observatory was chosen. Ebro actually served as the model for the new observatory. Thus, in order not to interfere with the two already existing observatories, astronomy was excluded from its aims. As in Ebro, the subjects were those of solar activity, cosmic radiation and electromagnetic phenomena in the earth. Thus the observatory was to have three sections, astrophysics, electromagnetic phenomena and geophysics. In 1934 Puig traveled to France and Germany in order to select and acquire the first instruments. The first building elegant and of three stories, with beautiful arches, designed by the architect Jorge Mayol, was inaugurated in 1935. At the solemn inauguration

Figure 29. Observatorio de Física Cósmica de San Miguel, Argentina (at about 1970). The white building at the center is the first building built in 1935.

were present the Papal Nuntius, Felipe Cortesi, the President of Argentina Agustín P. Justo, the Ministers of Education and Agriculture and many other visitors. This was a very brilliant act, with the presence of a good representation of the high society of Buenos Aires, whose cars filled the grounds of the Observatory. Father General Wladimir Ledochowski sent a telegram of gratitude to the Consejo Nacional de Observatorios for entrusting the direction of the new observatory to the Jesuit Order.

The first Jesuits who collaborated in the Observatory were, besides Puig, Nilo Arriaga (1895-1973) (Subdirector in 1935-37), Juan Rosanas (1881-1955) who had worked in 1910 in Ebro Observatory and Joaquín Selles (1885-1946). The first instruments installed were those of the sections dedicated to the electromagnetic phenomena and meteorology. Two lines for measurements of telluric currents of 2 km in NS and EW directions with 10 connections were laid out, with an electrometer Labo-Gif (Paris). Other instruments were a Gerdien for measurements of atmospheric electricity, an ionization detector by Ebert, and a Wulf electrometer. The meteorological station was donated by the

Servicio Meteorológico Nacional of Argentina. Thus, work began on the observation of telluric currents, atmospheric electricity and radio propagation. In 1940 the Instituto Entomológico San Miguel for the study of insects was established on the grounds of the Observatory. In this institute Gregorio Williner and Juan Bridarolli sampled a rich collection of more than 250000 insects. The collection had been started in 1925 by Jesuits professors of natural science in the Seminario Metropolitano of Buenos Aires.

In 1943 Puig returned to Spain and was succeeded by the Argentinean, Juan Antonio Bussolini (1905-1966) , who gave very great impulse to the scientific work of the observatory. Solar studies began in 1950 with photographic observations of sunspots by means of a Zeiss telescope. In 1964 a telescope with Lyot filter and two radiotelescopes of 5 m diameter were installed and observations were made of solar activity and radiation in the H-alpha line, and the 408 and 9700 MHz radio emissions. Activity at the observatory was at a high with Carlos Esponda (1917-) and Arturo Yriberry (1906-1995) working on geophysical problems, such as telluric currents, and Arriaga in solar physics. Arriaga was Subdirector from 1943 to 1969 and Director Emeritus after that year. In 1957 the observatory participated actively in the Argentinean programs of the International Geophysical Year. This period was the most fruitful of the Observatory owing to the able direction by Bussolini. A year after his death in 1967 a new spectroheliograph and radiotelescope were installed in a new building of three stories which had been designed by Bussolini, inspired by the Manila Observatory building, and research in solar physics took a new impetus.

In 1968 Mariano Castex was named Director. He launched a program of expansion of the observatory, which finally was proved to be very detrimental and contributed to its end as a Jesuit institution. New buildings were built, so that in 1970 there was a total of 13 buildings of various kinds. Between 1968 and 1969, the observatory was thoroughly renovated and modernized, especially in the line of solar physics research. An interesting addition was the establishment in the ground of the observatory of an institute for the study of the influence of radiation on animal reproduction, the Instituto Latinoamericano de Fisiología de la Reproducción (ILAFIR) which depended on the then Jesuit Universidad del Salvador. All these programs needed large funds, which the Jesuits could not cover. In order to obtain funds for the observatory, Castex, with official governmental support, created the Comisión Nacional de Estudios Geo-Heliofísicos (CNEGH), of which he was named President. This was a government commission, on which also other research institutions depended. In virtue of its connection with the CNEGH, the observatory changed its name to Observatorio Nacional de Física Cósmica. Many employees were hired and it was difficult to keep Jesuit control, although the observatory remained Jesuit property. Since the observatory depended on government funds for its operation, politics also played its part in the situation, creating serious problems. Finally Castex was removed from his post in 1970 and later left the Jesuit Order. From 1970 laymen were directors of the CNEGH and occupied the influential posts of academic and administrative subdirection of the observatory. Jesuits still figured as Directors, but its direction was only nominal. In 1975 the only Jesuit active in the observatory was Tomás Paneth (1926-). Finally, in 1977, the observatory was officially transferred to the Comisión Nacional de Investigaciones Espaciales, an institution dependent on the Argentinean Air Force.

Chapter 7

ASIA, AFRICA, AND AUSTRALIA: THE GREAT MISSION OBSERVATORIES (1814-2000)

With the exception of the observatories in Australia and India, which had characteristics akin to those in North America and Europe, observatories in Asia, the Middle East and Africa have some common features derived from their being in mission countries. These observatories were part of the apostolic work in non-Christian and mostly undeveloped countries. They were conceived of as an important witness of the work of missionaries in contributing to the progress of these countries and to the harmony between modern science and Christian faith. The scientific prestige of the observatories was considered to be an important factor in spreading the Christian message. In this sense, they were a continuation of the work of Jesuit scientists who in the 17th and 18th centuries had worked in observatories in China and India. Verbiest, one of the directors of the Imperial Observatory in Beijing had already expressed a similar view, saying that Christian faith entered China helped by the hand of astronomy (Chapter 1). In the 19th century and at the beginning of the 20th century, in many countries, Jesuit observatories were the only existing scientific institutions, and remained so for many years. In them, Jesuits laid the foundations for many state observatories and institutions which developed much later. For this reason, most of the Jesuit observatories in these countries covered a wide variety of subjects such as astronomy, solar physics, meteorology, seismology, and geomagnetism. In the practical aspects of these sciences these observatories offered, among other services, accurate time keeping, support of geodesic measurements and cartography, weather forecasts especially of hurricanes and typhoons, and evaluation of earthquake damage and assessment of the risk of earthquakes. These were invaluable services in countries to a great extent undeveloped. To these services we have to add the scientific training received by local personnel who collaborated in the observatories. In some countries those trained by the Jesuits were the first to receive any scientific formation at all. Thus Jesuits contributed in an important way to the scientific development of these countries. Among the observatories established in such countries the four great mission observatories, namely, Manila, Zikawei, Tananarive, and Lebanon excelled by their importance.

Manila

The founding of the Manila Observatory took place in 1865. This was the year in which Francisco Colina (1837-1893), a teacher of mathematics, began meteorological observations with rather rudimentary equipment in the secondary school founded by the Spanish Jesuits in Manila, Philippines, in

1860. It was known first as the Ateneo Municipal and later as the Ateneo de Manila. Thus, the first name of the observatory was Observatorio del Ateneo Municipal. This was one of the first observatories established in the Far East, antedating those of Batavia (Jakarta) in 1866 and Tokyo in 1875. Colina, the first Director, was helped by two young Jesuits, Juan Ricart (1838-1915) and Jaime Nonell (1844-1922). The same year, 1865, a violent typhoon produced heavy damage in Manila. The observations made by the observatory were published in the newspapers and attracted public attention. From different sections of Manila society, especially businessmen, merchants and mariners, funds were collected, so that new instruments were soon installed. From that date on meteorological observations were carried out at regular intervals.

Francisco Faura (1840-1897) arrived from Spain in 1867 and enthusiastically took charge of the observatory. He actually can be considered as the true founder of the observatory. In 1869 a Secchi meteorograph was acquired for the observatory and was set up by Faura although it came without instructions and was a very complex instrument. In 1870 an earthquake felt in Manila led Ricart to build two pendulums, horizontal and vertical, which two years later recorded the destructive earthquake of 1872. With a small telescope and other instruments, Faura, Ricart and Nonell traveled to Mantawalok, a small island of the Celebes archipelago, to observe the solar eclipse of 1868. After many difficulties, and little co-operation from the Spanish Colonial Government, the three young Jesuits boarded the British warship HMS Serpent, which took them to Mantawalok where they joined a Dutch team from the University of Utrecht. Despite the poor instruments used, they obtained interesting results, which were sent to the Director of the Observatory of Madrid and to Secchi in Rome. They were published in the Bulletin of the Observatory of the Collegio Romano. This may be considered the first scientific project carried out in the observatory by which it became known to the scientific community. Faura returned to Spain in 1871 and prepared himself for his future work, spending some time in 1877 at the Collegio Romano with Secchi and, the following year, at Stonyhurst with Perry. He also visited several observatories in Italy and made contact with the Italian seismologists Timoteo Bertelli, Filippo Cecchi and Michele de Rossi. Faura returned to Manila in 1878 and became Director of the observatory for nearly twenty years until 1897. With Faura as Director the observatory began more regular work. Though its main section was meteorological with very comprehensive equipment, including a Secchi meteorograph, it had also a magnetic section with an Elliot magnetometer and Dover inclinometer, calibrated at Stonyhurst by Perry. The seismological section housed a horizontal seismograph which recorded on a smoked glass plate, as well as a vertical pendulum, for visual observation. There was also a meridian telescope by Dollond among its

Figure 30. Federico Faura, Director of Manila Observatory, 1878-1897 (MO).

instruments. The observatory soon established relations and exchange of data with the observatories of the Far East, especially with the Jesuit observatory of Zikawei in China.

In 1879 Faura began the study of typhoons, or "baguios", the name given to them in the Philippines. He was able to forecast the arrival in Manila of a serious typhoon on July 7, 1879 and announced it to the public. This was the first official storm warning in the Far East and its value was immediately recognized. The same procedure was repeated on November 19 of the same year for another typhoon passing over Manila. Two days before it arrived Faura issued his warning and precautions were taken, such as closing the harbor. Places which received the warning in time to take precautions suffered little damage. The success of these forecasts raised the prestige of the observatory and its subsequent warnings were always taken into consideration. Faura contacted Viñes in Belen and soon realized that typhoons in the Pacific and hurricanes in the Caribbean were the same type of phenomenon. In 1882, Faura published his first study on typhoons, with emphasis on their precursory phenomena. The same year, based on the ideas proposed by Viñes, Faura developed an aneroid barometer adapted for the detection of approaching typhoons. On July 18, 1880 an earthquake violently shook Manila and was recorded by the instruments of the observatory. Its records were the first of an earthquake in the Philippines and brought the observatory to the attention of seismologists, among them John Milne with whom Faura began to correspond. In 1881 new seismological instruments made in Italy replaced the first set. These were a Cecchi horizontal seismograph, especially designed for the observatory at Faura's request, two microseismographs by Rossi and Cecchi, and a vertical seismoscope (tronometer) by Bertelli. Thus the seismological station was equipped with the best instruments available at that time. The observatory was first located on a tower of the main building of the Ateneo and underwent several renovations.

In 1884, the observatory acquired official status as the headquarters of the Meteorological Service of Philippines, created by a royal decree of the King of Spain, Alphonse XII. The Service was formed by thirteen stations dependent on the observatory, which was now renamed "Observatorio Meteorológico de Manila". Faura became the first Director of the Meteorological Service and all the observatory personnel were officials. The following year, the observatory assumed responsibility for official timekeeping, marking the correct time at noon by lowering a ball on a mast on the roof of the building; this was followed by a cannon shot in the harbor of Manila. The observatory also inaugurated a system of visual signs on the harbor to inform mariners of severe storm weather. In 1887, the Meteorological Service was formed by 10 meteorological stations on the island of Luzon, which sent observations by telegraph three times a day. Other stations were added later and in every one of them six observations were made every day, and more often during typhoon season. The observatory also received observations from stations in China, Japan and Indochina.

The old building of the observatory was inadequate for the importance that it was acquiring. As a consequence of the damage suffered in the earthquake of 1880, a new building was built as an annex to a new Jesuit school to the south of Manila for training school teachers. The new observatory, with ample space for offices, library and a tower for meteorological instruments, was inaugurated in 1886. The meteorological instrumentation was completed with Sprung-Fuess and Richard barographs, thermographs, heliographs, a Thompson electrometer by Mascart, etc. Two young Jesuits joined the observatory, at this time, Alphonse Renkin (1851-1886) and Martín Juan (1850-1888); both of them soon died. Juan had prepared for work in the observatory in the field of geomagnetism, spending a year with Perry at Stonyhurst and some time in the magnetic observatory of Montsouri, France, with Prof. Moureau. He traveled to Manila with new magnetic instruments that he had acquired in France and he took charge of the magnetic section of the observatory in 1887. From April to July 1888, Juan carried out, together with the English Jesuit, John Doyle (1862-1918), a magnetic field survey on various islands of the archipelago. His sudden death during the survey left the work incomplete.

In 1888 Ricardo Cirera arrived at the observatory and took charge of the magnetic section after Juan's death. He brought new magnetic instruments, among them Mascart variometers with photographic recording of the type used in the magnetic station of Parc Saint Maur in Paris. Between 1890 and 1893 Cirera completed the magnetic survey begun by Juan with an extensive program of measurements on different islands of the Philippines archipelago and on the coasts of China and Japan. In this work he was helped by Jovellanos and Miguel Saderra Mata (1852-1938). He published the results of the survey in 1893 along with the first magnetic map of Philippines. In 1894 Cirera left Manila for Spain to become the founder and first director of the Ebro Observatory (Chapter 4). He was succeeded in the magnetic section by Doyle and Miguel Saderra-Masó (1865-1939), who had arrived in Manila in 1890. In 1891, Saderra-Masó reorganized the seismological section and provided new impetus for its work. In 1895 he published the first work about the earthquakes in the Philippines, which was followed by other studies about the seismicity and tectonics of the region. At this time the observatory consisted of three sections: meteorology, terrestrial magnetism, and seismology.

Faura's health deteriorated, and his last days were saddened by the execution of his friend, Jose Rizal, the father of Philippines independence. In 1894 José Algué (1859-1930) arrived in Manila and occupied, first the post of Subdirector in charge of the section of meteorology, and later in 1897 he succeeded Faura as Director. Algué had studied at Georgetown University and collaborated with Hagen in the observatory. He had also spent time with Viñes in Havana. From 1884, it was intended to add an astronomical section to the observatory. Finally in 1897, after many delays, a especial building was inaugurated; it had a rotating dome of 10 m diameter, where a Merz equatorial telescope of 48 cm aperture was installed. Other astronomical instruments were also acquired, for example, a meridian circle for the official time service. The astronomical section was established, working in collaboration with Georgetown and Stonyhurst. Work in the observatory was recognized in international scientific circles and in 1896 its collaboration was asked for by the international project on cloud observation. Algué developed further and perfected the instrument created by Faura for the detection of typhoons. The instrument he called "barocyclonometer," consisted of a modified aneroid barometer with an adjustable scale to enable it to be used in a wide area of the Pacific, and an indicator which showed wind directions. This simple instrument, marketed in 1905, soon became very popular with Pacific mariners,

Figure 31. José Algué, Director of Manila Observatory, 1926-1948 (MO).

allowing them to determine distance and orientation from the typhoon center with relative accuracy. It became standard equipment on ships in East Asian waters for many years to come.

In 1898 Algué, using the observations made in the observatory from its foundation, published, *Baguios o Ciclones Filipinos. Estudio Teórico-Práctico*, the first book on typhoons in the Philippines. This book first appeared in Spanish, followed by an abridged edition in French, and later, in 1904, an edition in English. Algué identified the zones of origin and average trajectories of typhoons. He discovered two basic types: trajectories of parabolic shape that moved around the annual center of high pressure in the North Pacific in a clockwise direction, and a second type of storm moving in a linear path westward from the Philippines to decay over southern China. In 1900 Paul Bergholz, director of Bremen Observatory, Germany, published under his own name what was really a German translation of Algué's book. This was recognized in 1903 in the journal *Nature* by R. H. Scott who made the revision of the English version of Bergholz' book. Bergholz was not satisfied with his appropriation of Algué's book, but in England also constructed an instrument under his name which was an exact copy of Algué's barocyclonometer. Bergholz's book and instruments were extensively used in German ships.

In 1898, after the Spanish-American war, United States authorities occupied the Philippines and the observatory ceased to be an official Spanish institution. After the occupation of Manila, Admiral George Dewey had a dinner and a lengthy interview with Algué aboard his flagship Olympia. He praised the work of the observatory, promised to recommend the support of the U. S. Government, and requested that the observatory resume typhoon warnings. The situation changed the following year, when William Dorbeck, Director of Hongkong Observatory, accused the Jesuits in Manila of being lacking in scientific training and making, through the Spanish Consul, alarming forecasts of typhoons in Hongkong. On the basis of this information, the U. S. Secretary of Agriculture, James Wilson, forbade the observatory to make warnings of typhoons outside the area of the Philippines. Dorbeck's animosity against the observatory of Manila and also against the Jesuit observatory in Zikawei, China, had been clear since his arrival at the Hongkong Observatory in 1883. The strong protests of newspapers, businessmen, merchant marine and naval commanders in Manila and Hongkong, and of the Hongkong government itself, convinced the American authorities to lift the prohibition.

In 1901 the U. S. Administration, recognizing the work of the Observatory, entrusted to it the direction of the newly created Philippines Weather Bureau, and appointed Algué as Director. The meteorological service was established with nine first class, 25 second class and 17 third class stations, though not all of them were actually installed. Thus the observatory acquired again an official status, now under the United States administration. To highlight the work which had been done in the past and to help the United States authorities, the observatory prepared two long reports on the Philippines, with titles *El Archipiélago Filipino* and *Atlas de Filipinas*, which were published by the U. S. Government in 1901. In this new situation funds were made available to the observatory to improve its instrumentation. The seismological section installed new seismographs of the Vicentini, Omori and Wiechert type, and another station was installed in 1909 in Baguio with Vicentini and Omori seismographs. After the eruption of the Taal volcano in 1911, another seismological station was installed in Ambulong with Vicentini and Agamemnone instruments. Three years later another seismographic station was installed on the island of Guam. A further year later, another station was opened in Butuam, in the northern part of Mindanao. Because of the noise in the magnetic equipment, the geomagnetic station was moved in 1910 to Antipolo, near Manila, a quieter site. Between 1905 and 1920 other Spanish Jesuits were incorporated at the observatory. They had received their scientific training in American universities. In 1906, Juan Comellas (1863-1942), who had studied at Georgetown University with Hagen, took charge of the astronomical section an appointment that lasted twenty years. The following year José Coronas (1871-1938) arrived; he had received his training in Saint Louis University and took charge of the meteorological section. In 1915

Miguel Selga (1879-1956) arrived; he had studied astronomy at Harvard University and had worked at the observatories of Lick and Lowell. Selga succeeded Algué as Director in 1926 and was the last Spanish Director. Saderra Masó continued to be in charge of seismology until 1928, when he returned to Spain.

From 1920 American Jesuits began to replace the Spanish in the observatory. In 1928 William C. Reppeti (1884-1966), who had obtained his doctoral degree at Saint Louis University under Macelwane, took charge of the seismological and magnetic sections. Other Americans Jesuits joined the observatory at this time, among them , Bernard F. Doucette (1895-1974) and Leo Welch (1907-1990) in 1933, and Edmund J. Nuttall (1899-1972) in 1934. Between 1934 and 1938, Lejay, Director of Zikawei Observatory, carried out several gravity field surveys in different islands of the archipelago. The seismological section was fully renewed in 1930, with the installation in Manila of a set of three component Galitzin-Wilip seismographs, a Wiechert in 1931 in Baguio, and another Wiechert in 1939 in Tagaytay near the Taal volcano. The seismological section consisted of six stations in Manila, Butuan, Ambulong, Tagaytay, Baguio and Guam, which were in operation until the occupation of the Philippines by the Japanese army. In this prewar period the observatory, owing to its official character, grew to be a large institution with nearly two hundred persons working as technicians and in the administration in the central offices of Manila and in the secondary stations. The number of secondary meteorological stations in the different islands eventually numbered nearly three hundred. The Philippines Weather Bureau, directed by the observatory, was, then, a large and complex organization.

Charles E. Deppermann (1889-1957) who had obtained his doctoral degree at Johns Hopkins University and who had worked for some time at the Lick Observatory and the U.S. Weather Bureau, arrived at the observatory in 1926. He soon became the leading figure in the meteorological work of the observatory. He was named Subdirector in 1932 and later Director from 1945 to 1957. He continued Algué's classic work on Philippine typhoons, applying for the first time the modern ideas of frontology to tropical storms and introducing the idea of air mass analysis in reference to their genesis and path. He had learned this approach to weather analysis from the Bjerknes school on visiting in 1932 the Geophysical Institute of Bergen and the Meteorological Office of Oslo, Norway. Deppermann was one of the few early meteorologists who made a detailed effort to calculate radial and tangential wind velocity components of typhoons. He presented a theoretical model of typhoons that treated the tropical storm as a Rankine vortex with two rings of convection. He studied the differences between tropical and extratropical fronts and the motions of cirrus as prognostic of the movements of cyclones. Most of his contributions were included in his two works, *Outline of Philippine Frontology*

Figure 32. Charles E. Deppermann, Director of Manila Observatory, 1948-1957 (MO).

and *Some Characteristics of Philippine Typhoons*, published in 1936 and 1939. In 1951 H. Riehl in *Compendium of Meteorology* commented that, apart from Deppermann, researchers of his time in the study of typhoons were satisfied with the application of simple hydrodynamics.

In 1942, with the occupation of Manila by the Japanese army, all work in the observatory ceased and the following year the building was used as a military hospital. All American Jesuits were taken to concentration camps, and only Selga who was still the Director, remained free, Spain being a neutral country. In 1945 when the American troops were close to Manila, the observatory was destroyed by the Japanese. On the 9th of February four Japanese soldiers poured gasoline on the floor of the different rooms and on the astronomical dome and set fire to them. The dome, all the instruments, and the large library, with more than 20 000 volumes, were lost in the fire. Five days later incendiary bombs destroyed part of the building of the Ateneo where the offices of the meteorological service were located. Selga, his eyes filled with tears, contemplating the ruins of what had been a magnificent observatory and paraphrasing the Spanish poet Rodrigo Caro, wrote: "These, my soul, which you see now as fields of solitude, mounds of rubbish, were, once upon a time, a famous observatory" (Estos, alma, ¡Ay dolor! que ves ahora, campos de soledad, montón de escombros, fueron un tiempo cúpula famosa).

Soon after the war there were plans to reconstruct the observatory. In 1946 the Philippines acquired total independence through the Philippines Rehabilitation Act, though its autonomy had been considerable since 1934. The U.S. Government decided to reconstruct all branches of the Philippines Weather Bureau on a scale far surpassing the former service. The Jesuit observatory had no place in this plan. The new Philippine Government considered that it was time to have a weather bureau run entirely by its own scientists. Thus the Manila Observatory lost, for the second time, its privileged official status. However in 1947, the Jesuits decided to continue with the observatory, but to abandon the meteorological section, since this was already adequately covered by the new Philippines Weather Bureau. It was decided to start with seismology, and Deppermann was named Director in 1948. He traveled to Saint Louis to organize the new observatory in collaboration with Macelwane. In 1951 three Sprengnether seismographs were installed in Baguio, where the new offices of the observatory were established. Another program related to atmospheric studies, but not directly with weather, was the study of the ionosphere, which began in the same year with the installation of an ionospheric sounder. James J. Hennessey (1909-1987) joined the observatory in 1951 and succeeded Deppermann in 1957. In 1952 the new observatory buildings in Mirador, Baguio, were inaugurated with space for new programs in astronomy and solar physics. In connection with the new programs in ionosphere studies and the program in solar physics Vicente Marasigan (1918-) came to the observatory in 1952 with studies at the universities of Fordham and Georgetown. He was followed in 1957 by Richard A. Miller (1917-1974) who had studied at Fordham and at the University of Michigan. The observatory actively participated in the International Geophysical Year in 1958.

In 1962 the main part of the observatory was moved to Loyola Heights, Quezon City. A new building for ionospheric studies and solar physics was built, while the magnetic and seismographic stations remained in Baguio. During this time Sergio Su (1922-), Francis Glover (1925-) and José Dacanay (1925-) joined the staff. The seismological section grew with the development of three stations in Manila, Baguio and Davao and two new stations of the WWSSN network being at Baguio and Davao. New magnetic equipment was also installed in Baguio and Davos. The greatest advance was made in the solar physics program. In 1963, a new building of three stories was built for a telescope of 9.2 cm aperture for solar photography and for a spectroheliograph telescope equipped with a Lyot filter. In 1968 another building for a Razdow solar refractor telescope of 23 cm aperture was constructed. This instrument

Figure 33. Buildings of Manila Observatory at Quezon City, about 1960 (MO).

was part of a global network of nine similar instruments for monitoring solar flares. From 1965 the observatory has run four different programs, seismology, ionospheric studies, atmospheric electricity and solar physics.

As with other Jesuit observatories the importance of Manila Observatory began to decline in the 1970s. In 1967 Victor Badillo (1930-), after studying physics at Saint Louis University, joined the Observatory and succeeded Hennessey in 1972. That year Heyden, the last director of Georgetown Observatory, arrived in Manila and continued his work in astronomy until his death. Heyden had been appointed to Manila, but the beginning of the war and afterwards his work in Georgetown had kept him in the United States. After the closing of Georgetown, already 65 years old, he decided to dedicate the last years of his life to Manila Observatory. Work during this time was centered mainly on solar terrestrial relationships, the ionosphere and seismology. During the 1980s the situation remained stable, but no new Jesuits came to the observatory. By 1991 Badillo and Su were the only Jesuits working in the observatory. In 1993 Daniel J. McNamara (1939-) was named Director. The survival of the observatory was sometimes questioned, but it was maintained. Manila Observatory has returned to its roots in recent years. In cooperation with various Government agencies, such as the Weather Bureau (PAGASA) and the Department of Environment and Natural Resources, the observatory has been officially designated as the Climate Change Information Center for Philippines. In this role it is assisting the country to understand the issues of climate change and its consequences for the environment. To some degree traditional work in seismology and in the upper atmosphere has been continued, but this will be integrated into the climate change studies. The solar program has been discontinued. The spectroheliograph has been adapted as a differential absorption spectrograph to study atmospheric pollution.

Zikawei, Lukiapang, and Zose

The second Jesuit observatory in the Far East was the work of French Jesuits in China. In 1843 the Holy See entrusted the Jesuits with a mission in the Chinese Province of Kiangnan, whose capital was Nanjing. The first three French Jesuits arrived the same year in China, bringing a small telescope and other scientific instruments with them. Though they dedicated themselves primarily to pastoral work, they began to make some astronomical and meteorological observations. Evidence of these early astronomical observations is a manuscript found in the Archives of the French Jesuit Province with the title: *Tables astronomique pour la latitude et le meridien de Nankin, 1844*. In 1845, Claude Gotteland (1803-1856), Mission Superior, took the first steps to create a meteorological observatory. In a letter about his intentions for the mission, he added at the end "and besides these bigger works, I would wish a small observatory" (Et auprés de ces grands établissements, je voudrais un petit observatoire). There is no doubt that, in this project, there was clear influence of the memory of the great Jesuit astronomers such as Schall, Verbiest, Koegler and Hallerstein, who were Directors of the Imperial Observatory in Beijing from 1644 to 1773 (chapter 3). About 1865, Henri Le Lec (1832-1882) began the earliest meteorological observations made by Jesuits in China in the 19th century near Shanghai, in the region of Tong Kadou. Finally in 1871, with the encouragement of the Jesuit Mons. Adrien Languillat (1808-1878), Apostolic Vicar of Zhili, an observatory was established in Zikawei (Xujiahui), near Shanghai, where the Jesuits had established a boarding college, a seminary, a house for Jesuit students and two orphanages. The first director of Zikawei Observatory was Augustin Colombel (1833-1905) who began meteorological observations in December 1872. The observatory building of one story with a central tower of two stories was not begun until the following year. Regular observations, made every four hours according to the procedure established in the Observatory of Montsouris in France, began in 1874 and also the publication of a monthly bulletin in the same year. After the meteorological station established in the Russian Legation in Beijing in 1844, and the network of stations of the Chinese Maritime Customs about 1860, Zikawei was one of the oldest observatories in China.

Figure 34. Main Building of Zikawei Observatory, Shanghai, about 1923 (APFSJ).

Perry and Secchi, who had encouraged the establishment of the observatory, insisted on the need to find the right director. Marc Dechevrens (1845-1923), who had studied physics in Paris and had been a physics teacher in the colleges of Vannes and Vaugirard, was the person chosen and he was appointed Director. Before boarding the ship for the long journey to China, he spent three months at Stonyhurst where from Perry he learned about the work of an observatory and obtained some training in meteorology and geomagnetism. Dechevrens began as Director in 1876 and soon gave the observatory a solid scientific standard and he established international contacts. He may be considered to be the true founder of the observatory, which began with two sections of meteorology and magnetism. The observatory had very complete meteorological instrumentation having a Secchi Meteorograph and magnetic instruments for absolute measurements and variometers with photographic recording. Magnetic instruments acquired with Perry's help were installed in a separate small octagonal building. This was the first magnetic station in China and the only one for many years. The observatory had also a small telescope and a theodolite, used for time determination.

In 1879 a typhoon produced heavy damage in Shanghai and was the occasion for Dechevrens to bring the observatory to public attention with the publication of a detailed study of the nature and effects of this storm. From that year, the observatory, though it remained a private institution, was associated officially with the Maritime Department of Chinese Customs. The study and forecast of typhoons became its main task. These severe storms periodically devastate the coasts of China from July to September. The collaboration with the customs administration, the mail and telegraph services, and the shipping companies allowed Zikawei to establish a meteorological service with connections with all the other meteorological stations of the Far East. From 1882, the observatory began sending all newspapers a daily bulletin with weather information and forecasts for the following day. In 1890 the observatory began making weather maps for the whole coast of China and started issuing meteorological warnings by telegraph. Dechevrens proposed to the British authorities of Hongkong a detailed plan of collaboration of all the meteorological stations in China, centralized in Zikawei, which was not accepted. In 1883 a meteorological station had been established in Hongkong, whose first Director, William Dorbeck, tried to sidestep the work of Zikawei and began to create problems for Dechevrens. He even tried to interfere in the relations between Zikawei and Manila. We have already seen how Dorbeck intervened with the U. S. authorities against Manila Observatory. However, despite these problems, the work of Zikawei was always appreciated in the Crown Colony. In 1922, the Chamber of Commerce of Hongkong insisted with the Authorities that the same publicity was to be given to the forecasts from Zikawei, given their reliability, as well as to those of its own service.

Collaboration with the Chinese Customs in Shanghai and the support of its Director, Robert Hart, led in 1884 to the connection of the observatory with the harbor by telegraph and to the installation there of a mast, which, by means of visual signs, warned mariners of the weather conditions, especially during the typhoon season. The code, consisting of ten symbols, designed by Dechevrens, and later perfected by his successors, was adopted by all of the maritime stations dependent on the administration of the Chinese Maritime Customs. From 1896 the telegraph companies had carried all communications between Zikawei and the different harbors on the coast of China. In 1884, a time keeping service was also established at the observatory for the setting of ships chronometers. The exact time at noon was marked by the lowering of a ball at a mast in the observatory, followed by a cannon shot at the harbor, and from 1909, was also marked by a luminous sign. In September 1844, two admirals from England and Germany, accompanied the French Consul to Zikawei to learn about the accuracy of its time keeping. After watching Dechevrens' patient astronomical observations for some time, they were very impressed and convinced of their precision. They agreed to adjust the chronometers of their ships by the signals from the observatory.

Figure 35. Zose Astronomical Observatory, Shanghai, about 1923 (APFSJ).

In 1885 Dechevrens began the study of the observational and practical aspects of typhoons, and initiated the collection of the first statistically reliable data on the vertical temperature distribution pertinent to the structure of cyclones. He also studied the theoretical aspects of these storms. He presented his own theory of cyclones, suggesting they embodied a whirl-generator of turbulence in elevated regions of the atmosphere which attracted to its center the masses of air below its own level. From this level air was dispersed horizontally and, at a certain distance, descended to return horizontally to the center of the cyclone. Dechevrens observed these patterns of air circulation at the estuary of the Yangtze River. The vortices of this turbulence were located at a height of 500 m to 7000 m. This theory conflicted in some of its aspects with the later and accepted, thermal theory of cyclones. He also pointed out the importance of horizontal winds in the high atmosphere and the form of clouds at heights of between 5 km and 10 km, related to changes in the path of typhoons. In 1886, he observed the inversion of temperature along the vertical axis of cyclones and the changes of humidity. Together with Faura and Algué, he was among the first to study the nature and characteristics of typhoons.

Figure 36. Marc Dechevrens, Director of Zikawei Observatory, 1876-1887 and of the Observatory of the Island of Jersey, 1893-1923 (APFSJ).

Dechevrens also for many years studied in detail the vertical component of wind velocity. For this purpose he designed a special anemometer, which he called a "clino-anemometer." During the World Fair in Paris in 1889, this instrument was installed on top of the Eiffel Tower. Dechevrens returned to France in 1887 due to health problems and, as we have seen, established an observatory on the island of Jersey, where he continued his studies of the vertical component of wind velocity (Chapter 4).

Stanislas Chevalier (1852-1930) arrived in China in 1883 and succeeded Dechevrens in 1887. Between 1897 and 1898 he carried out an important cartographic study of the upper part of the Yangtze River. During his long river expedition he made more than 1200 astronomical observations. He carefully determined the geographic coordinates of 50 locations of the most important towns, which were then used to plot the position of the remaining sites. The final product was 64 maps of 1/25000 scale, for which he was awarded the Golden Logerot Medal by the Societé Geographique of Paris. In 1900 Chevalier took the initiative in establishing a true astronomical observatory to complement the work of Zikawei. This astronomical observatory was built on the Zose Hill (Sheshan), 25 km southwest of Zikawei, with a large dome of 10 m diameter where a double telescope with equatorial mounting by Gauthier of 40 cm aperture and 6.8 m focal length was installed. This was the same type of instrument as was used in the international program Carte du Ciel (see Chapter 4, Vatican Observatory). Chevalier was director of Zose for 28 years. He made a large number of observations of stars, sunspots and the orbits of asteroids and their perturbations which were published in 17 large volumes. Zose operated autonomously from its inauguration in 1901.

In 1896 Louis Froc (1859-1932) became Director of Zikawei, for 36 years until his death. Between 1899 and 1902 a new and larger building was built at Zikawei for the observatory with three stories and a central tower of seven stories. In 1904 the section of seismology was inaugurated with the installation of two horizontal Omori seismographs, donated by the Japanese Government. Instrumentation was improved in 1909 with a Wiechert horizontal seismograph of 1000 kg mass, in 1915 with a vertical Galitzin seismograph and in 1932 with a three component set of Galtzin-Wilip electromagnetic seismographs, the best available at that time. Zikawei was the first seismographic station in China. With the installation of these instruments the observatory established its five sections of meteorology, astronomy, magnetism, seismology and the time service. Froc perfected the system of visual signs to warn mariners of weather conditions used in the harbor of Shanghai and from 1918, in most of the important harbors of China. Zikawei sent weather information by telegraph to more than 20 harbors. From 1915, at Shanghai harbor, the imminent arrivals of typhoons were signaled by a cannon shot. Froc's work on typhoon forecasting and his accurate warnings merited him the name of "Père des Typhons," among the pilots and sailors of the China Sea. In 1906 Dorbeck, Director of the Hong Kong Observatory, motivated by his animosity against the Jesuits, prevented the publication in the Colony of the forecast of a typhoon made by Froc two days in advance, with the result of the loss of many lives and ships. Froc published in 1920 an atlas with the paths of 620 typhoons which had occurred between 1893 and 1918. During this period Zikawei received telegrams with weather information from a large number of stations in the Pacific, from Siberia to the Philippines, which allowed Froc to draw and publish accurate daily weather maps.

In 1908 the geomagnetic section, which had been established in 1877, was moved to Lukiapang, about 30 km from Zikawei, owing to the installation of electric traction streetcars in Shanghai, which disturbed the magnetic measurements. In Lukiapang four small buildings were built with non-magnetic materials for the installation of absolute measurement instruments and for variometers with photographic recording. In 1932 new magnetic instruments were installed in Zose, and for two years the observations at Lukiapang were also continued. After 1934 magnetic observations were done only at Zose and

Lukiapang was closed. From 1908, the observatory had a general director and three directors of the sections of meteorology, seismology and the time keeping service (Zikawei), astronomy (Zose) and magnetism (Lukiapang and Soze). These sections functioned with great autonomy and sometimes with little coordination, despite the existence of a general director of the observatory. The greatest part of the work of the observatory was routine observation, weather analysis and forecasts, especially during typhoon season, and some very limited research on sunspots, stellar spectra and microseisms. Among the Jesuit collaborators, few remained at the observatory for a very long time. In 1899 Edward Pigot arrived; he stayed until 1907, when he returned to Australia where he founded Riverview Observatory, as we shall see later. The only Japanese Jesuit to work in an observatory was Paul Tsuchihashi (1866-1965), who was at Zose from 1906 to 1910 working on astronomy. He worked, especially, on the orbit of asteroids, and he was a professor of mathematics at the Jesuit Aurora University in Shanghai. He returned to Japan where he was professor of mathematics and Chinese literature and later, in 1939, Rector of the Jesuit Sophia University in Tokyo. Henri Gauthier (1870-1949) and Joseph de Moidrey (1858-1936) came to the observatory in 1907; Gauthier was in charge of the meteorological section and de Moidrey of the magnetic. In 1920 Ernesto Gherzi (1886-1976) arrived at Zikawei. He was an Italian who had entered the French Novitiate with the desire to be sent to the China Mission. He worked first in meteorology with Froc, eventually in 1940, becoming the last Director of Zikawei. From 1924 to 1946 Edmund de la Villemarqué (1881-1946) was at Zose and became astronomy section director from 1940. Louis Gauchet (1870-1951) and Maurice Burgaud (1884-1977) arrived in 1924. Gauchet collaborated with de la Villemarqué on the study of the variation of latitude, stellar fields and the orbits of asteroids.

In 1931, Pierre Lejay (1898-1958) was appointed General Director of the three observatories. He greatly advanced the science of the observatories, especially in atmospheric physics and gravimetry. He had finished his doctoral degree at the University of Paris in 1926 with a thesis on the propagation of electric disturbances in the atmosphere. From 1922 to 1926 he had collaborated at the Observatory of Paris with the Service Meridién et de l'Heure. In 1927 Lejay, recently arrived in Zikawei, offered a rather negative evaluation of the work being done at the observatory and was very critical of its future. He pointed out the lack of originality and the low scientific standards of the research being carried out, its isolation and the fact that no native Chinese had been assimilated into the scientific personnel. In 1932 as Director he installed new instruments to study solar radiation, such as Angstrom and Moll pyroheliometers and, in 1935, a heliospectrograph. He created a section of atmospheric physics, where some of the earliest studies of the ozone content in the atmosphere were made in 1935. In 1938 an ionospheric sounder was installed, opening a line research on the physics of the upper atmosphere.

Figure 37. Group of the personnel of Zikawei Observatory with occasion of the International Project on Longitudes in 1926. Jesuits are: First row (seated) Ernesto Gherzi, Stanislas Chevalier, Edmund de la Villemarqué, Edward Pigot (Director of Riverview Observatory, Australia). Second row (standing) Maurice Burgaud, Pierre Lejay. Gaston Fayet (Director of Nice Observatory) is seated between Gherzi and Chevalier (APFSJ).

He made some modifications in the sounder which increased the efficiency of the reflections; and he studied the variation in height of the ionized layers of the atmosphere and their reflecting characteristics. Together with the ionospheric studies his other most important work was on the study of the Earth's gravity field. In 1933 he developed a gravity meter, together with Fernand Holweck, based on an inverted pendulum, with a sensitivity 200 times greater than that of the existing instruments and easy to transport (Holweck-Lejay gravity meter). Between 1933 and 1936 Lejay carried out many gravity field surveys in China, Indochina, the Philippines, Malaysia and Indonesia with two of these instruments. In 1939, Lejay returned to France where he was mobilized. After the end of the war he remained in France, and took many official scientific positions such as the presidency of the Société des Radioélectriciens in 1946 and in 1952 the presidency of the International Union of Scientific Radioelectricists. He was elected a member of the Académie des

Sciences and Officier of the Légion d'Honneur. He was greatly esteemed by his colleagues, who respected his enthusiasm in defending his ideas and French science with tenacity and the benevolent authority with which he was imbued. He died suddenly at sea during a journey to America; an appropriate end for a man who was the son of an admiral and a brother of Navy officers.

In 1926, Zikawei and Zose participated in the international project for the revision of longitude proposed in 1925 by the International Commission of Longitudes and led by General Gustave Ferrié. The aim of the project was to measure with great accuracy, down to milliseconds of arc, the longitude in three observatories, approximately at the same latitude and separated in longitude by 120 degrees, by comparing astronomical observations and time signals transmitted by radio. The three stations chosen were the observatories of San Diego, Algiers and Zikawei. For this purpose Gaston Fayet, Director of the Nice Observatory, stayed at Zikawei from September to November 1926. Lejay, Chevalier, Gauchet, de la Villemarqué and Burgaud participated in the observational work of different aspects of the project. This was, practically all of the personnel of the observatory with the exception of Gherzi and de Moidrey, who continued the routine observational work in meteorology, seismology and magnetism. For this project new radio receivers and antennas were installed under Lejay's supervision. The result of this very laborious work was the determination of the longitude of Zikawei in relation to Greenwich, with an accuracy of 5 milliseconds; and in relation to San Diego with an accuracy of 3 milliseconds (about 1.5 meters). One of the aims of the project was to detect the variations in relative longitude of the three observatories at the time of the observations in order to verify the theory of continental drift proposed by Wegener. However, owing to the short duration of the project, only three months, with this accuracy this could not possibly be done. Relative motion of continents is only one to eight centimeters per year.

In 1927, during armed confrontations in China between nationalists and communists, the observatory was protected by French soldiers and did not suffer damage. The work of the observatory continued and in 1933 the weather service to maritime navigation was extended also to air navigation. In 1937 Japan declared war on China. The following year Japanese troops arrived in Shanghai and, when war was declared between Japan and the United States, the Japanese occupied the whole city. Gherzi used his Italian nationality to protect the observatory and maintained good relations with the Japanese authorities. From 1939, after Lejay's departure to France, the three observatories of Zikawei, Lukiapang and Zose functioned autonomously with three directors, namely, Gherzi, Burgaud and de la Villemarqué. Lejay nominally retained the title of director, but he never returned to China. During the war the meteorological service was closed resuming its work in 1946 in collaboration with the new Central Weather Bureau of China. During these last years of the observatory, its key figure was Gherzi, who kept the meteorological and seismological sections in operation. He worked on the climatology of China using all of the observations of the observatory, and on the nature of typhoons, continuing the work of Dechevrens and Froc. He tried to find a correlation between the heights of the ionospheric layers and the occurrence of storms and he explored the possibility of using this information for weather forecasting. Gherzi was also, in 1924, one of the first to investigate the relationship between microseismic noise and oscillations in the atmospheric pressure, research he continued for many years.

After the triumph of Maozedong's Communist troops, the new Communist government occupied the observatory in December 1949, allowing the Jesuits only half an hour to leave the premises. All of their notes and books were confiscated. The President of the Academia Sinica, whom Gherzi had protected during the Japanese occupation, led the communists into the observatory. Gherzi was then in the Philippines and could not return to China. He always regretted the loss of all of his scientific papers. Gherzi spent four years in Macao organizing the Meteorological Service at the request of the Portuguese Colonial Government. In 1955, after a short stay in Saint Louis, he moved to the observatory of Jean de Brebeuf in Montreal, Canada. Burgaud was held in prison until 1955, when he was expelled from China and he went to the Observatory of Tananarive, where he continued working until his death. The Zose observatory, still using the old telescope, was reopened in 1962 with the name of Shanghai Astronomical Observatory dependent on the Chinese Academy of Science. After the cultural revolution, new instruments were installed in Zose in 1979 and astronomical observations resumed. The Zikawei building is at present used as the headquarters and administrative offices.

Antananarive

The most important Jesuit observatory in Africa was established in Madagascar by French Jesuits. We have seen the great effort French Jesuits made at Zikawei Observatory in China; a similar effort was made in the observatories established in Madagascar and Lebanon. Madagascar was practically closed to European influence from 1828 until 1861, when King Radama II allowed Europeans to be established again in the capital Tananarive (Antananarive). French Jesuits of the Province of Toulouse then founded their first mission. They were soon aware of the culturally backwardness of this large African island. Scientific work was non-existent. There were no modern maps. No geodetic work or cartographic surveys had yet been made nor meteorological, magnetic and seismological observations had been done. The Jesuits thought that part of their mission work had to be to try to remedy this

Figure 38. The first Observatory of Tananarive, Madagascar, about 1890.

situation. The first meteorological observations were begun in Andohalo by Jean Laborde (1819-1884) from 1864 to 1878. These are probably among the first meteorological observations ever made in Africa. In 1878 the French Consul, Alfred Grandidier, asked Jesuits to make meteorological observations in Tananarive. These were undertaken in Mahamisina by Augustin Debosc (1824-1885) from 1880 to 1882. The other scientific work needed was cartography, since no reliable maps of Madagascar existed. Désiré Roblet (1828-1914), was the first to begin this work in 1873 making topographic surveys of the island. He was an indefatigable explorer and geographer, a true pioneer of the early cartography of Madagascar. He also continued the meteorological observations at Andohalo in 1887 and 1888. After the first French-Madagascan war in 1884, the French presence in Madagascar became firmly established. The General Representative of France in Madagascar, M. le Myre de Vilers, conceived the idea of creating an observatory in Tananarive and interested Monsignor Cazet, Vicar Apostolic of Madagascar. One of the reasons behind this idea was that France did not have then any observatory in the southern hemisphere. Le Myre de Vilers convinced the French government and the Académie des Sciences, who in 1887 gave their approval to the project. As we have seen in other cases, he thought that the Jesuits were the only ones who could carry out this project and made contact with them. French Jesuits accepted this responsibility for the project and appointed Elie Colin (1852-1923) to supervise this work. He was still a student of theology in Uclés, Spain, and had been a science teacher in the colleges of Bordeaux, Sarlat and Monaco. After he had finished his studies of theology, Colin spent some time with Perry at Stonyhurst and at the Observatoire del Bureau des Longitudes in Montsouris, at the Bureau Central Météorologique of France and at the Observatoire Astronomique. He elicited the collaboration of all these institutions and acquired scientific equipment for the new observatory, such as two telescopes with equatorial and meridian mounting, chronometers, theodolites, and magnetic and meteorological instruments.

Colin arrived in Tananarive in 1889. He searched for a suitable site for the observatory and finally a site was selected on the hill of Ambohidempona at 120 m height above the plateau and 1402 m above sea level (ASL). This was at the time the highest astronomical observatory in the world. The building had the form of a T with a long section of 32 m length and an octagonal central body with a large dome of 8 m diameter. It had besides three other smaller domes at the three ends of the building . On top of the hill, the observatory with its four domes presented a beautiful sight from the capital. Instruments brought from France were installed and astronomical, meteorological and magnetic observations began in 1889. Astronomical observations were also used for the support of the geodetic work and for the time keeping service which was established in the observatory.

The first phase of the observatory, under Colin's direction, between 1889 and 1895, enjoyed the support of Madagascan Royal Family. The King Ranavalomanjaka III gave it the title of Royal Observatory. In 1893, owing to health problems, Colin returned to France and was replaced temporarily by other Jesuits. In 1894 relations between France and the Madagascan government, under Queen Ranavalona II, were broken. The following year all missionaries were forced to abandon Tananarive. The observatory remained under the care of two native assistants. Later a search was made for arms supposed to be hidden in the observatory and part of the instruments were taken to the Royal Palace. The following year French troops occupied part of the island. As they advanced toward Tananarive, the Government of Madagascar destroyed the observatory, which they considered to be in a strategic position which dominated the capital. The beautiful domes, part of the walls and the instruments which were still there were wantonly destroyed or carried away. French troops finally entered the capital and Madagascar was made a French Protectorate. Queen Ranavalona II remained as a figurehead for a couple of years. Finally, in 1897, General Joseph Gallieni assumed full powers as Governor General.

Figure 39. Désiré Roblet and Elie Colin, Director of the Observatory of Tananarive, 1889-1893 and 1896-1923, at their geodetical work in Madagascar.

In 1896 Colin, who learned in France of the destruction of his observatory, returned to Madagascar. He found that only a few of the instruments were still in working condition; most were lost or severely damaged. In 1898 Colin began the reconstruction of the observatory. Many difficulties were encountered in the work and funds were not easily available. The new observatory, finished in 1902, was built on the ruins of the old one, on a more modest scale, with only one dome, where the Eichen equatorial telescope, which had been repaired, was installed. Magnetic and meteorological observations were resumed, but astronomical work was practically discontinued some years later. Colin installed a network of meteorological stations along the island initially with 11 stations and later 25 stations. The observatory was put in charge of the Colonial Meteorological Service until 1920, when this work passed to the Service des Travaux Publiques and later to the Direction de Services Maritimes. The French colonial government entrusted to the observatory the observation and forecast of cyclones, providing special funding for this purpose. Thus Tananarive was the fourth Jesuit observatory involved in the study and forecast of cyclones after Belen, Manila and Zikawei. Though not so frequent as in the Caribbean and the Pacific, cyclones which have originated in the Indian Ocean, in some cases continued on a westward path to Madagascar, where, at times, they caused serious damage. Large earthquakes are not frequent in Madagascar; nevertheless in 1899 the observatory installed a Cecchi seismograph, one of the earliest in Africa.

An important part of Colin's work was his participation in the geodetic and cartographic work carried out by the French colonial administration. Together with Roblet, who was already nearly seventy years old, he measured with great accuracy a base line of 1800 meters in the plane of Maharema, which was used to measure the distance of 5571 m between Ambohimahavony and Ialamalaza. Then he established a triangulation network of 76 stations over a distance of 211 km from Tananarive at 1400 m altitude (ASL) to the coast. Between 1896 and 1906 Colin and Roblet carried out geodetic work in the region of Imerina with a triangulation network covering an area of 32000 square kilometers. Colin was associated as geodesist with the French Army, collaborating in an official manner with the Generals Voyron and Gallieni. For this work Colin and Roblet received the Herbert-Fournet Prize of the Société de Géographie de Paris in 1898. Later, in 1903, Colin received the Gay Prize of the Académie des Sciences and the Prize of the Golden Medal of Louis Bourbannaud for his geodetic work and he was named Officier de la Legion d'Honneur in 1921.

Charles Poisson (1882-1965) was appointed Director in 1923 after Colin's death and occupied this position for forty-two years. He had been an officer in the French Navy and had participated in First World War as a commander in the marine infantry. He always retained traces of his military bearing and his commanding attitude. In 1927 he renewed the seismographic section with two horizontal Mainka seismographs of 450 kg mass, donated by the Académie des Sciences of Paris. Between 1932 and 1936 the magnetic section was also renewed with a La Cour magnetograph which had been loaned during the program of the International Polar Year and with an inductometer to measure the magnetic inclination. In 1931 the meteorological section was incorporated into the official Meteorological Service of Madagascar and the observatory concentrated on seismological and magnetic work. From 1926 several Jesuits joined the observatory at different stages, so that there were always at least two Jesuits working there. Work with Poisson was not always easy since he did not easily accept the changes in the observatory suggested by younger Jesuits. One of these collaborators was Louis Cattala (1904-1963), who went to France in 1928 and studied in the Institut de Physique du Globe of Paris and of Strasbourg and in the Universities of Lille and Louvain. He returned to the observatory in 1943 and was in charge of the magnetic section, for which he

Figure 40. Charles Poisson, Director of the Observatory of Tananarive, 1923-1965.

had acquired new instruments. He began to carry out magnetic and gravity survey field work. He had mistakenly expected soon to be appointed director. His differences with Poisson increased and in 1949 he left the observatory and dedicated himself to gravity and magnetic field work, participating in the magnetic map of Madagascar and teaching sciences in the seminary at Tananarive.

In 1947 the observatory established a collaboration with the Institut de Recherche Scientifique of Madagascar, which made possible a renewal of part of the equipment. In 1955 Burgaud, after being expelled from Zikawei and having spent some time in prison in China, arrived at the observatory and took charge of the magnetic section. The situation of the observatory, however, was declining owing to the aging personnel and instruments, so that Jesuits began to contemplate its closure. In 1956, the Mission Superior, Jacques Tiersonnier (1914-), gave a very pessimistic evaluation of the situation of the observatory. He said of Poisson, already seventy four years old, that he was truly the old captain resolved to sink with his ship (c'est vraiment le vieux capitain résolu à sombrer avec son navire). He insisted that the observatory should collaborate more in the International Geophysical Year (IGY) and incorporate native personnel, but Poisson was opposed to both suggestions. Lejay, who was consulted, also saw the situation as no longer sustainable and proposed that the observatory should be transferred to the colonial administration. The 1958 participation of the observatory in the IGY program was very mediocre and its continuation was clearly problematic. The process was accelerated by the independence of Madagascar in 1960. Owing to lack of funds and personnel the observatory was transferred in 1965 to the University of Madagascar, but still under Jesuit direction. Finally in 1967 the transfer was complete. Three Jesuits, Louis de Laitre (1923-), the last director, Burgaud and Jean Coze (1924-1975) continued working in the Observatory until 1969.

Boroma, Zumbo, Bulawayo, and Addis Ababa

Early meteorological observations in Africa were also carried out by Jesuits from 1883 in the mission post of Boroma, some 16 km from the village of Tete, near the Zambezi River, in the present Mozambique, close to the frontier with Zambia. They were made over 10 years by the Hungarian Istvan Czimmermann (1849-1893). In 1890 another Hungarian Jesuit, Laszlo Menyhart (1849-1897), arrived at the mission. He had been a science teacher in Kalocsa College, Hungary, and was familiar with the work of its observatory. Although his primary scientific interest was in botany and mineralogy, in 1893 he took charge of the meteorological station and became interested in meteorology. In the same year he established a second station in Zumbo, a few kilometers north of Boroma. The station of Boroma had three thermometers, a thermograph, a barograph and psicometer, which had been donated by Cardinal Haynald, Archbishop of Kalocsa and founder of its observatory. The observations at Boroma and Zumbo were systematized and published by Fenyi in the Bulletins of the Observatory of Kalocsa (Chapter 4). Fenyi considered these stations in Africa as part of his observatory and worked with their data. After Menyhart's death in 1897, observations were continued until 1905, when, despite Fenyi's efforts to maintain the stations in operation, they were closed. Not much is known about the reasons for their closing, but it would have to be due to changes in the direction of the Mission.

The Zambezi Mission was entrusted to the English Jesuits in 1879. In 1896 they founded St. George's College in Bulawayo. The first idea of establishing an observatory dedicated to astronomy and meteorology in Rhodesia came from Victor Nicot (1858-1935), a missionary in Macloutsie, Matabeleland, who in 1897 began meteorological observations at St. George's College and published the results in the *Zambesi Mission Record*. In 1896 Alphonse Daignault (1850-1938) tried to interest Cecil Rhodes in the creation of an observatory in the then Rhodesia. At that time there were none there and only two in South Africa in Cape Town and Lovedale. Although the idea was accepted, some years were still to pass until in 1901, Richard Sykes (1854-1920), the Mission Superior, began preparations for the setting up of an observatory in Bulawayo and appointed Edmund Goetz (1865-1933) as Director. Nicot participated in the setting up of the observatory; he insisted on the importance of meteorological and magnetic observations. Goetz, a native of Alsace, had joined the Jesuits in England and was a physics and mathematics teacher from 1887 at St. Aidan's College, and later in the Island of Jersey. In 1901 he finished his Licenciate of Science in the University of Paris with certificates in astronomy, geography, physics and geology. In 1902 he visited Stonyhurst; and spent eighteen months at Georgetown Observatory with Hagen, collaborating in the observation of variable stars. Goetz arrived in Bulawayo in 1903 with the astronomical, meteorological and magnetic instruments he had acquired in France and England. The observatory was inaugurated in 1903 with two buildings, the main one for astronomy and meteorology and a small one for magnetic observations. The astronomical section had two small telescopes. In 1913 a larger telescope with equatorial mounting by Cooke of 16 cm aperture was installed in a rotating dome. For the work in meteorology it had very complete equipment, namely, a Tonnelot barometer, thermometers, barographs, pluviometer, etc. For magnetic observations, it had a Nalder unifilar magnetometer and a Dover inclinometer. Goetz soon began a program of regular meteorological, magnetic and astronomical observations.

Goetz´s first scientific work was in astronomy, collaborating with Hagen in the observation of variable stars. His most important contribution, however, was the study of the climatology of Zimbabwe, especially, its rainfall pattern. He also made early magnetic observations. In 1909 and 1914 he carried out two magnetic field surveys; the first with a profile from Broken Hill (Kabwe, Zambia) to the "Star of the Congo Mine", in the Congo region, and the second in Barotseland (Zambia) covering a distance of more than 200 miles from Kazungula to Lealui. Goetz was active in the scientific circles of the then Rhodesia and was a member of the Rhodesian Scientific Association. In 1926 the Jesuits left Bulawayo and the observatory was transferred to the Colonial government. Some of the instruments were taken to St. George's College in Salisbury where Goetz taught mathematics and French. The telescope was installed there in a dome and it is still being used by the students. The observatory continues today with the name of Goetz Observatory.

The last observatory directed by Jesuits in Africa was a state observatory entrusted to and directed by the Jesuits in Ethiopia. The origin of this observatory was in the recommendation of the Scientific Committee of the International Geophysical Year (IGY), which in 1955 asked for the installation of a magnetic station near the magnetic equator and suggested as a possible location a site in Ethiopia near Addis Ababa. This recommendation was taken up by the University College of Addis Ababa, which had been entrusted to the Canadian Jesuits by the Ethiopian Government of Haile Sellassie in 1945. The observatory began operations in 1958, in time to participate in the programs of the IGY. Its director was the French Canadian Jesuit Pierre Gouin (1917-). The observatory had very complete magnetic instrumentation; three Askania QHM (quartz horizontal magnetometer), a Ruska earth inductor inclinometer, Chasselon Fennel D-magnetometer and three-component Ruska variometers with photographic recording. The observatory also had meteorological and seismological stations, the latter with three-component short period Willmore seismographs. In 1962 a seismological station of the global network WWSSN was installed.

During the 20 years under his direction, Gouin carried out important work in geomagnetism. He collaborated in the study of the temporal variations of the magnetic field and in the study of magnetic storms with the French Jesuit, Pierre Noel Mayaud (1923-) who worked at the Institute de Physique du Globe of Paris. Between 1958 and 1962 Gouin completed a magnetic map of Ethiopia. He also worked in seismology, in the study of the seismicity of Ethiopia and its tectonic and geologic characteristics. Between 1963 and 1975 he directed several gravity field surveys and the observation of solar eclipses in 1959 and 1961. Gouin was director until 1978 when, after an absence abroad, the new Government, which had overthrown Haile Selassie in 1974, denied him an entry permit to the country. This was the end of the Jesuit participation in the observatory. Gouin returned to Canada and joined the Observatory of Jean de Brébeuf in Montreal.

Ksara, Lebanon

In 1843 French Jesuits of the Lyon Province established a mission in Syria, and in 1875 founded the University of St. Joseph in Beirut. From this university there arose, in 1903, the initial idea of establishing an observatory in Lebanon. Bonaventure Berloty (1856-1934) was selected for this job. Berloty had obtained his doctorate in science in the University of Paris in 1886. To prepare for his work in the new observatory he spent some months in 1905 and 1906 at Ebro Observatory in Spain and at Stonyhurst in England. Berloty arrived in Beirut in 1906 and selected a location in Ksara in the Bekaa Valley about 40 km from Beyrouth, at 960 m ASL, for the new observatory. The first section installed was the astronomical section with a telescope with equatorial mounting by Bardoux of 11 cm aperture, another of 9.5 cm aperture and a meridian circle of 4.8 cm aperture. In 1908 the magnetic section was set up with Mascart variometers. In 1909 the meteorological building was built and in 1910 the seismological section began with the installation of two Mainka horizontal seismographs of 130 kg mass, which were lent by the International Association of Seismology. Thus by 1910 the four sections of astronomy, magnetism, meteorology and seismology, were working regularly. The publication of the meteorological bulletin began also the same year. From 1907, several Jesuits joined the observatory, among them, Joseph Brachet (1882-1904) and Charles Combier (1880-1950). In 1913 new astronomical instruments, among them a telescope of 16 cm aperture and a heliograph, were acquired. A tower of 12 m height was built to support a rotating dome of 5 m diameter to house the telescopes. Work was finished in October 1914, the same month that the Ottoman Empire entered the First World War on the German side. Since Lebanon was part of the Ottoman Empire, French Jesuits had to leave the observatory. The observatory was sacked several times, first by the Turkish army who carried away part of its instruments, and, after being abandoned, its buildings were reduced to ruins.

At the end of the war in 1918, Berloty returned to Ksara and the work of reconstruction began. Funds were made available by the Government of France, which had established a protectorate over Syria. New buildings were built that incorporated a new rotating dome where a new equatorial telescope of 20 cm aperture was installed. New seismological and magnetic equipment were also installed, a Mainka horizontal seismograph of 435 kg mass and Mascart variometers with photographic recording. From 1919 Berloty collaborated in the measurement of the geodetic network of Lebanon and Syria carried out by the French Army. He made repeated observations of star transits with a meridian telescope to establish the geographical coordinates of Ksara which were then used as the fundamental point for the geodetic network. As part of the geodetic work, Jesuits of the observatory also participated in the measurement of two base lines of 12 km and 13 km in 1920 and 1923, the first in Lebanon near Ksara and the second in northern Syria.

In 1920, after taking part in the war and being decorated with the Légion d'Honneur, Combier returned to the observatory. He was appointed Director in 1925. The same year George Horan (1876-1936), an Englishman who had joined the French Jesuits with the idea of going to the China mission, joined the observatory. He had been first in Zikawei and was sent to Lebanon, because of health problems. He stayed until 1934, working in the observatory mainly on instrumentation and radio communication. The contributions of the observatory to meteorology ensured that, when, in 1921, the Meteorological Service of Syria was created by the Office National Météorologique of France, its direction was assigned to the director of the observatory, who at that time was Berloty. The Service consisted, then, of 14 stations distributed through Syria and Lebanon. In 1934, after Berloty's death, problems arose in regard to the direction of the Service. The intention was to transfer it to the Meteorological Station in Beirut, but, finally, it stayed with the Ksara Observatory. The direction of the observatory by Combier, who did not enjoy the prestige of Berloty, was at times questioned. New Jesuits arrived during the 1930s, including Jean Rey (1896-1979), who had studied in the Institut de Physique du

Globe in Paris and worked in meteorology and magnetism; Henri Destezet (1872-1957), who worked in seismology; and Jean Delpeut (1892-1977), who took charge of the magnetic section. New instruments were installed, such as a vertical Galitzin seismograph in 1935 and a meridian telescope of eight centimeter aperture. In 1926 and 1930 the observatory participated as one of 40 stations in the International Project of Longitude. After Berloty's death in 1934 the astronomical section became virtually inactive.

After the independence of Lebanon and Syria in 1944, the government of Lebanon assumed part of the costs of the observatory. The government of Syria lost all interest in the observatory, and Ksara limited its meteorological observations to the territory of Lebanon. Combier's efforts for joint climatological studies of Lebanon and Syria did not bear fruit. He had made an interesting study of the sandstorms which he could not manage to complete. He died in a car accident in 1950 and was succeeded by Jacques Plassard (1911-), who was the last Jesuit director of the observatory. Plassard had begun his scientific career in astronomy in 1940 in the observatory of Lyon working with Jean Dufay and obtained his doctoral degree in 1949 from the University of Lyon with a work on stellar spectroscopy. He tried to apply his work in Ksara to the study of the spectrum of the night sky. He planned to revive the work of the observatory concentrating on geophysics. In 1954 Plassard collaborated with the French geophysicist Pierre Stahl in the observations of gravity with several field surveys of Lebanon. The earthquake of 16 March 1956, which caused casualties and heavy damage in Lebanon, drew the attention to the observatory and to the need for renewing its instruments. Barthelemy Kogoj (1897-1982), who had arrived at the observatory the year before and was in charge of seismology, installed new more sensitive Grenet-Coulomb seismographs.

In 1962 Lebanon created its Conseil National de la Recherche Scientifique (CNRS) on the model of that of France and asked for the collaboration of Ksara. In 1966 the CNRS of France stopped its funding of the observatory and from 1968 funding was received from the Lebanese CNRS. Plassard quit his position in the French CNRS in order to work in the Lebanese administration. He thought that this was the best way to fulfill his missionary work in the service of Lebanon, though he would have preferred to keep his ties with the French CNRS as well. Between 1967 and 1979 an extensive program of synthesis of the long series of geophysical data collected by the observatory in the Lebanon region was carried out, in which all the Jesuits participated. In this work, directed by Plassard, Rey took part in climatology, Delpeut in magnetism and Kogoj in seismicity. The outcome was reflected in several publications, among them a climate atlas, a catalogue of earthquakes, a study of seismicity, a study of the gravity field and several pluviometric and gravimetric maps. In 1974 the advanced age of Delpeut, Rey and Kogoj, all of whom were over seventy-five years old, and the fact that no younger Jesuits were being made available for this work led to the recommendation of the closing of the observatory. The four Jesuits working at the observatory were aware of having accomplished their particular work as missionaries in Lebanon and that the time had come to pass the responsibility to the people they had come to help. This was finally done five years later, in 1979, and all the equipment was transferred to the CNRS of Lebanon. Thus, after 72 years of work, Ksara ceased as a Jesuit observatory. Some years later the observatory was practically destroyed during the Lebanese war which especially affected the Bekaa Valley.

Calcutta

One of the most influential Jesuit educational institutions in India is St. Xavier College, founded in 1860 at Park Street in Calcutta, and which since 1862 has been associated with the University of Calcutta. In 1866 Eugene Lafont (1837-1908), a Belgian, arrived in India. He began to teach science and to carry out meteorological observations at the College. In 1867, he detected a cyclone nearby and forewarned the Government Meteorological Department. In 1870 a Secchi meteorograph was installed and a bulletin began to be published. These were the modest beginnings of the meteorological observatory. Lafont, who had a striking personality, in 1868 began to hold public lectures on scientific subjects, which attracted large audiences. He soon became known in scientific circles in India and was a strong promoter of scientific education. In 1871 Lafont became Rector of St. Xavier College where he fostered scientific education, which at that time in India was at a very low ebb. As a way of reversing this situation he collaborated with Mahendralar Sarkar in the founding of the Indian Association for the Cultivation of Science.

On the 8th of December, 1874 a transit of Venus over the solar disk took place. We have seen that Perry was in charge of the British expedition to the Kerguelen Islands to observe this phenomenon (Chapter 4). For the same purpose the Italian astronomer Pietro Tacchini, Director of the Palermo Observatory, organized an Italian expedition to India. Secchi was supposed to have taken part in this expedition, but a deterioration in his health did not allow him to travel. Tacchini invited Lafont to take part in the observations which were made in Madhupur, north of Calcutta. Tacchini impressed by the clarity of the Indian sky, encouraged Lafont to create an astronomical and solar observatory in St. Xavier College. In 1877, with Tacchini's collaboration, Lafont installed a Steinheil telescope of 23 cm aperture with equatorial mounting. The same year an equatorial telescope by Merz of 18 cm aperture was also acquired. The latter telescope was equipped with a Browning spectroscope equivalent to ten prisms of sixty degrees. He also acquired a smaller equatorial telescope of 7.5 cm aperture, a meridian circle and a celostat with double mirror by Gauthier. For the telescopes a rotating dome of 7 m diameter was built on the roof of the college. In 1878 a third floor was added to the college building and the observatory was moved to a new separate building. Lafont succeeded in collecting the necessary funds from different sources, among them the British Colonial Administration and the Asiatic Society of Bengal. Alphonso de Penaranda (1834-1896) participated in the installation; he was a professor of mathematics and astronomy, who became director of the observatory and continued the astronomical observations for many years.

After the installation of the telescopes, the observatory worked mainly on solar observations, namely sunspots and prominences. Other astronomical observations were also carried out. Personnel from the observatory also participated in some scientific expeditions. In 1890, led by Penaranda, they went to Bhagalpur, Bihar and Meghna, East Bengal, for the observation of two annular solar eclipses. In 1898 a group of eleven Jesuits under the direction of Constantin de Clippeleir (1843-1907) and Victor de Campigneulles (1847-1917) observed the solar eclipse of 22 January in Dumraon, Bihar. They took with them two telescopes and a variety of other instruments and published their results. In 1901 a small group went to Padang, Sumatra to observe another solar eclipse. From 1869 Edward Francotte (1843-1923) was in charge of the meteorological observations and later, between 1895 and 1924, became formally director of the meteorological section of the observatory. Francotte, who worked at the observatory for 35 years, compiled and analyzed the meteorological observations from 1868 to 1918. From 1889, the direction of the observatory was divided. From that time on there was a director for meteorology and another for astronomy and solar physics. Work at the observatory began to decline after about 1920. The meteorological section was closed in 1925, two years after Francotte's death. Solar observations were continued until 1939, when the observatory was finally closed.

Riverview, Australia

One of the few Jesuit observatories in the southern hemisphere is the Riverview Observatory of St. Ignatius College in New South Wales, Australia. Its founder was Edward F. Pigot (1858-1929), an Irishman, who had been a medical doctor before he joined the Jesuits. In 1899 he had been appointed to Zikawei Observatory, but owing to health problems went to Australia in 1907. He visited Manila Observatory and conceived of the idea of establishing an observatory at St. Ignatius College. In 1907 Pigot began meteorological observations, and two years later installed horizontal and vertical Wiechert seismographs. A year later he installed Mainka horizontal seismographs. Thus the observatory began meteorological and seismological observations. Owing to its location the seismographic station soon became known and its observations were often sought and much appreciated by seismologists around the world. At that time, there were very few seismographic stations in the southern Pacific region. In 1923, the famous Japanese seismologist F. Omori was visiting Riverview when the large Tokyo earthquake took place. He looked at the recordings without knowing that the earthquake had destroyed the capital city of his homeland. Jesuit interest in solar observations could not fail to find a place at Riverview, and Pigot observed several solar eclipses, in 1910 in Brunei Is., in 1911 in Tonga and in 1922 in Queensland. He carried out a program of observations of solar radiation between 1925 and 1929. In 1922 a telescope of 15 cm aperture was installed, which allowed some more ambitious astronomical projects. In 1926 Pigot began a joint collaboration with Bosscha Observatory of Lembang, Java, in the observation of variable stars. We have seen that this had been a research topic which had attracted the interest of many Jesuits, especially of Hagen in Georgetown and the Vatican, of Hisgen and Esch in Valkenburg, and of Goetz in Bulawayo. In 1927 Pigot using pendulums began a program of recording earth tides in a mine in Cobar, NSW, which did not in the end give the expected results.

In 1929 Pigot was succeeded by William J. O'Leary (1869-1939), who, as we have seen, had been in charge of the seismological station at Rathfarmham Castle, Dublin from 1911 (Chapter 4). In 1933, he was joined by Daniel O'Connell (1896-1982) who had collaborated with him in Ireland, and afterwards, from 1923 to 1926, was a physics teacher at St. Ignatius College and succeeded him in 1938. O'Leary and O'Connell continued the program of observations of variable stars with a new telescope and two photographic cameras lent by Lembang Observatory. O'Leary was a very able instrumentalist. In 1934 he developed an instrument to compare photographic plates of stars, which he called Blink Comparator, in order to detect variable stars. He also built several clocks, one of them being sent to Georgetown Observatory. The seismological station was upgraded in 1941 with a three-component electromagnetic seismographs built in Australia according to the blueprints of Galitzin. War had prevented buying the original instruments which were made in Estonia. In 1952 O'Connell was appointed Director of the Vatican Observatory and was succeeded by Thomas Noel Burke-Gaffney (1893-1958) who had been working at the Observatory from 1946. He worked in observational seismology and made the observatory one of the world's first-class seismological observatories. His success as a seismologist came primarily from his remarkable insight in reading and analyzing seismic records. In 1962 a seismographic station of the global network WWSSN was installed.

Figure 41. Edward F. Pigot, Founder and Director of Riverview Observatory, Australia, 1907-1929

In 1965 Lawrence A. Drake (1931-) was named Director, though soon after he was absent from the observatory until 1972, because of his doctoral studies in seismology at the University of California, Berkeley. He had been appointed to the observatory in 1951. In 1972 the meteorological section was closed, except for the rainfall observations which were continued till 1978. Since 1972 the Bureau of Mineral Resources, later the Australian Geological Survey Organization (AGSO), has helped to finance the observatory. Drake advanced the seismological studies and lectured in geophysics at nearby Macquarie University. In 1992 he retired from Macquarie University and traveled to Bolivia to become Director of Observatorio San Calixto in La Paz (Chapter 6). From 1992 to 1996 Luke Tyler, a lay geography teacher at St. Ignatius College, took care of the observatory. Because AGSO was interested in the continuation of the seismographic station it continues to the present finance its operations and collects and stores the seismograms. A Jesuit, Michael Hansen, a teacher at St. Ignatius College, was in charge of the observatory from 1997 to 1999.

PART II. JESUIT OBSERVATORIES (1814-2000)

1. EUROPE

OSSERVATORIO DEL COLLEGIO ROMANO, ROME, ITALY

1824-1879
Astronomy, Meteorology, Geomagnetism, Solar Physics

The observatory was established in 1787 after the suppression of the Society of Jesus at the time the college was run by the diocesan clergy. Its first director was Giuseppe Calandrelli. The observatory was located on a tower on the right side of the front of the college building and had astronomical and meteorological instruments. When in 1824 the college was returned to the Jesuits, Dumouchel was named Director of the Observatory. In 1840 de Vico gave a great impulse to the observatory and installed a new Ertel meridian circle. In 1848 Jesuits were expelled from Rome and returned a year later. In the mean time de Vico has died and Secchi was named director. He gave the observatory an international name. In 1853 new buildings were built on top of the roof of Saint Ignatius Church adjoining the college. A new Merz equatorial telescope was installed on a rotating dome. Secchi and his collaborators worked on physical astronomy, specially, on spectra of stars, and solar physics. In 1858 magnetic equipment was installed and work in geomagnetism was added to astronomy and meteorology. In 1870 Jesuits were banished from Rome, but Secchi continued as director of the observatory until his death in 1878. The following year Jesuits were expelled from the observatory which continued in operation until 1923.

Directors

Dominique Dumouchel	1824-1840
Francesco de Vico	1841-1848
Pietro Angelo Secchi	1850-1878
Gaspare S. Ferrari	1878-1879

Collaborators

Luca Boccabianca	
Victor della Rovere	1840-1848
Benedetto Sestini	1840-1848
Bernardino Gambara	
Paolo Rosa de Conti	1858-1868
Enrico Cappelletti	1859
Nazareno Mancini	1867-1870

Giovanni Egidi	1871
Felice Ciampi	1879

Publications

Memorie intorno ad alcune osservazioni fatte alla specola del Collegio Romano, 1838-1845
Memorie dell' Osservatorio del Collegio Romano; 1851-1863.
Bulletino Meteorologico dell' Osservatorio del Collegio Romano. I-XVI, 1862-1877.

Buildings and instruments

The first observatory was installed in a square tower, built in 1878, on the eastern corner of the main front of the college. Secchi built a new observatory on the roof of Saint Ignatius Church that was inaugurated in 1853. It consisted of a circular room of 7.7 m diameter with a rotating dome which housed the Merz telescope on a massive granite pier. An elliptical room was also built for the Ertel meridian circle and other minor telescopes. These two rooms were built on top of two of the large columns designed to support the dome of the church that was never built. The Cauchoix refractor was installed in a smaller room. Other rooms were used as library and workrooms for the astronomers. The magnetic equipment was installed in the western end of the roof in a room built with non-magnetic materials.

Astronomy
1824: The observatory began with some instruments left from the old observatory, namely, Reichenbach transit, Dollond telescope with equatorial mounting and Bellet meridian circle and two sidereal clocks.
1825: Cauchoix telescope with equatorial mounting of 14 cm aperture, Gambey teodolite, and Breguet chronometer.
1843: Ertel meridian circle.
1853: Merz telescope with equatorial mounting of 24.5 cm aperture and 4.33 m focal length, Dent chronometer and Pons clock.

Meteorology
1824: Complete meteorological station with thermometers, barometers, Robinson anemometer, Kreil thermograph, balance barograph, etc.
1867: Secchi Universal Meteorograph.

Magnetism
1853: Grubb bifilar magnetometer, Wiber-Jones unifilar magnetometer, Lloyd vertical magnetic balance, Barrow inclinometer and declinometer,

Bibliography

Anonymous, 1850. *Memorie dell Osservatorio dell' Università Gregoriana in Collegio Romano diretto dai PP. della Compagnia di Gesu.* Tipografia. delle Belle Arti, Rome.

Anonymous, 1855. Il nuovo osservatorio del Collegio Romano. *Civiltà Cattolica*, serie II, 9, 125-128.

A. Secchi, 1856. *Descrizione del nuovo Osservatorio del Collegio Romano.* Tipografia delle Belle Arti, Rome, 36 pp.

A. Secchi, 1859. *Memorie del Osservatorio del Collegio Romano* (description of the meteorological and magnetic equipment). Tipografia delle Belle Arti, Roma.

A. Secchi, 1877. *L'astronomia in Roma nel Pontificato de Pio IX.* Tipografia della Pace, Rome. 50 pp.

C. Andre, G. Rayet and A. Angot, 1878. *L'astronomie practique et les observatoires en Europe et en Amerique,* 5 partie, Observatories d'Italie. Gauthier-Villar, Paris (Observatoire du Collége Romain, 110-143).

P. de Vregille, 1906. Les observatoires de la Compagnie de Jésus au début du XXe siècle. *Rev. des Questions Scientifique*, 59 (3 serie, vol 9), 10-72; 493-579 (Observatoire du Collége Romain, p. 577).

S. Maffeo, 2001. *La Specola Vaticana. Nove Papi, una Missione.* Pubblicazioni della Specola Vaticana, 398 pp., 99 Figs.

OBSERVATORY OF STONYHURST COLLEGE, LANCASHIRE, ENGLAND

1838-1974.
Astronomy, Meteorology, Geomagnetism, Solar Physics, Seismology

The observatory was established in 1838 with meteorological instruments and in 1845 the first telescope was installed. In 1846 Weld began the real scientific work and installed in 1858 the magnetic equipment. A new building for astronomy was built in 1867 and a better telescope was added. The most brilliant time was under Perry's directorship from 1868 to 1889. Research was done mainly on terrestrial magnetism, solar physics and astronomy. Perry made the observatory nationally and internationally known. After his death in 1889, the observatory continued scientific work, especially on solar and stellar spectrography, by Sidgreaves and Cortie. In 1924 seismographs were installed. After 1930 work began to declined and finally in 1947 the observatory was officially closed, though meteorological observations continued. An attempt to reopen the observatory in 1958 dedicating it to geomagnetism did not last long. In 1974 the observatory was definitely closed. In 1980, the astronomical building was restored and a telescope installed for use of the college students and local amateur astronomers.

Directors

Henry McCann	1842-1845
Joseph Howell	1845-1846
Alfred Weld	1846-1851
James Clare	1852-1856
Alfred Weld	1856-1860
Stephen J. Perry	1860-1862
John E. Moore	1862.
Walter Sidgreaves	1863-1868
Stephen J. Perry	1868-1889
Walter Sidgreaves	1889-1919
Aloysius L. Cortie	1919-1925
Edward O'Connor	1926-1932
James P. Rowland	1932-1947
James Lawrence	1947-1960
James Worthy	1960-1974

Collaborators

Joseph Howell	1845-1848
Richard Vaughan	1850-1857
Joseph Hostage	1867-1876
William Carlisle	1870-1875
Joseph Howett	1870-1879
William McKeon	1876-1881
Joseph Rooney	1880-1893
Henry Macklin	1930-1939

Publications

Results of Meteorological Observations 1848-1859
Results of Meteorological and Magnetic Observations, 1860-1886
Results of Meteorological, Magnetical and Solar Observations, 1888-1896
Results of Meteorological and Magnetical Observations, 1897-1911

Results of Meteorological, Magnetical and Seismological Observations, 1912-1918
Results of Meteorological and Magnetical Observations, 1919-1920
Results of Geophysical and solar Observations, 1921-1942.

Buildings and instruments

The first building for meteorological observations was built in 1838 in the college garden in the form of an octagonal room of 6.6 m diameter with four prolongations in cross in N-S and E-W directions. In this building the meteorological and first astronomical instruments were installed. In 1867, at a certain distance, the place for the absolute magnetic measurements was located. In 1867, in the SW corner of the garden, the astronomical building was built, consisting of a circular room of 8 m diameter with a metallic rotating dome and a pier for the main telescope, with two additional small rectangular rooms used for other instruments and work space. In 1868 an underground room was excavated for the magnetic variometers and their recorders. In 1909 the seismographs were also installed there.

Astronomy
1845: Telescope with equatorial mounting of 10.16 cm aperture and meridian circle, of 7.6 cm aperture, both by Jones.
1850: Newtonian reflector telescope of 19 cm aperture by Wright and Cassegrain reflector of 24.1 cm aperture with altazimuth mounting by Carey.
1858: Newtonian reflector telescope by Carey of 17.8 cm aperture, a gift of Sir Radcliffe.
1867: The main telescope with equatorial mounting of 20.3 cm aperture and 3.45 m of focal length by Napier and Carey with lens by Troughton and Simms and finder by Cooke.
1893: The main telescope was upgraded with a new lens of 38 cm by Howard Grubb,
Dublin (Perry Memorial Telescope).
1903: A small Jones equatorial telescope of 10 cm aperture, a telescope of 12.5 cm aperture by Clarke and two transit telescopes by Carey of 6.6 cm aperture.
1932: Alvan Clark refractor of 18.6 cm aperture.

For the study of solar and stellar spectra:
1870: Two Browning spectroscopes, the largest with 4 prisms by Troughton and Simms. Browning photographic spectrograph with 6 prisms supplemented with half prism by Christie-Hilger and net by Rowland. Hilger photographic spectrograph. Spectroscopes and spectrographs were adapted to the telescopes of 20 cm and 38 cm.

Other equipment included Hilger solar prisms, Troughton and Simms and Dolland and Carey micrometers, chronometers, chronographs, sidereal clocks and other complements.

Meteorology
1838: A complete meteorological station with thermometers, anemometers, pluviometers and barometers.
1866: The Board of Trade provided meteorological equipment of direct recording, namely, thermographs, barographs, pluviographs, anemographs and a recorder of the intensity of solar radiation. Later three aneroid barometers, maxima and minima thermometers and two pluviometers.

Magnetism
1863: Barrow unifilar magnetometer for absolute measurements, declinometer and inclinometer of Kew type
1866: A set of variometers with photographic recording for Z, H and D components.
1957: A set of La Cours variometers.

Seismology
1909: Milne horizontal seismograph.
1924: Milne-Shaw horizontal (NS,EW) seismographs.

Bibliography

W. Sidgreaves, 1866. Stonyhurst Observatory. *Letters and Notices* 3, 259-262.

Anonymous (possibly of Perry), 1868. Stonyhurst Observatory, 1839-1868. (Manuscript) 25 pp. SJBPA (Society of Jesus British Province Archive)

S. Perry, 1870. Stonyhurst Observatory. *Monthly Notices Roy. Astr. Soc.* 30, 94-98.

C. André and G. Rayet, 1874. *L'Astronomie practique et les observatoires en Europe et Amerique.* 1er partie, Angleterre (Observatoire du Collége de Stonyhurst 147-150). Gauthier-Villars, Paris.

S. Perry, 1877. The Stonyhurst Observatory. (Manuscript) 31 pp. SJBPA

A. Hewitson, 1878. *Stonyhurst College, present and past.* Chronicle Office, Preston, (Ch. X) 147-155.

S. J. Perry, 1880. L'Observatoire de Stonyhurst. *Annales de la Soc. Scient. de Bruxelles* 4, 147, 281-302.

S. J. Perry, 1882. History of Stonyhurst Observatory (1838-1881). (Manuscript) 15 pp. SJBPA.

G. Gruggen and J. Keating, 1901. *Stonyhurst, its past history and life in the present*. Kegan, Trench and Trübner, London, (Ch. XIII, The Observatory) 238-245.

P. de Vregille, 1906. Les observatoires de la Compagnie de Jésus au début du XXe siècle. *Rev. des Questions Scientifique* 59, 10-72; 493-579.

W. Sidgreaves, 1909. Stonyhurst College Observatory, Landcashire. A brief sketch of its history and work. (Manuscript) 10 pp. SJBPA.

J. Rowland, 1920. Private journal on matters relating to Stonyhurst observatory from Oct 27, 1919. (Manuscript) 16 pp. SJBPA.

Anonymous, 1932. The story of a famous observatory. Astronomers of Stonyhurst. *The Meccano Magazine*, February, 102-103, 150.

B.G.S., 1948. The Stonyhurst Observatory. *The Stonyhurst Magazine*, 29, 145-147.

P. J. Treanor, 1948. Stonyhurst College Observatory. *Nature* 161, 285.

Anonymous, 1957. The Observatory. *The Stonyhurst Magazine* 32, 445-448.

Anonymous, 1961. Stonyhurst College Observatory. *The Stonyhurst Magazine* 34, 96-97.

Anonymous, 1961. Stonyhurst Observatory. *Letters and Notices* 66, 169-171.

F. O'Reilly, 1980. Reopening of the Dome Observatory - Sunday 1st June 1980. *The Stonyhurst Magazine* 41, 269-272.

F. O'Reilly, 1991. The Perry Memorial Telescope. *The Stonyhurst Magazine* 47, 94-96.

T. E. Muir, 1992. *Stonyhurst College*, 1593-1993. James and James, London, (Excursus: The Stonyhurst Observatories 1838-1992) 122-124.

J. Bommer, 1995. Stonyhurst College Observatory. (Manuscript) 12 pp.

OBSERVATOIRE DU COLLEGE NOTRE DAME DE LA PAIX, NAMUR, BELGIUM

1838-1936

Astronomy, Meteorology

The observatory was first part of the laboratory of physics installed in 1838 in the College of Notre-Dame de la Paix in Namur by Antoine Maas. It included two telescopes one by Dolland and the other by Merz of 9 cm aperture. In 1857 a meteorological station was installed with which V. Van Tricht made observations in 1880. In 1883 a tower was added to the college building with a dome for a new telescope by Frères Henry, Paris of 20 cm aperture and 1.20 m focal length. The observatory continued until 1940 when it was closed

Bibliography

D. Lucas y F. Willaert, 1912. L'éclipse de soleil du 17 Avril 1912 observée au Laboratoire du Collège N.-D. De la Paix a Namur. *Rev. des Questions Scientifiques*. 72. 187-225.

A. Allart, 1915. Chronicon VIII (1909-1915) Collegium Namur Reginae Pacis. Sur l'eclipse de soleil du 17 du Avril 1912, 147-181. (Manuscript) Archives of Collège N.D. de la Paix, Namur.

H. Tihon, 1999. Notes sur les observatoires du Collège Notre-Dame de la Paix, Namur. (Manuscript) 1 p.

OBSERVATORIO DEL COLEGIO DE S. MARCOS, LEON, SPAIN

1860-1868

Meteorology

In 1860 an observatory with all basic instruments for meteorological observations was installed at the Jesuit Faculty of Philosophy and Theology in Leon, Spain. Its observations were sent to Instituto Central Meteorológico, Madrid and published in the Anuario Estadístico from 1865 to 1867. Directors were Francisco Vinader from 1860 to 1862 and Antonio Cabré from 1863 to 1868. The observatory was closed in 1868, together with the faculty when Jesuits were expelled from Spain.

Bibliography

Anonymous, 1866. *Resumen de las observaciones meteorológicas recogidas en el Observatorio del Colegio de S. Marcos, Leon. 1865 seguidas de los estados correspondientes a los años 1862, 1863 y 1864.* Miñón, Leon.

HAYNALD OBSERVATORY, KALOCSA, HUNGARY

1879-1950

Astronomy, Geomagnetism, Meteorology, Solar Physics

The observatory was founded by the initiative of the Archbishop of Kalocsa, Cardinal Lajos Haynald in the Jesuit College, Stephaneum Gymnasium. Built on the roof of the college the observatory was inaugurated in 1879. The first director was Braun who installed a Merz telescope with equatorial mounting attaching to it a Hilger spectroscope for the study of the solar crown, sunspots, prominences and faculae. The observatory had a complete meteorological station. From 1885 to 1913 Fényi gave the observatory its greatest fame with his work on solar physics. After him the observatory began to decay and in 1950 was confiscated by the Communist government.

Directors

Karl Braun	1879-1884
Adolf Hüninger	1884
Gyula Fényi	1885-1913
Theodor Angehrn	1913-1947
Mátyás Tibor	1948-1950

Collaborators

Adolf Hüninger	1880-1911
Johann Schreiber	1891-1902
Riegle Sándor	1897-1913

Publications

Publikationen des Haynald Observatoriums, Heft I-XVII (1896-1949)

Buildings and instruments

The observatory was built on the second floor of the college. It consisted of five rooms one of them for the library and two circular rooms with conical roofs, one of 4 m diameter and the other of 3 m for the large and small telescopes. Two of the rooms had apertures in NS and EW direction for the meridian circle and an astronomical teodolite. In the west corner of the building there was a tower with a geodesic pier used also for magnetic measurement. The meteorological instruments were installed on the roof.

Astronomy
1878: Merz telescope with equatorial mounting of 18 cm aperture and 2.20 m focal length. Merz telescope of 8 cm aperture. Cooke transit telescope of 5.8 cm aperture
Hilger spectroscope with 6 prisms. Spectroscopes by Merz, Häntsch and Browning with 2 poliprisms by Amici. Zöllner astrophotometer and Glan and Vogel spectrometer. Three chronometer and a chronograph by Mayer and Wolf.

Meteorology
1878: Thermometers, barometers, etc. A nephoscope and an instrument of own design to record electric discharges in storms, "Gewitterregistrator" by Schreiber and Fényi.

Magnetism
1878: Dover magnetometer, declinometer, and inclinometer.

Seismology
1920? Wiechert horizontal seismograph

Bibliography

C. Braun, 1886. *Berichte von dem Haynaldschen Observatorium zu Kalocsa in Ungarn.* Aschendorffschen Buchhandlung, Munster, 45 pp.

P. de Vregille, 1906. Les observatoires de la Compagnie de Jésus au début du XXe siècle. *Rev. des Questions Scientifiques* 59 (3 serie, vol 9), 10-72; 493-579.

J. Fényi, 1906. Die Sternwarte in Kalocsa, das Haynald observatorium. *Nachr. der öster-ungarische Provinze* 3, 14-23.

B.O. Kelényi, 1920. *Geschichte der ungarischen Astronomie.* Stephaneum Nyomda R.T., Budapest, 86-89.

T. Angehrn, 1928. *A Haynald Observatorium.* Árpád-könyvek, Kalocsa.

I. Mojzes, (ed.) 1986. *A Kalocsai Haynald Observatórium története.* Orzágos Müszaki I.K.K., Budapest, 139 pp.

M. Tibor, 1991. Die Tätigkeit des Haynald Observatoriums. (manuscript), 4 pp.

OSSERVATORIO METEOROLOGICO TUSCULANO, MONDRAGONE, ITALY
1879-1944
Meteorology, Seismology

The observatory was established in 1879 in the Collegio di Mondragone, near Frascati, founded in 1865 in the old Villa Borghese. The observatory had very complete meteorological instrumentation with Fortin barometer, pluviometer, Robinson anemometer, heliograph, thermometers, ozonometers, etc. In 1888 a seismometer built by Egidi was installed and in 1908 Cancani and Agammenone seismoscopes, donated by the Uficio Centrale de Meteorologia e Geodinamica of Rome. Directors were science teachers in the school and changed every two or three years. The observatory was closed in 1944.

Directors

Felice Ciampi	1879-1889
Giuseppe Giovenale	1889-1891
Giovanni Egidi	1891-1894
Giuseppe Giovenale	1894-1901
Carlo Rinaldi	1901-1905
Vittorio Bovini	1905-1907
Ercole Iannelli	1907-1908
Giuseppe Preti	1908-1909
Ercole Iannelli	1909-1910
Filippo Rossetti	1910-1912
Otho Werner	1912-1913
Erasmo Blasio	1913-1915
Giacomo Morissey	1915-1916
Oreste Mirri	1916-1917
Luigi Danese	1917-1919
Angelo Tom,	1920-1922
Dinus Dippi	1923-1925
Guilielmo Miserville	1926-1928
Guilielmo Como	1929-1930
Vicentius Monachino	1931-1933
Giovanni Scarpellino	1934-1936
Humberto Damiani	1937-1938
Ignatius Ruggeri	1939-1941

Demetrio Russos 1942-1944

Bibliography

F. Ciampi, 1880. *Notizie intorno all'Osservatorio Meteorologico Tusculano.* Tip. delle Scienze Matemat. e Fisiche, Rome, 19 pp.

P. de Vregille, 1906. Les observatoires de la Compagnie de Jésus au début du XXe siècle. *Rev. des Questions Scientifiques* 59, (3 serie, vol 9) 10-72; 493-579.

L. Danese, 1918. *L'Osservatorio Meteorologico Tusculano. 1868-1918 Memorie storiche e descrittive.* Manuzio, Rome, 41 pp.

OSSERVATORIO SUL GIANICOLO, ROME, ITALY

1880-1913
Astronomy

The observatory was installed in 1880 in the Villa Cecchina, on the slope of the Gianicolo Hill by Ferrari after the confiscation by the state of the observatory of the Collegio Romano. A dome was built to house the telescopes. The observatory was associated with the Gregorian University where Ferrari was professor of astronomy. He was succeeded in 1895 by Müller, also professor of astronomy in the university until the observatory was closed in 1913.

Directors

Gaspare S. Ferrari	1884-1894
Adolph Mueller	1895-1913

Buildings and instruments

The building was a small dome that housed the telescopes. The main instrument was a Merz equatorial telescope of 23 cm aperture and 3.82 m focal length. Later a smaller telescope was added of 10.8 cm aperture and 1.50 m focal length.

Bibliography

G. Ferrari, 1882. Il nuovo Osservatorio Astronomico Privato sul Gianicolo. *Pont. Universita Gregoriana, Bull. Meteorologico,* 21, 4, 1-3.

OBSERVATORIO METEOROLOGICO DEL COLEGIO DEL JESUS, TORTOSA, TARRAGONA, SPAIN

Meteorology
1880-1905

The Colegio Máximo (Faculty of Philosophy and Theology) of the Province of Aragón was established in Tortosa, Tarragona in 1880. The same year began the meteorological observations of temperature, atmospheric pressure and rainfall, from 1886, air humidity, cloudiness and wind direction and from 1892 wind velocity and evaporation. Observations were made by Jesuit students under the direction of the professors of science, Antonio Canudas till 1891 and Ramón Faura till 1905. The observatory was closed when the faculty was moved to Roquetas and Ebro Observatory was established. The records of the observatory from 1880 to 1905 are kept in Ebro Observatory.

OBSERVATORIO METEOROLOGICO DEL COLEGIO DE LA GUARDIA, PONTEVEDRA, SPAIN

1880-1916
Meteorology

The Colegio Apostol Santiago, a Jesuit secondary school, was established in 1872, near La Coruña and moved in 1875 to La Guardia, Pontevedra near the Miño River. The meteorological observatory began in 1880. It was installed in an hexagonal building in the gardens of the school. It had very complete equipment, namely, two maxima and minima thermometers (Walferdin and Rutherford), Bellani thermograph, psicrometer, Tonnelot pluviometer, Eon Fils barometer, evaporimeter, etc. Observations were made twice a day and sent to Instituto Central Meteorológico in Madrid and published in its annual bulletin since 1883. In 1908-1911 the observatory was moved to a tower built in the south front of the school building. Merino, Director from 1888 to 1906, published several climatological studies of the region, based on the observations of the observatory. The observatory was closed in 1917 when the school was moved to Vigo.

Directors

Ramón Martínez	1880
Sandalio García	1801-1884
Juan B. Rojas	1885-1887
Baltasar Merino	1888-1906
Julián Zabala	1907
Lucio Rodríguez	1908
Eduardo Fernández Regatillo	1909-1913

Bernardino Urra 1914-1915

Publications

Colegio de la Guardia. Observaciones Meteorológicas. (several bulletins from 1881 to 1897).

Bibliography

B. Merino, 1891. R*esumen de las observaciones meteorológicas efectuadas en el Colegio de la Guardia de la Compañía de Jesús, en el decenio 1881 a 1890.* Rafael Marco y Viñas, Madrid

B. Merino, 1893. *Estudio sobre las borrascas en la costa occidental de Galicia.* Tipografía Gallega, Tuy, 59 pp.

B. Merino, 1894. *Un estudio acerca del agua meteorológica.* Tipografía Gallega, Tuy, 70 pp.

B. Merino, 1893,1894, 1897. *Las observaciones meteorológicas de los años 1891 a 1896. Observatorio Meteorológico del Colegio de la Compañía de Jesús en La Guardia (Pontevedra).* Tipografía Gallega, Tuy.

E. Rivera, 1993. *Colegio Apóstol Santiago. Historia de una larga peregrinación.* (El P. Merino, maestro y climatólogo*).* Colegio Apóstol Santiago, Vigo, 1993

E. Rivera, 1995. Baltasar Merino S.J. Un grande Naturalista no Baixo Miño. *Monografía da Associación Naturalista Baixo Miño*, 5, 1-21.

OBSERVATORIO ASTRONOMICO Y METEOROLOGICO DEL COLEGIO DE OÑA, BURGOS, SPAIN

1882-1967
Meteorology, Astronomy

In 1882 a meteorological station was installed in the Jesuit Faculty of Philosophy and Theology, established in 1880 in an old Benedictine Monastery in Oña, Burgos. Observations were sent to the Instituto Central Meteorológico, Madrid and published in its annual bulletin until 1931. The directors were professors of science at the Faculty of Philosophy. In 1906 a telescope with altazimuth mounting of 20.3 cm aperture and 3 m focal length was installed in a rotating dome on the roof of the building. From that year the observatory was named Observatorio Astronómico y Meteorológico, but astronomy never progressed. In 1931 the faculty was closed and opened again in 1942 after the Spanish Civil War. Meteorological observations continued until the faculty was moved to Bilbao in 1967. In this period no directors were assigned to the observatory.

Directors

Ramón Martínez	1882-1885
Eleuterio Fernández	1886
Melchor Delgado	1887-1890
Bonifacio Fernández Valladares	1891-1899
Fernando Gutiérrez del Olmo	1900-1903
Pedro Fernández	1904-1906
José María Ibero	1907
Ignacio Cantarell	1908-1909
Fernando Gutiérrez del Olmo	1911-1915
Emiliano Echaguibel	1916, 1918
José María Aldasoro	1917, 1919
Emiliano Echaguibel	1918
José María Aldasoro	1919-1931

Publications

Observaciones Meteorológicas hechas en el Colegio Máximo de la Compañía de Jesús en Oña. Imprenta Privada del Colegio, Oña, Burgos. Monthly volumes between 1885 and 1895 and annual volumes from 1906 to 1920.

Bibliography

B. Fernández Valladares, 1885. *Observaciones Meteorológicas del Colegio de Oña*. Imprenta del Colegio, Oña, Burgos.

B. Fernández Valladares, 1896. *Noticia del clima de Oña sacada de las observaciones hechas desde Febrero de 1882, en que se fundó el Observatorio hasta 1895*. Imprenta del Colegio, Oña, Burgos.

Anonymous, 1906. *Observaciones del eclipse total de Sol del 30 de Agosto de 1905 hechas por los Padres de la Compañía de Jesús en el Colegio de Oña.* Imprenta Privada del Colegio, Oña.

METEOROLOGICAL OBSERVATORY OF SACRED HEART SEMINARY, GOZO, MALTA

1882-1909
Meteorology

The observatory was established in 1882 in the Sacred Heart Seminary, a diocesan seminary entrusted to the Jesuits in 1866 in the Maltese island of Gozo. The observatory was installed on the roof of the Seminary and had a very complete meteorological instrumentation, namely, Fortin barometer, aneroid barometer, anemograph, thermograph, piscrometers, thermometers, pluviometers, etc. The observatory was associated with the Rete Meteorologica Italiana, and sent monthly bulletins to other observatories among them the U.S. Weather Bureau. Directors were the science teachers in the seminary. Jesuits left the seminary in 1909 and the observatory was closed shortly after.

Directors

Filipo Borrello	1883-1884
Giuseppe Micali	1885
Felix Lo Re	1886
Angelo Basile	1887
Aloisius Grech	1888
Angelo Basile	1889
Salvatore Curmi	1890
Angelo Basile	1891
John Edge	1892-1893
Manuel Grima	1894
Salvatore Curmi	1895
Aloisius Albanese	1896
Aloisius Briffa	1897
Aloisius Albanese	1898
Giuseppe Billota	1899
Salvatore Mammana	1900-1901
Alban Robinson	1902-1903
Francesco Longhitano	1904-1908

Bibliography

P. de Vregille, 1906. Les observatoires de la Compagnie de Jésus au début du XXe siècle. *Rev. des Questions Scientifiques* 59 (3 serie, vol 9), 10-72; 493-579.

J. Bezzina, 1983. The Meteorological Observatory at Gozo Seminary 100 years after its foundation. *Gozo Year Book 1983*, 19-23.

J. Bezzina, 1983. The Gozo Seminary- A Historical Note. *Sacred Heart Seminary Directory, 1983-84*. pp. 4-5.

METEOROLOGICAL OBSERVATORY AT SAINT IGNATIUS COLLEGE, MALTA

1883-1907
Meteorology

The observatory was established in 1883 in St. Ignatius College in Malta, founded in 1877 by English Jesuits. Observations were published in the Bulletin of Stonyhurst Observatory. The observatory was closed together with the college in 1907.

Directors

James Scoles	1884-1893
Joseph Dobson	1894-1907

Bibliography

F. Edwards, 1985. *The Jesuits in England from 1580 to present day*. Burn and Oates, Wellwood, 333 pp.

OBSERVATORIO METEOROLOGICO DEL COLEGIO DE ORDUÑA, VIZCAYA, SPAIN

1883-1931.
Meteorology

In 1883 a meteorological station was installed in the Colegio de Nuestra Señora de la Antigua in Orduña, Vizcaya. Its directors were the science teachers in the school. Observations were sent to the Instituto Central Meteorológico, Madrid and published in its annual bulletin. The station was closed in 1931

Directors

Hilario Retolaza	1883-1885
Baltasar Merino	1886
Efrén Astudillo	1887-1902

Daniel Sola	1894-
Saturnino Gutiérrez	1898
José Julián Larrazabal	1899-1900
Carlos Varona	1902-1904
Manuel Guinea	1905-1908
Federico Rodríguez	1909-1910
Hermógenes Basauri	1911-1916
Florentino Legarreta	1917-1920

OBSERVATOIRE DE LOUVAIN, BELGIUM

1884-1970
Astronomy

The observatory was established in 1884 in the Jesuit Faculty of Philosophy and Theology in Louvain, dedicated to teaching and students demonstrations. It was located in a tower with a rotating dome built on top of the building of the faculty which housed an equatorial telescope of 15 cm aperture and a meridian circle of 7.5 cm aperture. The observatory had auxiliary equipment and was used mainly for student demonstrations. The observatory is listed in the catalogue of Lancaster, *Liste General de Observatoires et Astronomes* of 1888. In 1929 the faculty was moved to Eegenhoven and the observatory was also installed there. In 1958 the faculty was moved again to a new building in Heverlee, outside Louvain, where a new observatory was built. Jesuit student used the observatory until about 1970. From that time the observatory is used sporadically by students of the University of Louvain.

Directors

Joseph Delsaulx	1884-1887
Julien Thirion	1887-1918
Henri Dopp	1922-1928
August Grégroire	1929-1940

OBSERVATORY OF THE KOLLEGIUM STELLA MATUTINA, FELDKIRCH, AUSTRIA

1889-1936
Meteorology, Astronomy

The origin of the Observatory in the College Stella Matutina, in Feldkirch, Austria can be placed in the meteorological observations which began in 1889. Later the meteorological station was moved to a nearby location in Tisis. The observatory had a Merz equatorial telescope of 15 cm aperture and instruments for spectroscopy. Province Catalogues do not mention the names of directors, but only those of astronomy teachers. The observatory was used only for teaching purpose. In 1936 the College was taken over by the Nazi government. Returned to the Jesuits after the war the college was closed in 1979.

Bibliography

P. de Vregille, 1906. Les observatoires de la Compagnie de Jésus au début du XXe siècle. *Rev. des Questions Scientifiques* 59 (3 serie, vol 9), 10-72; 493-579.

SPECOLA VATICANA, VATICAN, CASTELGANDOLFO, ITALY AND TUCSON, ARIZONA

1890 -
Astronomy

The observatory was founded in 1890 by Pope Leon XIII with name Specola Vaticana directly dependent on the Holy See. The first three directors were not Jesuits. The observatory was installed in the Vatican in the Tower of Winds with a dome for the two Merz telescopes and had also sections of meteorology and magnetism. In 1891 the observatory began its participation in the Carte du Ciel program. The first Jesuit director was Hagen named in 1906. He concentrated in astronomy and suppressed the meteorological and magnetic sections. A new Merz telescope of 40 cm aperture was installed in 1909 in a tower of the Leonine wall. In 1932 under Stein's direction the observatory was transferred to Castelgandolfo in the papal palace and gardens of the Villa Barbarini. There two domes were built and a new Zeiss equatorial telescope was installed. The new observatory was inaugurated in 1935 and it was entrusted officially to the Society of Jesus by the Pope Pius XI. The observatory continued the work in the program of Carte du Ciel, in the Atlas of Variable Stars and other astronomical projects. In 1949 a very complete spectrographic laboratory was installed for work on spectra of stars. For this purpose a new telescope was acquired in 1957. The bad conditions for astronomical observations in Castelgandolfo motivated the signing of agreement in 1980 with the University of Arizona to use its telescopes in Tucson. In 1993 a new telescope called VATT (Vatican Observatory Advanced Technology Telescope) was installed on Mount Graham, Arizona. From this date on all observations and research work are done in Arizona. At present about 13 Jesuits are working at the Observatory.

Directors

Francesco Denza	1890-1894	Barnabite
Giussepe Lais	1894-1898	Oratorian
Angel Rodriguez de Prada	1898-1906	Agustinian
Johann G. Hagen	1906-1930	
Johan Stein	1930-1951	
Daniel O'Connell	1952-1970	
Patrick Treanor	1971-1978	
George Coyne	1978-	

Collaborators

Ramón Puigrefagut	1932-1934
Alois Gatterer	1933-1953
Oscar Probst	1934
Johann Beckers	1934-1935
Antonio Fusarelli	1935-1937
Karl Treusch	1935-1978
Josef Junkes	1935-1977
Peter Albert Zirwes	1935-1968
Antonio Pignatelli	1936-1944
Mátyás Tibor	1937-1940
Mathew Timmers	1938-1954
Vinzenz Frodl	1940-1947
Walter Miller	1946-1955
Ernst Salpeter	1946-1976
Jules de Kort	1947-1958
Roger Le Claire	1950-1952
Aloysio Puhl	1955-1983
Florent Bertiau	1958-1974
Martin F. McCarthy	1958-
Omer Van de Vijver	1969-1971
Emiel de Graeve	1972-1983
Juan Casanovas	1976-
Edmund Benedetti	1978-1986
Pierre Cardon de Lichtbuer	1978-1981
Christopher Moss	1979-1986
William R. Stoeger	1979-
Richard Boyle	1981-
Christopher Corbally	1983-
Sabino Maffeo	1985-
Gustav Teres	1988-1993
Saverio Corradino	1990-

Guy J. Consolmagno 1993
Andrew P. Whitman 1997
Richard Murphy 1998
José Funes 2000

Publications

Pubblicazioni della Specola Vaticana 1891-1905. (7 vols.)
Miscellanea Astronomica (Ossia Raccolta di articoli pubblicati dalla Specola Vaticana in diversi giornali scientifici). 1920-1958.
Ricerche Astronomiche edite dal Direttore. Ricerche Minori della Specola Astromica Vaticana. 1939-1979 (9 vols.)
Miscellanea Astronomica della Specola Vaticana (1931-1956) (2 vols.)
Annual Report. 1963-1969
Vatican Observatory Publications. 1970-
Studi Galileiani, 1981-
Physics, Philosophy and Theology. 1988.
Scientific Perspectives on Divine Action, 1993-2000 (4 vols)

Buildings and instruments

The observatory was first established in 1890 at the Vatican in the Tower of Winds where a rotating dome was built for the telescopes and meteorological and magnetic instruments were installed. In 1891 new telescopes for the program Carte du Ciel were installed in a dome of 8 m diameter on the Tower of S. Giovanni, part of the Leonine Wall. In 1909 another dome of 8.8 m diameter was built on another tower of the Wall and the Tower of Winds was abandoned.

In 1932 the Observatory was moved to Castelgandolfo. In 1935 the installation was finished; two rotating domes of 8.5 m and 8 m diameters were built on top of the roof of the Papal Palace. Offices and library were installed in the fourth and fifth stories. In 1941 the astrograph of the Carte du Ciel with its dome was also moved to Castelgandolfo.

In 1993 a building with a rotating dome was built on Mount Graham near Tucson, Arizona where the new VATT telescope was installed.

Astronomy

1890: Two Merz telescopes, one of 10.2 cm aperture and equatorial mounting and the other of 10.6 cm aperture and altazimuth mounting and a meridian circle by Stark. 1891: Double photographic equatorial telescope by Gilon, Paris, for the Carte du Ciel program, of 33 cm aperture and 3.43 m focal length housed in a rectangular metal tube.

1893: Henry and Gauthier heliograph with Janssen design of 14 cm aperture and 2.15 m focal length which produced a solar image of 28 cm diameter.

1909: New Merz equatorial telescope of 40 cm aperture and 6.1 m focal length, a Graff photometer, a Danjon interferometer, a double astrograph by Zeiss consisting in a refractor of 40 cm aperture and 2.4 m focal length and a reflector with a parabolic mirror of 60 cm diameter and equivalent focal length of 8.2 m.
1933: A complete spectrographic laboratory with two prisms of 61.2 cm diameter and angles of 4 and 8 degrees, a Komess instrument for coordinates measurements, a Hartmann spectrocomparator, Askania microphotometer, GH Steinheil, Zeiss and Halle spectrographers and other equipment
1957: Schmidt reflector telescope by Cox, Hargreaves and Thomson, London with a parabolic mirror of 98 cm diameter, correcting plate of 65 cm and 2.4 m focal length. The telescope was adapted with 3 prisms for stellar spectroscopy.
1950-1980: Several auxiliary instruments and spectrographs by Jarrell-Ash, Boston, and from Bausch and Lomb, Rochester
1960: Computer center with IBM1620 computer. Upgraded the following years, 1978: IBM5100 computer, 1990: VAX3100 computer.
1985-1995: Telescope of new design VATT (Vatican Observatory Advanced Technology Telescope) with a principal mirror of 1.83 m diameter and secondary of 38 cm resulting in a system of 16.47m focal length, installed in Mount Grahan, Arizona.

Bibliography

F. Denza, 1891. *Cenni storici sulla Specola Vaticana.* Pubblicazioni della Specola Vaticana, Fascicolo II, 88 pp.

P. de Vregille, 1907. L'Observatoire du Vatican et le P. Hagen. *Lettres d'Ore.* Supp. 1, 8-14.

J. Stein, 1908. I restauri della Specola Vaticana. *Riv. di Fisica e Scienze Naturali (Pavia)* 108, 5-20

J. Hagen, 1923. The Vatican Observatory. *The Observatory* 46, 54.

J. Stein and J. Junkes 1952. *La Specola Vaticana nel passato e nel presente.* Specola Vaticana, 68 pp.

J. Stein and J. Junkes, 1952. *Die vatikanische Sternwarte in Vergangenheit und Gegenwart.* Typ. Poliglotta Vaticana, 72 pp.

P. Treanor, 1969. *The Vatican Observatory.* Specola Vaticana, 34 pp.

S. Maffeo, 1991. *In the service of nine Popes. 100 years of the Vatican Observatory.* Vatican Observatory, 241 pp.

Anonymous, 1995. *Vatican Observatory. Annual Report* 1994. Vatican Observatory Publications, 31 pp.

S. Maffeo, 2001. *La Specola Vaticana. Nove Papi una Missione*. Publicazioni della Specola Vaticana, 398 pp.

OBSERVATORY OF THE SEMINARY OF OUDENBOSCH, HOLLAND

1891-1929
Meteorology and astronomy

The seminary of Oudensbosch was entrusted to the Society of Jesus in 1878. In 1891, the observatory was built in a tower with a dome where an equatorial telescope of 8 cm aperture was installed in the new building of the seminary. The same year began the meteorological observations which were sent to the Meteorological Institute in Utrecht. At the seminary there were professors of astronomy until 1915 who took care of the observatory, among them Alphonse Renkin, Victor Becker, from 1878 to 1898 and Henri Bolsius from 1901 to 1915. Jesuits left Oudenbosch in 1929.

Bibliography

P. de Vregille, 1906. Les observatoires de la Compagnie de Jésus au début du XXe siècle. *Rev. des Questions Scientifiques*, 59 (3 serie, vol 9), 10-72; 493-579.

Letter of de G. J. Adriaansen (18-5-1992).

OBSERVATOIRE DE ST. LOUIS, JERSEY ISLAND, UNITED KINGDOM

1893-1979
Meteorology, Geomagnetism, Seismology

The observatory was founded in 1893 by Dechevrens, former director of Zikawei Observatory in the Island of Jersey, in the English Channel, where in 1880 the Maison Saint Louis, for philosophy studies of French Jesuits had been established. The observatory had a building for recorders, offices, library and a meridian circle. The main structure was an iron tower of 50 m high, where a special anemometer for the measurement of horizontal and vertical components of the wind, designed by Dechevrens, was installed. In 1897 began the magnetic observations of inclination and declination and in 1936 a seismograph was installed. In 1923 several instruments were sent to Zikawei and in 1929 the tower was dismantled. The observatory, directed by Rey, continued its observations during the Second World War until 1979, although the Jesuit school of philosophy had been moved to France in 1954. In 1979, the observatory was transferred to the Jersey Meteorological Department and is still in operation.

Directors

Marc Dechevrens	1893-1923
Christian Burdo	1923-1933
Charles L. Rey	1934-1979

Buildings and instruments

The observatory consisted of a one-story building for offices, recorders, library and telescopes and an iron tower of 50 m height where the anemometers were installed.

Meteorology
Universal anemometer (Dechevrens), Robinson anemometer, Jordan heliograph, Benson nephososcopy, Tonnelat barometer, Richard and other barographs, thermometers, psicrometers, etc.

Astronomy
Gautier meridian circle of 5 cm aperture and telescope of 11 cm aperture

Magnetism
1897: Dover declinometer and inclinometer.

Seismology
1936: Mainka horizontal seismograph

Bibliography

M. Dechevrens, 1890?. Project de fundation d'un observatoire météorologique à Jersey. (manuscript) 6 pp. APFSJ (Archives Province de France Societatis Jesu)

M. Dechevrens, 1904. *Bulletin des Observations Magnétiques et Meteorologiques, Résumé de 10 années, 1894-1903*. Vol 9.

P. de Vregille, 1906. Les observatoires de la Compagnie de Jésus au début du XXe siècle. *Rev. des Questions Scientifiques* 59 (3 serie, vol 9), 10-72; 493-579.

M. Dechevrens, 1923. Etude du vent à Jersey (dans la Manche), vingt années d'observations (1895-1914) à l'Observatoire Saint Louis. *Memoirs de l'Office National Meteor. de France* 1, 3.

C. Rey, 1960. The Observatory; Maison Saint Louis. *Bull. de la Soc. Jersiaise*, 17, 313-320.

C. Cornet y J. Rabourdin. 1981. Charles Rey (1897-1981) *Compagnie-Courrier de la Province de France*, 149, 115-116.

F. Le Blancq, 1994. One hundred years of weather recording at the Maison St. Louis Observatory, Jersey, 1894 to 1994. (manuscript). 12 pp.

STERNWARTE DES IGNATIUSKOLLEGE, VALKENBURG, HOLLAND

1896-1941
Astronomy and meteorology

In 1894 the Ignatiuskolleg of philosophy and theology of German Jesuits was established in Valkenburg, Holland after their expulsion from Germany. The observatory was installed in 1896 in a tower with a dome of 5 m diameter, built in the part of the building dedicated to the faculty of philosophy. It was dedicated to astronomy and its main instrument was a telescope with equatorial mounting. The main aim of the observatory was to serve as an aid to the science courses, although observational work was also carried out. The most active directors were Hisgen, Baur and Esch, who carried out observations of variable stars in collaboration with Hagen. Other professors of science also participated in the observatory. In 1901 a meteorological station was installed which sent the observations to the Royal Meteorological Institute of Holland in Utrecht. The observatory was closed in 1941 when the college was occupied by SS German troops, after the invasion of Holland in the Second World War.

Directors

Johann Springer	1896-1897
Joseph Hisgen	1898-1902
Alfred Baur	1902-1904
Michael Esch	1905-1909, 1919-1938
Hermann Schmitz	1939-1940

Publications

Beobachtungen veränderlichen Sterne. Sternwarte des Ignatiuskollegs, Valkenburg, vol. 1-7. 1930-1937

Buildings and Instruments

The observatory was built in 1896 at the same time as the college with a tower of 30 m height where a rotating dome of 5 m diameter was installed. The dome housed a telescope of equatorial mounting of 23 cm aperture and 2.80 m focal length by Saegmueller and lens by Clacey, Washington. Other instruments were a telescope of altazimuth mounting of 15.3 cm aperture by Floess, Vienna, a Briethaupt theodolite, chronographs and spectrographs. In 1901 a complete meteorological stations was installed

Bibliography

P. de Vregille, 1906. Les observatoires de la Compagnie de Jésus au début du XXe siècle. *Rev. des Questions Scientifiques*, 59 (3 serie, vol 9), 10-72; 493-579.

A. Feder, 1919. Wissenschaftliches Arbeiten. In *: 25 Jahre Ignatiuskolleg Valkenburg 1894-1919.* Herder, Freiburg.

M. Esch, 1919. *Beobachtungen veränderlichen Sterne. Sternwarte des Ignatiuskollegs, Valkenburg, Niederlande*, Veröff. N. 1.

H. J.M. Keulen, 1996. *Van Jesuitencollege tot Academie voor Bewustzijnsont wikkeling. En Historische en Heemkundige Studie in en ron het Geudal Jarrboek 1996*. Valkenburg. 185-224.

OBSERVATORIO METEOROLOGICO UNIVERSIDAD PONTIFICIA DE COMILLAS, SANTANDER, SPAIN
1900-1967
Meteorology

In 1900 a meteorological station was established in the Seminario and later Universidad Pontificia de Comillas in Santander (northern coast of Spain). Directors were professors of science of the faculty of philosophy. The observations were discontinued during Jesuits expulsion from Spain, from 1931 to 1940 and continued afterwards until 1967. Directors were assigned only in the first period. Observations were sent to the Servicio Meteorológico Nacional.

Directors

Antonio Suárez	1900-1903
Julián Zabala	1904
Santiago Viña	1905
Carlos Varona	1906
Antonio María Lozano	1907-1908
Lucio Rodríguez	1909-1912
Pedro Fernández	1913-1914
Antonio María Lozano	1915-1931

OBSERVATORIO DE CARTUJA, GRANADA, SPAIN

1902-1975
Seismology, Meteorology, Astronomy

The observatory was founded in 1902 in the Jesuit Faculty of Theology and Philosophy in the neighborhood of Cartuja, Granada. The observatory had three sections, namely, astronomy, meteorology and geodynamics. The section of astronomy had a good telescope with equatorial mounting and other equipment; the meteorological section had very complete equipment and the geodynamic section several seismographs. The three sections functioned with certain autonomy and had, sometimes, different directors. Due to its location in one of the most seismic active regions of Spain, the most important section was that of seismology, especially, during the time when Navarro Neumann was director between 1907 and 1931. He modernized the equipment with mechanical and electromagnetic seismographs built by himself. After the expulsion of the Jesuits from Spain in 1931, the observatory was operated by the state. After the Spanish Civil War in 1940, Jesuits came back to the observatory which was directed by Due Rojo until 1967. Vives tried to give a new impulse to astronomy installing a telescope in Sierra Nevada at 2506 m altitude, but he left the Observatory in 1971. The same year the Observatory was transferred to the University of Granada.

Directors

Ramón Martínez	1902-1904	
José Mier y Terán	1904-1908	astr. 1907 meteor.
Ramón Martínez	1904-1906	meteor and seism.
Ricardo Garrido	1909-1912	astr.
Manuel M. Sánchez Navarro-Neumann	1907-1920	seism.

Mariano Ayala	1908-1909	meteor.
Juan Murillo	1910-1911	meteor.
M. M. S. Navarro-Neumann	1913-1914	
Agustín Moral	1915-1916	meteor.
Ginés Yañez	1917	meteor.
Federico Savoie	1918	meteor.
Rafael Barraquer	1919	meteor.
Ricardo Garrido	1920	meteor. and astr.
Manuel Grund	1921-1922	
M. M. S. Navarro-Neumann	1923	
M. M. S. Navarro-Neumann	1924-1931	seism.
Cándido Guerrero	1924-1928	meteor.
Joaquín Meseguer	1929	meteor.
Juan Murillo	1930-1931	meteor.
Antonio Dué Rojo	1940-1966	
Teodoro Vives	1967-1971	

Collaborators

Luis Hurtado	1907-1948
Luis López	1907-1910
Antonio Sola	1913-1931
Manuel Merlo	1966-1971

Publications

Boletín Mensual del Observatorio de Cartuja, 1903- 1930

Buildings and instruments

The building, inaugurated in 1902, has a plant in form of a cross with a circular room in the middle with a rotating dome of 8 m diameter where the equatorial telescope and the seismological instruments were installed. The east room was for the meridian circle and the west one for the meteorological instruments. The northern front of the building was embellished with a classical ornamentation of Doric order. The astronomical observatory in Sierra Nevada was built in 1966 in the upper flank of Veleta Peak at an altitude of 2506 m. This is a rectangular building with two stories and two domes one for the Cassegrain telescope.

Astronomy

1902: Telescope with equatorial mounting of 32 cm aperture and 5.35 m focal length and Meridian Circle both by Mailhat, Paris. Grubb telescope of 15 cm

aperture and 2.30 m focal length. Two portable telescopes. Spectrograph by Littrow. Heliostat by Siberman.
1966: Reflector Cassegrain Telescope of 32 cm aperture donated by the Observatory of Georgetown University and installed in Sierra Nevada.

Meteorology
1902: Thermometers and Richard thermograph, Ducretet barometer, anemometer with Richard recorder, hygrometers, etc.

Seismology
1902: Vicentini horizontal seismographs, NE 305 kg. Vicentini vertical pendulum 45
kg mass; Stiattesi horizontal seismographs NS and EW, 208 kg mass.
1908: Wiechert horizontal seismograph, 200 kg mass
1908-1924: Instruments built by Sánchez Navarro-Neumann:
1908-1924: Horizontal seismographs (Cartuja Vertical) 280 kg mass
1909-1919. Horizontal seismographs of 340 kg (Cartuja Bifilar).
1920: Horizontal seismograph of two components (Wiechert type) of 3000 kg mass and magnification of 1000, period 5.5 s (Berchmans).
1924: Electromagnetic horizontal seismograph, (Galitzin type), with photographic recording with 7.5 kg mass and 30 s period with galvanometer Deprezt-D'Arsonval of 19 s. (Javier). Two similar seismographs of horizontal component (Canisio). A vertical electromagnetic seismograph of 12 s period (Belarmino).
1949: Horizontal mechanical seismograph of own construction (Cartuja Máximo) with 4500 kg mass, 15 s period and 500 magnification.

Bibliography

J. Granero, 1902. El Observatorio Astronómico, Geodinámico y Meteorológico de Granada. *Razón y Fe* 3, 222-225.

J. Granero, 1902. Observatorio de Granada. Sección Geodinámica. *Razón y Fe* 3, 512-520.

R. Martinez, 1902. Observatorio de Granada. Sección Meteorológica. *Razón y Fe* 4, 478-490.

J. Granero, 1903. Observatorio de Granada. Sección Astronómica. *Razón y Fe* 5, 339-347.

G. Gerland, 1904. Erdbebenbeobachtungen in Spanien. *Beiträge zur Geophysik* 6, 508-542.

J. Mier y Terán, 1905. *Eclipse total de sol del 30 de Agosto de 1905*. López Guevara, Granada, 114 pp.

P. de Vregille, 1906. Les observatoires de la Compagnie de Jésus au début du XXe siécle. *Rev. des Questions Scientifiques*, (3 serie, vol 9) 52, 10-72; 493-579.

R. Garrido, 1909. Les observations solaires à l'Observatoire Astronomique de Cartuja (Grenade). *Ciel et Terre* , 1-8.

M. Sánchez Navarro-Neumann, 1912. Diez años de actividad de la Estación Sismológica de Cartuja (1903-1912). Asocc. Española para el Progreso de las Ciencias 3, 245-260.

M. Sánchez Navarro-Neumann, 1924. Les séismographes de la Station séismologique de Cartuja (Granada). *Bull. U.G.G.I., Section de Séismologie*, 4, 119-131.

M. Sánchez Navarro-Neumann, 1928. La Estación Sismológica de Cartuja (Granada) y su labor científica (1903-1928). *Razón y Fe* 82, 59-74.

M. Sánchez Navarro-Neumann, 1932. La station séismologique de la Compagnie de Jesus a Cartuja, Espagne. *Rev. des Questions Scientifiques* 102, 247-253.

A. Due Rojo, 1941. El Observatorio de Cartuja, Granada. In: *Dios y la Ciencia*, Facultad de Teología S.J., Granada, 217-228

A. Due Rojo, 1953. El cincuentenario del Observatorio de Cartuja. *Urania* 234, 67-81.

J.M. Quintana González, 1989. La astronomía en Andalucía. *Rev. Española de Física* 3, 3-13.

OBSERVATORIO METEOROLOGICO DEL COLEGIO DE NUESTRA SEÑORA DEL RECUERDO, CHAMARTIN, MADRID, SPAIN

1902- 1931
Meteorology.

In 1902 a meteorological observatory was established, in the Colegio de Nuestra. Señora del Recuerdo, Chamartin, Madrid. From 1903 to 1906, the observatory was the central station of a network of meteorological stations in the Jesuit colleges of Puerto de Santa María, Málaga, Orihuela, La Guardia,

Oña, Orduña, Comillas, Gijón and Villafranca de los Barros. The observatory published a joint monthly bulletin. After 1906 when the network ceased operating, the observatory continued as an individual meteorological station until 1931. Directors were only assigned till 1924.

Directors

José A. Pérez del Pulgar	1901-1904
Juan Oliva	1905-1906
Francisco Ruipérez	1907-1908
Miguel Martínez Ropero	1909
José M. Loubiere	1910-1911
Juan Murillo	1912-1914
Ginés Yañes	1915-1916
Diego González	1917
Joaquín Cordero	1918-1920
Julio Vertiz	1921
Juan Cavestany	1922-23
Antonio Dué	1924

Publications

Boletín del Observatorio Meteorológico del Colegio de Nuestra Señora del Recuerdo, Chamartín de la Rosa, Madrid. (Monthly from 1903 to 1906).

Bibliography

J. Aguilera y Osorio, 1930. *Memoria del Colegio de Na. Sra. del Recuerdo en sus bodas de oro (1880-1930)*. Madrid, p. 72.

OBSERVATORIO DEL EBRO, ROQUETAS, TARRAGONA, SPAIN
1904-
Geomagnetism, Solar physics, Electricity, Meteorology, Seismology

The observatory was established in 1904 in the Faculty of Philosophy and Theology in Roquetas, Tarragona, near the Ebro River. Cirera, the first director oriented the new observatory to the study of the relationship between solar activity and terrestrial magnetism and electricity. The observatory was built on a hill near the faculty building and inaugurated in 1905. Ten different buildings were built for the different sections. In 1913 the observatory acquired a separate identity with its own foundation. From the beginning the observatory had very complete instrumentation in the sections of geomagnetism, solar physics, atmospheric and terrestrial electricity, meteorology and seismology. In

1914 Rodes gave a great impetus to the observatory and made it known in international circles. In 1939, after the Spanish Civil War, Romañá reorganized the observatory and installed new equipment. The observatory continued working in geomagnetism and solar physics, and kept renewing the instrumentation installing the first ionospheric equipment in Spain. Instrumentation kept being upgraded and, in the last years, lay scientists began to be incorporated. In 2001 a foundation has been created to ensure the continuity of the observatory in which several institutions are present including the Society. The same year a layman was named director. At present only three Jesuits actively work at the Observatory. It is not probable that active Jesuit presence will be continued.

Directors

Ricardo Cirera	1904-1920
Luís Rodés	1920-1939
Antonio Romañá	1939-1970
José O. Cardús	1970-1985
Luís F. Alberca	1985-2001

Collaborators

Mariano Bacells	1904-1908	heliophys.
Bienvenido Clusella	1904-1915	meteor.
Luís Gravalosa	1904-1907	seism.
Juan Forcadas	1904-1927	
Ludwig Dressel	1905	elect.
Etienne Merveille	1905	astro. magnet.
Miguel Carreras	1906	
Juan García Mollá	1906-1908, 1915-1920	elect. meteor. subd.
Narciso Hornos	1906-1924	
José Lleonart	1907	subdir.
Alfred Baur	1908	seism.
Juan Rosanas	1909-1910	
José Ubach	1910-1911	magn.
Joaquín Pericas	1910-1919 1924	astr. seism.
José Múnera	1912-1914	seism.
Juan Ortega	1912-1914	magn. elect.
José Albiñana	1914-1916	subdir. seism.
Carlos Ubach	1915-1926	
Francisco Rubio	1916-1931	astr.
Pedro Trullas	1918-1926	seism.
Enrique de Rafael	1921	
Félix Pareja	1922	

Ramón Sostres	1923	
Ignacio Puig	1926-1934	subdir. elect. magn.
Francisco Benitez	1926-1954	
Zacarias Cremades	1928-1937	
Francisco de Paula Mas	1940	
Joaquín Lamolla	1942	
Juan Torrens	1943-1987	
Ramón Pahí	1946	
Eduardo Galdón	1951- 2000	subdir. elect. ionosf.
Ricardo Alegre	1956-1964	
Pascual Bolufer	1957-1962	magn. elect.
Pío José Villarroya	1958-1982	
Jaime Tilló	1962-1986	
Edmundo Benedetti	1963-1965	elect.
Juan Casanovas	1963-1967	seism. heliofis.
Antonio Echevarría	1968-1976	seism
Ernesto Sanclement	1970-	magn.

Publications

Boletín Mensual, vol. 1- 26 (1910-1937); vol. 30-, 1942-desde el v. 30, serie A, (heliofísica, meteorología, sismología), serie B, magnetismo.
Boletín de Ionosfera 1957-
Boletín de Heliofísica 1947-1970
Boletín de Sismología 1948-1951.
Miscelánea 1-41 (1947-1998)
Memorias 1-17 (1906-1995)

Buildings and instruments

The observatory was built on the hill behind the main building of the Faculty of Theology and Philosophy with separate buildings for each section. In 1904 there were 10 buildings for seismology, magnetism of absolute measurements, magnetism of variations, meteorology, electricity, aerology, astrophysics, solar physics, workshop and offices. The solar physics building has the form of a cross with two rotating domes, the first for the equatorial solar telescope and the second added in 1957 for the solar filter. In 1930 the larger Landerer building was added for library, museum and offices.

Solar Physics

1904: Mailhat solar telescope with equatorial mounting of 16.2 cm aperture. Meridian circle, Arago actinometer, Evershed spectro-heliograph, Rowland spectro-goniometer; Angström pirheliometer; Grubb celostat
1957: Solar radiotelescope of 1 m wave length.

1966: Lyot solar filter and Molol-Gorzinsky solarigraph.
1972: Solar radiotelescope of 6, 11 and 21 cm wave length.
2001: Zeiss-APQ 150/1200 Astrophysics solar telescope with equatorial mounting

Atmospheric electricity
1904 - 2 Thomson-Mascart electrometers, Kalocsa detector of electric discharges, Gerdien and Elster-Geitel instrument for atmospheric ionization.

Ionospheric Soundings
1954: Ionospheric sounder ST-35 (France)
1966: Ionospheric sounder JW5 Magnetic AB (Sweden).
1987: Digital ionospheric sounder Digisonde 256 (USA).

Telluric currents
1904: Two lines NS of 1280 m and EW of 1415 m with galvanometers Depres-d'Arsonval.

Meteorology
1904: Barometer Fortin and Tonnelot, baroragraphs Richard, pyroheliometer Angström, actinometer, anemometer, thermometers, etc.
The meteorological instruments were renewed several times. The station was operated in collaboration with the Instituto Meteorologico Nacional.

Magnetism
1904: Mascart variometers with photographic recording and direct reading (H, Z, D). Dover-Kew magnetometers for absolute measurements (H and D), Schulze-Potsdam induction inclinometer
1950: La Cour magnetometers (QHM and BMZ).
1954: La Cour variometers (H,Z,D).
1960: Ruska and Askania magnetometers.
1965: Bar fluxometer for magnetic pulsations
1966: Elsec proton magnetometer
1979: AMOS (Canada) automatic magnetic station
1992: Automatic magnetic stations by GEOMAG (France) andARGO (United Kingdom).

Seismology
1904: Vicentini seismographs, NE, 100 kg (2.3s) and Z, 50 kg (0.8s); Grablovitz seismographs NW and NE, 12 kg (13s).
1914: Mainka horizontal seismograph NS, 1500 kg (14.8s); EW, 157 kg (7.8s) and vertical pendulum NS, 316 kg (2.6s) (own construction)
1940: Mainka-Ebro NS and EW, 1500 kg (15.4s y 10.8s)
1966: Benioff short period seismographs ZNE (1-0.5s)

1968: Spregnether horizontal long period seismographs, NS and EW (15s-90s)
1974: Sprengnether vertical long period seismograph, Z (15-90s).
1987: SS Ranger (Geotechnical Co.) short period vertical seismograph (1s), installed outside the Observatory in collaboration with the Instituto Geografico Nacional.
1991: The short period station is completed with Kinemetric horizontal seismographs NS and EW.
1999: Broadband three component seismographs Streckeisen-Quanterra, NEZ.

Bibliography

R. Cirera, 1904. Observatorio de física cósmica del Ebro. *Razón y Fe* 9, 91-98

R. Cirera, 1904. Rapport succint sur l'Observatoire de l'Ebre. *Beiträge zur Geophysik* 6, 534-337.

G. Gerland, 1904. Erdbebenbeobachtungen in Spanien. *Beiträge zur Geophysik* 6, 508-542.

B. Berloty, 1905. L'Observatoire de l'Ebre, a Tortosa (Espagne). *Etudes* 104, 643-657.

R. Cirera, 1906. Noticia del Observatorio y de algunas observaciones del eclipse de 30 de agosto de 1905. *Memorias del Observatorio del Ebro* 1, 57 pp.

P. de Vregille, 1906. Les observatoires de la Compagnie de Jésus au début du XXe siécle. *Rev. des Questions Scientifiques*, 54 (3 serie, vol 9), 10-72; 493-579.

M. Balcells, 1907. Sección Astrofísica del Observatorio del Ebro. *Razón y Fe* 18, 220-233; 362-374.

M. Balcells, 1908. La observación solar. *Memorias del Observatorio del Ebro* no. 2. 143 pp.

E. Mervielle, 1908. La seccion magnétique. *Memorias del Observatorio del Ebro* No. 3. 75 pp.

R. Cirera, 1910. El Observatorio del Ebro. *Boletín Mensual del Observatorio del Ebro,* 1, 1, 9-14.

J. García Moll , 1910. La seccion eléctrique. *Memorias del Observatorio del Ebro.* No. 4. 123 pp.

P. Mezzetti, 1912. L'Osservatorio dell'Ebro e i nouvi metodi della fisica solare. *Riv. di Fisica, Matemática e Scienze Naturali* 145, 5-27.

I. Puig, 1926. La Sección Electro-meteorológica del Observatorio del Ebro. *Razón y Fe* 77, 411-423.

I. Puig, 1926. La sección geofísica del Observatorio del Ebro. *Razón y Fe* 77, 504-518.

I. Puig, 1926. L'Observatori de l'Ebre. *Ciencia* 3, 100-108; 4, 161-172.

I. Puig, 1927. *El Observatorio del Ebro. Idea general sobre el mismo*. Imp. Moderna del Ebro de Algueró y Baiges, Tortosa, 188 pp.

J.O. Cardús, 1983. El Observatorio del Ebro. *Volumen Conmemorativo del 75 aniversario del Observatorio del Ebro,* Roquetas, 9-19.

J. Batlló, 1995. L'Observatori de l'Ebre. *Revista de Fisica* 8, 41-46.

J. Batlló, 1995. Instruments i altres materials d'interés científic conservats a l'Observatori de l'Ebre. *Actes III Trobades d'Historia de la ciéncia i de la técnica als països catalan, SCHCT*, Barcelona, 523-531.

J. Batlló and A. Ugalde, 2000. Els sismògrafs de l'Observatori de l'Ebre. *Actes V Trobades d'Historia de la ciéncia i de la técnica als països catalan, SCHCT,* Barcelona, 161-165

J.J. Curto, J.O. Cardús, L.R. Gaya, E. Sanclement and J.M. Torta, 2000. Un segle d'instrumentació magnètica a l'Observatori de l'Ebre. *Actes V Trobades d'Historia de la ciéncia i de la técnica als països catalan, SCHCT* , Barcelona, 175-180.

L. R. Gaya and J. Batlló, 2000. La secció solar de l'Observatori d'Ebre. *Actes V Trobades d'Historia de la ciéncia i de la técnica als països catalan, SCHCT* , Barcelona, 191-196

J. G. Solé, D. Altadill, L.F. Alberca and E. Galdón, 2000. La secció inosfèrica de l'Observatori d'Ebre. *Actes V Trobades d'Historia de la ciéncia i de la técnica als països catalan, SCHCT* , Barcelona, 235-240.

OSSERVATORIO METEOROLOGICO DEL COLLEGIO PENNISI, ACIREALE, CATANIA, SICILY, ITALY

1915-1978
Meteorology

In 1915 a small meteorological observatory was installed in the Collegio Pennisi in Acireale, Catania, Sicily which lasted until the school was colsed in 1978.

Directors

Nuntius Longhitano	1915-1935
Fredericus Weber	1936
Vincentius Barcellona	1937-1953
Vincentius Schiliro	1954-1958
Giuseppe Damiani	1959-1978

SEISMOLOGICAL OBSERVATORY AT MUNGRET COLLEGE, LIMERICK, IRELAND

1908-1915
Seismology

Mongret College, near Limerick was founded in 1892. In 1908 in a small building a seismological observatory was established with two Bosch-Omori horizontal seismographs. In 1911 William J. O'Leary, a teacher of physics installed a seismograph of his own design consisting in an inverted pendulum with a mass of 600 kg suspended by three cables which recorded the two horizontal components on smoked paper. In 1915 he moved with the instrument to Rathfarnham Castle.

Bibliography

T. Murphy (1995). The seismology observatories of Mungret and Rathfarnham. (manuscript) 29 pp.

SEISMOLOGICAL OBSERVATORY OF RATHFARNHAM CASTLE, DUBLIN, IRELAND

1913-1968
Seismology

In 1913 the Faculty of Philosophy of Irish Jesuits was established in Rathfarnham Castle. In 1915 William J. O'Leary moved there from Mungret

College and installed the seismograph he has built there. Another instrument of the same type with 1500 kg mass and a natural period of 15 s was also installed. From 1929, several Jesuits were in charge of the observatory. In 1932 a horizontal Milne-Shaw seismograph was acquired with photographic recording. In 1935 Richard Ingram became director who was active in seismology and in 1950 installed a modern short period seismograph. The station was closed after his death in 1967.

Bibliography

R.E. Ingram and J.R. Timoney, 1954. Theory of an inverted pendulum with trifilar suspension. Geophysical Bull. School of Cosmic Physics, Dublin Institute for Advance Studies, 9, 3-8.

Anonymous, 1960. Rathfarnham Castle. Seismological Observatory. Irish Jesuit Yearbook 1960, p. 93.

Anonymous, 1973. A last look at Rathfarnham. (manuscript) p. 50-53.

T. Murphy, 1995. The seismology observatories of Mungret and Rathfarnham. (manuscript), 29 pp.

SEISMOLOGICAL STATION OF THE COLEGE-SEMINARY OF TRAVNIK, BOSNIA AND HERZEGOVINA

1924-1927
Seismology

In 1924 a seismological station was installed in the College-Seminary of Travnik in Bosnia and Herzegovina which had been founded by Jesuits in 1882. The station consisted in a horizontal Conrad seismograph of 23 kg of mass. The first director was Peter Gärtler a mathemátics and science teacher. In 1927 he was succeded for one year by Stephan Pesko. The station was closed in 1928.

2. NORTH AMERICA

GEORGETOWN UNIVERSITY OBSERVATORY, WASHINGTON, D.C.

1844-1972
Astronomy, Seismology

The astronomical observatory of Georgetown University was founded in 1844. The first telescopes were a transit by Ertel, a meridian circle by Simms and a telescope with equatorial mounting of 12 cm aperture by Troughton and Simms. In 1888 Hagen became Director and in 1891 installed a larger telescope of 30.5 cm aperture by Fauth in a new dome and a new building was built. Hagen worked mainly on the observation of variable stars. In 1893 Fargis, Algué, Hedrick and Rigge worked on the new field of photographic star observation. In 1909 Tondorf installed the seismological station which became part of the JSS and later of the JSA. The station was enlarged in 1912 and 1925 with new instruments. In 1962 a WWSSN station was installed. The sections of seismology and astronomy functioned with certain independence. In 1947 Heyden established a graduate program in astronomy and research projects in solar spectrography and radio astronomy. The observatory was closed in 1972.

Directors

Astronomy

James Curley	1841-1888
John G. Hagen	1888-1906
John Hedrick	1906-1916,
Peter Archer	1915-1924
John Gipprich	1925
Edward C. Phillips	1926-1928
Paul McNally	1928-1948
Francis J. Heyden	1948-1972

Seismology

Francis A. Tondorf	1910-1930
Frederik Sohon	1929-1970
Bernard McConnell	1970-1972

Collaborators

Francisco de Vico	1848
Benedetto Sestini	1849-1869
Angelo Secchi	1849-1851

William Rigge 1890
George A. Fargis 1890-
José Algué 1893-1894
Joseph Cohalan
Edward Powers
Lawrence C. McHugh
Roger LeClaire
Martin McCarthy
George Coyne

Buildings and instruments

The building of the observatory, finished in 1844, had a central body with three stories, a dome of 6 m diameter and two lateral bodies of one story. In the central dome the equatorial telescope was installed and in the other two the meridian circles. In the second story was the library. A second building was added in 1894 where the zenithal telescope and the equatorial of 12 cm were installed. The seismographs were installed in an underground vault near Maguire Hall.

Astronomy

1844: Meridian circle of 10 cm aperture by Troughton and Simms; transit telescope of 22 cm aperture by Saegmueller.
1849: Telescope of 12 cm aperture with lens by Troughton and Simms and equatorial mounting by Merz and Mahler.
1891: A reflector telescope was adapted to a photochronograph of own design.
1893: Telescope with equatorial mounting of 30.5 cm aperture by Fauth.
1944: Equipment was restored and a photometer was added to the equatorial telescope

Seismology

1910: Wiechert horizontal and vertical seismographs of 80 kg mass,
1912: Bosch-Omori horizontal seismographs of 25 kg and 200 kg; Wiechert horizontal and vertical seismograph of 200 kg mass; Mainka horizontal seismograph of 135 kg mass.
1923: Three component Galitzin electromagnetic seismographs (ZNE).
1962: WWSSN station with short and long period three components seismographs .

Publications

Annals of the Astronomical Observatory of Georgetown College
Georgetown University Seismological Bulletin. vol 1, 1915

Bibliography

J. Curely and J. Roothan, 1843-1884. Letters. *Archivum Romanum Societatis Jesu* (ARSI). Provincia Maryland (1007-III-4, Registri 1)

J. Curley, 1852. *Annals of the Astronomical Observatory of Georgetown College*. 215 pp.

C. André and A. Angot, 1877. *L'Astronomie practique et les Observatoires en Europe et en Amerique*. Part 3: Etats Unis d'Amerique (Observatoire de Georgetown College). 3, 20-24.

J.G. Hagen, 1891. *The photochronograph and its application to star transits*. Georgetown College Observatory, Washington, 76 pp.

W. F. Rigge, 1904. Jesuit astronomy. Part II, the restored Society. *Popular Astronomy* 12, 303-310, 375-385.

P. de Vreguille, 1906. Les observatoires de la Compagnie de Jésus au début du XXe siècle. *Rev. des Questions Scientifiques* 59 (3 s. 9) 10-77, 493-579

J. S. O'Connor, 1932. The Georgetown Seismological Observatory. *Earthquakes Notes* 3, 3-5.

T.D. Barry, 1940. A century of astronomy. The Georgetown University Observatory (abstract). *Bull. Am. Ass. Jesuit Scientists*. 18, 20.

J. B. Macelwane, 1941. American Jesuits in Science. *Thought* 16, 122-133.

J. B. Macelwane, 1950. The Georgetown Seismological Observatory, Washington, District of Columbia. In: J.B. Macelwane (ed.), *Jesuit Seismological Association 1925-1950*. Saint Louis University. 237-257.

F.J. Heyden, 1975. *The beginning and end of a Jesuit Observatory (1841-1972)*. Pub. of Manila Observatory, 24 pp.

R. E. Curran, 1993. *The bicentennial history of Georgetown University. Vol. I, From academy to university 1789-1889*; Vol II, The second century 1889-1989. Georgetown University Press, Washington.

C.H. Geschwind, 1998. Embracing science and research. Early twenty-century Jesuits and seismology in the United States. Isis 89, 27-49.

SAINT LOUIS UNIVERSITY GEOPHYSICAL OBSERVATORY

1860-

Seismology, Meteorology

A small astronomical observatory was established in Saint Louis University between 1855 and 1888 and some meteorological observations were made from 1835 to 1841. A complete meteorological observatory was established in 1860 and closed in 1871. In 1909 the observatory was installed in definite form with meteorological and seismological instruments by Goesse and Rueppel. The seismological station consisted of a Wiechert horizontal seismograph as part of the JSS program. In 1925 Macelwane took charge of the observatory, created the Department of Geophysics with a graduate program. He reorganized the JSS with the new name Jesuit Seismological Association (JSA) and established its central station in Saint Louis. Between 1928 and 1938 seismographic stations were installed in Florissant outside St. Louis, Little Rock, Arkansas and Cape Girardeau, Missouri. In Macelwane's time Saint Louis University became a prestigious center of seismographic research. The seismographic part developed further with the installation of a WWSSN station in Florisant, and the renewal of instrumentation. In 1964 the department was reorganized as the Department of Earth and Atmospheric Sciences in a new building with graduate programs in geophysics and meteorology. In 1973 a network of 32 stations was installed in the region of New Madrid, Missouri and in 1991 a network of broad-band seismographic stations. Stauder ceased as director in 1988, he was the last Jesuit to occupy this post. The observatory continues today without Jesuit presence.

Directors

John B. Goesse	1911-1920
George E. Rueppel	1920-1923
James B, Macelwane	1925-1956
Victor J. Blum	1956-1962
William Stauder	1962-1988

Collaborators

William C. Reppeti	1926-1928
Joseph S. Joliat	1927-1933
George J. Brunner	1931-1949
Henry F. Birkenhauer	1936-1944

Jesús Emilio Ramírez	1937-1941
Patrick A. Heelan	1949-1952
Agustín Udías	1960-1963
Luis Fernández	1962-1968
Gabriel Leblanc	1962-1966

Buildings and Instruments

Offices and workrooms were at university buildings. In 1944 they were moved to the building of the Institute of Technology. In 1964 the Department of Earth and Atmospheric Sciences occupied a new building. Meteorological instruments were installed on the roof of the university buildings. Seismographs were first installed in underground vaults at the basement of Du Burg Hall and at the basement of the Gymnasium. In 1928 an underground vault was built on the grounds of the Jesuit Novitiate in Florissant, outside Saint Louis. Seismographs were also installed at several sites outside St. Louis, namely, Cape Girardeau and Little Rock in 1949, at five sites in the New Madrid area in 1960, at Cathedral Cave in 1991, and in the Upper Mississippi Valley in 1998.

Meteorology

A complete meteorological station was installed on the roof of the University main building from 1860 to 1876 and again in 1909. In 1964 it was moved to the new building of the Department.

Seismology

1919: Wiechert horizontal seismograph of 80 kg mass installed in the basement of DuBourg Hall.
1928: Three components Galitzin-Wilip seismograph (ZNE) and two horizontal Wood-Anderson (NE), installed at Florissant, outside Saint Louis and two Wood Anderson (NE) in the Gymnasium vault.
1930: Wood-Anderson seismographs (NE) at Little Rock, Arkansas.
1938: Long period horizontal Sprengnether seismographs in the Gymnasium vault and Wood-Anderson seismographs (NE) at Cape Girardeau, Missouri
1949: Sprengnether seismographs at Little Rock and Cape Girardeau
1960: Network of 5 stations in Mew Madrid region, Missouri
1962: WWSSN seismographic station at Florissant.
1972: WWSSN moved to French Village
1973: Network of 32 stations of short period in New Madrid region.
1991: Broad band seismographic station Cathedral Cave (CCMO).
1998: Upper Mississippi Valley Seismic Network consisting of several stations with broad-band and short-period seismographs.

Bibliography

J.B. Macelwane, 1926. The new seismographic station of Saint Louis University. *Physics Bull.* 6, 5-7.

J.B. Macelwane, 1926. The Jesuit seismographic stations in the United States and Canada. A retrospect. *Bull. Seis. Soc. Am.* 16, 187-193.

J. B. Macelwane, 1950. The Saint Louis University Stations. In J. B. Macelwane (ed.), *Jesuit Seismological Association, 1925-1950.* Saint Louis University. 134-180.

Anonymous, 1955. Meteorology at St. Louis University. *Bull. Am. Meteor.* Soc. 36, 358-359.

H.F. Birkenhauer, 1956. Father Macelwane and the Jesuit Seismological Association. *Earthquakes Notes* 27, 12-13.

B. J. Mitchell, 1995. A brief history of SLU geosciences. (manuscript), 2 pp.

B.J. Mitchell, J.S. Chester, F.M. Chester, R. B. Herrmann, E.J. Haug, C.J. Ammon, C.B. Belt, K.M. Chauff, G. M. Smith, 1995. Geoscience Newsletter, *The Department of Earth and Atmospheric Sciences.* Saint Louis University, 21 pp.

C.H. Geschwind, 1998. Embracing science and research. Early twenty-century Jesuits and seismology in the United States. Isis 89, 27-49.

OBSERVATORY AT CAMPION COLLEGE, PRAIRIE DU CHIEN, WISCONSIN

1881-1888

Astronomy

In 1881 Hagen installed a small astronomical observatory at Campion College, Prairie du Chien, Wisconsin, a boarding high-school for boys. It began with two telescopes of 7.5 cm aperture and focal lengths of 96 cm and 55 cm. In 1883 a small building with a dome of 4 m diameter was built. Hagen used the observatory for observations of variable stars and it was closed after his departure in 1888.

Bibliography

W.P. Faherty, 1941. John Hagen eminent European astronomer sojourn in Wisconsin. *Wisconsin Magazine of History* 25, 178-186.

CREIGHTON UNIVERSITY OBSERVATORY, OMAHA
1885-1932
Astronomy

In 1885 a small astronomical observatory was established in Creighton College (later Creighton University) in Omaha, Nebraska. The building of the observatory was finished in 1885. It was a small one story building with a rotating dome which housed a telescope of 12.5 cm aperture with equatorial mounting by Steward. There were also a meridian circle of 7.5 cm aperture and a telescope with altazimuth mounting both by Steward, and chronometers, chronographs, theodolites and other equipment. In 1904 and 1908 two spectroscopes by Steinheil and Zoellner were acquired. The observatory was used for teaching purpose and research. W. Rigge was director from 1897 to 1927. In 1900 it was considered as one of the 61 observatories in the United States. After Rigge's death interest decayed and operations ceased after 1932.

Directors

Joseph F. Rigge	1885-1894
William F. Rigge	1897-1927
William C. Doyle	1928-1930
John Markoe	1931-1932

Bibliography

P. de Vregille, 1906. Les observatoires de la Compagnie de Jésus au début du XXe siècle. *Rev. des Questions Scientifiques*, 59 (3 serie, vol 9), 10-72; 493-579.

W. F. Rigge, 1911. The Observatory. *Creighton Chronicle* (February), 1-11.

Anonymous, 1985-86. Fr. Rigge's 1910 encounter. *Window Magazine, Creighton University* (Winter issue) 25.

J. M. Scott, 1990. Fr. Rigge and the case of the confirmation shadow. *Window Magazine, Creighton University*. (Fall issue)

SAINT MARY'S COLLEGE OBSERVATORY

1899-1910

Astronomy, Seismology

In 1899 a small astronomical observatory was established in Saint Mary's College, in the Jesuit Faculties of Theology and Philosophy. A telescope of 10 cm aperture with equatorial mounting was installed in a dome as well as a transit telescope of 5 cm aperture. The observatory was used for student demonstrations in the classes of astronomy. Some of the professors of astronomy between 1902 and 1909 were James C. Daly, Jerome Tenk and Thomas MacCourt. In 1909 a Wiechert seismograph was installed, as part of the JSS program, but it was damaged in a flood the same year.

Bibliography

P. de Vregille, 1906. Les observatoires de la Compagnie de Jésus au début du XXe siècle. *Rev. des Questions Scientifiques*, 59 (3 serie, vol 9), 10-72; 493-579.

J. B. Macelwane, 1950. The seismological station at Saint Mary's College, Kansas. In J. B. Macelwane (ed.) *Jesuits Seismological Association, 1925-1950*. Saint Louis University, 181-184.

RICARD MEMORIAL OBSERVATORY, SANTA CLARA UNIVERSITY, SANTA CLARA, CALIFORNIA

1895-1967

Meteorology, Astronomy, Seismology

In 1895, Ricard established an astronomical and meteorological observatory in Santa Clara College (now Santa Clara University) in the town of Santa Clara, California. In 1900 a telescope with equatorial mounting of 20 cm aperture by Alvan Clark and a second telescope of 10 cm aperture together with other astronomical instruments and a complete meteorological station were installed. Ricard dedicated himself mainly to meteorology and weather forecasts. Two Wiechert seismographs were installed in 1909. In 1928 two new buildings were built, one with a large dome of 15 m diameter for a large reflecting telescope with a 150 cm mirror which was never installed. The same year an underground vault was built for the seismographs. New seismographs were installed in 1930 and a new telescope by Alvan Clark in 1941. After Ricard's death in 1930, interest for astronomy fell off except for a brief interval under Weber in 1950. In 1960 the observatory began to be used for student demonstrations. In 1991 after extensive renovation the observatory was opened for use by students under faculty supervision.

Directors

Jerome S. Ricard	1895-1930
James B. Henry	1930-1931
Raymond J. Buclet	1931-1932
Albert J. Newlin	1932-1941 (no S.J.)
John A. Weber	1941-1959
Bernard R. Hubbard	1959-1962
Alexis I. Mei	1962-1966

Buildings and instruments

The original building was a modest wooden structure where instruments were kept and were taken out for observations. In 1921 the California Knights of Columbus raised funds for a new observatory building finished in 1928. The observatory has two buildings one with a large dome of 15 m diameter. At the same time an underground vault for the seismographic station was built. Building and dome were renovated in 1991.

Astronomy

1900: Refractor telescope with equatorial mounting of 20 cm aperture by Alvan Clark, and a telescope of 10 cm aperture.

1941: Telescope with equatorial mounting of 41 cm aperture by Alvan Clark

Seismology

1907: Seismometer of own construction

1909: Wiechert vertical and horizontal seismographs of 80 kg mass

1930: Three component set of Galitzin-Wilip electromagnetic seismographs (ZNE)

1931: Two horizontal Wood-Anderson torsion seismographs

Bibliography

P. de Vregille, 1906. Les observatoires de la Compagnie de Jésus au début du XXe siècle. *Rev. des Questions Scientifiques* 59 (3 serie, vol 9), 10-72; 493-579.

E. Shipsey, 1950. History of the University of Santa Clara: Seismological Station, Santa Clara, California. In J. B. Macelwane (ed.), *Jesuit Seismological Association, 1925-1950.* Saint Louis University 191-215.

Anonymous, 1970. Ricard Memorial Observatory. (manuscript) 4 pp.

G. McKevitt, 1979. *The University of Santa Clara, a history 1851-1977*. Stanford University Press, (The Padre of the Rains, 221-229).

JOHN CARROLL UNIVERSITY OBSERVATORY, CLEVELAND

1896-1992
Seismology, Meteorology

In 1896 Odenbach established in Saint Ignatius College, Cleveland (today John Carroll University) a meteorological observatory and began his studies of atmospheric electricity. The station was completed in 1898, with the installation of a Secchi's Meteorograph and it received the name of Angelo Secchi Observatory. In 1900 Odenbach installed two seismographs of his own making, the first Jesuit seismographic station in North America. In 1911, he founded the Jesuit Seismological Service (JSS), a network of sixteen seismographic stations with uniform instruments, Wiechert horizontal seismographs of 80 kg mass, installed in Jesuit colleges and universities, fifteen in the United States and one in Canada with Cleveland as Central Station. Data were sent by the individual stations to the Central Station which published a bulletin and send information to the International Seismological Center in Strasbourg. The JSS lasted only one year and the stations continued independently until 1925 when the JSA was organized. The meteorological station was closed in 1950. The seismographic station was renovated in 1946 with Sprengnether electromagnetic seismographs installed in an underground vault in the basement of Rodman Hall. New instruments were acquired in 1983 up to eleven seismographs of short and long period with visible recording. The station was closed in 1992.

Directors

Frederik L. Odenbach	1895-1933
Joseph S. Joliat	1933-1947
Henry Birkenhauer	1947-1962
Edward J. Walker	1962-1983 (no S.J.)
W. Richard Ott	1983-1992

Bibliography

P. de Vregille, 1906. Les observatoires de la Compagnie de Jésus au début du XXe siècle. *Rev. des Questions Scientifiques* 59 (3 serie, vol 9), 10-72; 493-579.

H. F. Birkenhauer, 1950. John Carroll University Seismological Observatory. In. J. B. Macelwane (ed.). *Jesuit Seismological Association 1925-1950.* Saint Louis University 73-90.

C.H. Geschwind, 1998. Embracing science and research. Early twenty-century Jesuits and seismology in the United States. Isis 89, 27-49.

MARQUETTE UNIVERSITY OBSERVATORY, MILWAUKEE, WISCONSIN

1909-1955
Seismology, Astronomy, Meteorology

In 1909 a small astronomical observatory was established at Marquette University, Milwaukee, Wisconsin. For this purpose a dome was built on the roof of Johnston Hall where two telescopes with equatorial mounting of 15.2 cm and 13.3 cm aperture were installed. The same year a seismological station was installed in the basement of the same hall with a horizontal Wiechert seismograph as part of the JSS program. In 1910 a meteorological station was installed. The telescopes were used by McGeary, the first director, to make observations of sunspots, comets, planets and stars, but the main use of the observatory was as teaching aid to students. In 1937 a new telescope of 13.3 cm aperture was installed. Interest in the observatory decayed soon. Carroll kept the seismographic station in operation until his death in 1955. The dome of the observatory was dismantled in 1983.

Directors

James L. McGeary	1909-1912.
Ignatius Kircher	1912-1916
John B. Kremer	1916-1927
Albert Poetker	1927-1931
Charles Heyden	1931-1932
Joseph F. Carroll	1932-1955

Bibliography

Anonymous, 1909. First quake record made in Milwaukee. *Free Press*, 2 Jan 1909.

Anonymous, 1909. Sun spots located by Father McGeary. *Sentinel*, 27 Sept 1909.

P. Timken, 1943. Fr. Carroll's seismograph records quakes here. *Marquette Tribune*, 25 Feb 1943.

J.F. Carroll, 1950. Marquette University Seismological Station, Milwaukee, Wisconsin. In J. B. Macelwane (ed.), *Jesuit Seismological Association 1925-1950*. Saint Louis University. 100-108.

Anonymous, 1955. Carroll dead; MU physicist, earthquake expert. *Milwaukee Journal*, 13 Dec. 1955.

CANISIUS COLLEGE SEISMOLOGICAL STATION, BUFFALO, NEW YORK

1909-
Seismology

The seismological station of Canisius College in Buffalo, New York was part of the JSS program and began in 1909-1910 with a horizontal Wiechert seismograph of 80 kg mass. The seismograph was originally installed in an old building of the College and was moved in 1913 to a new building. In 1932 a vertical Galitzin-Wilip seismograph was installed and in 1946 two horizontal Sprengnether seismographs of long period and a vertical Sprengnether seismograph of short period. In 1962 a new system of time control was installed. In 1963 all records were sent to the U.S. Geological Survey where they are kept. Directors were professors of science in the College. The station is still in operation.

Directors

John T. Vitek	1909-1910
William Repetti	1910-1911
Michael J. Ahern	1911-1914
John P. Delaney	1930-1938
Austin C. McTigue	1940-1960 (no S.J.)
James J. Ruddick	1960-

Bibliography

A. C. Mc Tigue, 1950. The Canisius College Station, Buffalo, New York. In J. B. Macelwane (ed.). *Jesuit Seismological Association, 1925-1950*. Saint Louis University. 42-60.

James J. Ruddick, 1985. Some further notes on the history of the Canisius College Seismological Station. (manuscript) 1p.

REGIS COLLEGE SEISMOLOGICAL STATION, DENVER, COLORADO

1909-
Seismology

In 1909 a with a Wiechert horizontal seismograph of 80 kg mass was installed in Regis College, Denver, in the basement of the main building, as part of the JSS program. The first director was Forstall who took care of the station for 39 years. In 1946 Sprengnether horizontal seismographs of long period and vertical seismographs of short period were installed. In 1962 three component short period seismographs were installed. In 1966 instruments were installed in a new underground vault with Helicorder visual recording donated by the USGS. Downey was the last Jesuit director. The stations continues in operation with a lay director.

Directors

Armand W. Forstall	1909-1948
Joseph V. Downey	1948-1989

Bibliography

J. V. Downey, 1950. Regis College Station, Denver, Colorado. In J. B. Macelwane (ed.). *Jesuit Seismological Association 1925-1950*. Saint Louis University 93-99.

J. V. Downey, 1948. The Rev. Armand W. Forstall, S.J. 1859-1948. *Bull. Seis. Soc. Am.* 38, 233.

SEISMOLOGICAL OBSERVATORY OF LOYOLA UNIVERSITY, NEW ORLEANS, LOUSIANA

1909-1985
Seismology

The Seismological Observatory of Loyola University, New Orleans began in 1909 as a station of the JSS program with horizontal and vertical Wiechert seismographs of 80 kg mass. In 1924 the instruments were moved to an underground vault in Bobet Hall. In 1946 two Sprengnether horizontal long period and a vertical short period seismographs were installed. Maring directed the station from 1938 to 1983. The station was closed in 1985.

Directors

Anthony L. Kunkel	1910-1922
Anthony J. Westland	1922-1923
Edward T. Cassidy	1923-1924
Oree L. Abell	1924-1938
Karl A. Maring	1938-1983

Bibliography

J. B. Macelwane, 1950. The seismological Observatory of Loyola University of the South New Orleans, Louisiana. In J. B. Macelwane (ed.), *Jesuit Seismological Association 1925-1950.* Saint Louis University 110-118.

HOLY CROSS COLLEGE SEISMIC STATION, WORCESTER, MASSACHUSETTS

1909-1934
Seismology

In 1909 as part of the JSS program a seismographic station was installed in Holy Cross College, Worcester with a horizontal Wiechert seismograph of 80 kg mass. The location of the instrument was changed several times until the seismograph was donated in 1934 to Weston Observatory.

Directors

Ferdinand Rousseau	1909-1910
Daniel P. Mahoney	1928-1929
Joseph Merrick	1930-1933
William Logue	1934

Bibliography

D. Linehan, 1950. Holy Cross College Seismic Station, Worcester, Massachusetts. In J. B. Macelwane (ed.), *Jesuit Seismological Association, 1925-1950.* Saint Louis University 258-264.

SPRING HILL COLLEGE OBSERVATORY, MOBILE, COUNTY, ALABAMA

1910-1989
Seismology, Meteorology

In 1910, as part of the JSS program, a seismographic station was established in Spring Hill College with a horizontal Wiechert seismograph of 80 kg of mass. The station was closed between 1935 and 1939 and opened again in 1941 with two horizontal McComb-Romberg seismographs. Between 1939 and 1947 a meteorological station was installed. In 1958 a graphic system of recording with electronic amplification and a vertical seismograph of own construction were installed. In 1962 a WWSSN station was installed. The station was closed 1989.

Directors

Cyril Ruhlmann	1910-1926
Karl A. Maring	1926-1929
Anthony J. Westland	1930-1935, 1941-1944
Walter J. Rheim	1945
Walter L. Furman	1946
Louis J. Eisele	1947-1988

Bibliography

W.J. Rhein, 1940. The Seismological Observatory at Spring Hill College. *Bull. Am. Ass. Jesuit Scientists* 18, 147-148.

J. B. Macelwane, 1950. The Seismological Observatory of Spring Hill College, Spring Hill, Mobile, Alabama. In J.B.Macelwane (ed.), *Jesuit Seismological Association 1925-1950.* Saint Louis University, 229-236.

L. J. Eisele, 1985. Seismology at Spring Hill College under the direction of L.J. Eisele. (manuscript) 10 pp.

FORDHAM UNIVERSITY SEISMOGRAPHIC STATION, NEW YORK
1910-1977
Seismology

In 1910 a seismographic station was established at Fordham University, as part of the JSS program, with a horizontal Wiechert seismograph of 80 kg mass installed in the basement of the administration building. In 1927 a small separate building was built for the station where new Galitzin seismographs were installed. Later more seismographs were acquired so that the station had very complete equipment. Directors changed every two or three years until in 1928 Lynch was appointed who occupied this post until 1977. After him the station was run for a short time by students and closed in 1980. In 1983 all seismographic records were donated to Lamont-Doherty Seismological Observatory of Columbia University, New York, where they are preserved.

Directors

Edward P. Tivnan	1910-1912
Clement Risacher	1912-1914
William C. Repetti	1914-1918
Joseph McAree	1918-1920
Joseph Lynch	1920-1923
Frederik Sohon	1923-1924
John S. O'Connor	1924-1926
John Tynan	1926-1928
Joseph Lynch	1928-1977

Buildings and instruments

The first instruments were installed in 1910 in the basement of the administration building. In 1927 a small building with an underground vault was built with funds from the family of William L. Spain, a student of the University. In 1931 instruments were moved to another new building with a larger underground vault.

1910: Wiechert horizontal seismograph of 80 kg mass.
1924: Two Milne-Shaw horizontal seismographs with photographic recording
1927: A set of three component Galitzin seismographs (ZNS)
1931: Two Wood Anderson torsion seismographs (NE)
1940: Short period Sprengnether seismographs (ZNE) and short and long period Benioff seismographs.

Bibliography

J. J. Lynch, 1932. Seismology at Fordham. *Bull. Am. Ass. Jesuit Scientists,*

J.J. Lynch, 1936. New seismic vault at Fordham. *Trans. Am. Geophys. Union* 17, 103.

J.J. Lynch , 1950. The Fordham University Station, New York, New York. In J. B. Macelwane (ed.), *Jesuit Seismological Association 1925-1950.* Saint Louis University. 119-133.

J.J. Lynch, 1970. *Watching our trembling earth for 50 years.* Dodd, Mead and Co. New York, 110 pp.

SEISMOLOGICAL OBSERVATORY OF SAINT BONIFACE COLLEGE, MANITOBA, CANADA.

1910-1922
Seismology

The only seismographic station of the JSS program in Canada was established in Saint Boniface College, Manitoba. It had the standard horizontal Wiechert seismograph of 80 kg mass. Its first director was Joseph Blain. The station was destroyed in a fire in 1922.

Bibliography

J.B. Macelwane, 1950. The Seismological Observatory of Saint Boniface College, Saint Boniface, Manitoba, Canada. In J. B. Macelwane (ed.), *Jesuit Seismological Association 1925-1950*. Saint Louis University. 134-136.

GONZAGA COLLEGE SEISMOLOGICAL STATION, SPOKANE, WASHINGTON

1910-1970
Seismology

The station of the JSS program was first installed at Gonzaga College with a Wiechert instrument by Eugene Bacigalupi. In 1930 the instrument was moved to the Jesuit school in Mount Saint Michael some 8 miles north of Gonzaga. In 1946 two Wood Anderson seismometers were installed. The station was operated by Jesuit professors and students.

Bibliography

F. J. Altman, 1950. The Jesuit Seismological Station at Spokane, Washington. In J. B. Macelwane (ed.), *Jesuit Seismological Association 1925-1950*. Saint Louis University. 216-228.

LOYOLA UNIVERSITY SEISMOLOGICAL STATION, CHICAGO, ILLINOIS

1912-1990
Seismology

The seismographic station of Loyola University in Chicago was established in 1912 as part of the JSS program. Originally it had the standard horizontal Wiechert seismograph of 80 kg mass installed in the basement of the Michael Cudahy Science Hall. In 1957 Donald Roll installed a set of three components short period Sprengnether seismographs and in 1983 changed the recording system. He was the last Jesuit director. The station was closed in 1990.

Directors

Thomas A. Kelly	1912-1916
George J. Brunner	1926-1931
Alphonse R. Schmitt	1931-1950
J. Donald Roll	1957-1987

Bibliography

A. R. Schmitt, 1950. The Loyola University Seismological Station, Chicago, Illinois. In B. Macelwane (ed.). *Jesuit Seismological Association 1925-1950.* Saint Louis University. 60-66

J. D. Roll, 1985. The Seismological Station at Loyola University Chicago since 1950. (manuscript) 1 p.

Anonymous, 1990. History of the Seismological Station of Loyola University of Chicago. (manuscript) 4 pp.

XAVIER UNIVERSITY SEISMOLOGICAL OBSERVATORY, CINCINNATI, OHIO

1925-1984
Seismology

The Seismological Observatory of Xavier University in Cincinnati was established in 1925 after the reorganization of the JSA. The first instruments were two horizontal Wood-Anderson seismographs installed in 1927 and a vertical Galitzin-Wilip in 1928. In 1949 the Wood-Anderson instruments were changed for a set of three component electromagnetic seismographs of short period of own construction. The station was especially active under Stechschulte's direction. The station was in operation until 1984 when it was closed

Directors

Vincent V. Herr	1926-1929
Joseph Wilczewski	1929-1931
Victor C. Stechschulte	1932-1956
Edward Bradley	1956-1984

Bibliography

V. C. Stechschulte, 1950. Xavier University Seismological Observatory. In J. B. Macelwane (ed.). *Jesuit Seismological Association 1925-1950*. Saint Louis University, 67-72.

E. Bradley and T. J. Bennett, 1968. Earthquake history of Ohio. *Bull. Seis. Soc. Am.* 55, 745-752.

WESTON SEISMOLOGICAL OBSERVATORY, BOSTON, MASSACHUSETS

1928-

Seismology, Geomagnetism, Astronomy,

The seismographic station in Boston College, Weston, was established in 1828 with the installation of two Bosch-Omori horizontal seismographs, a donation of Georgetown Observatory. The same year a telescope of 23 cm aperture was also installed. The astronomical section was enlarged in 1931 when a dome was built. In 1935 a new telescope of 38 cm aperture was installed. The astronomical section was closed in 1959. The seismographic station progressed specially under Linehan, who founded the Department of Geophysics in 1948. Seismic exploration equipment was acquired with the financial support from oil exploration companies. In 1949 a new building was built with 15 rooms for offices, laboratories and classes. Main work of the Observatory was shallow seismic exploration. In 1955 a network of 6 seismographic stations, which were later increased to 30, were established in the north-east region of the United States. Magnetic equipment was installed in 1957 with the occasion of the IGY (International Geophysical Year) with protons and rubidium magnetometer for absolute measurements and variometers. In 1962 a WWSSN seismographic station was installed. The observatory continued under Jesuit direction until 1993, when Skeehan the last Jesuit director retired. Today the observatory continues as an acknowledged research institute in seismology.

Directors

Astronomy

Thomas D. Barry	1940-1959

Seismology

G. A. O'Donnell	1935-1940
Michael J. Ahern	1940-1950
Daniel Linehan	1950-1974

James P. Skehan	1974-1993

Collaborators

Henry M. Brock	(astronomy)
John Blatchford	(astronomy)
Thomas D. Barley	(astronomy)
Sydney Judah	(astronomy)
T. H. Quigley	(seismology)
J. B. Doherty	(magnetism)
T. J. Smith	(seismology)
Francis J. Donehoe	(seismology)
John F. Devane	(seismology)
James P. McCaffrey	(seismology)

Buildings and instruments

The first building was a small one story with a dome for the first telescope. In 1931 a larger dome was built. The seismographic instruments were installed in the basement of one of the buildings of the College. In 1949 a separate building with 15 rooms was built for the Geophysics Department with space for offices, classes and laboratories.
1928: A small telescope of 23 cm aperture by Eastern Science Supply,
1935: Clark refractor telescope of 38 cm aperture with equatorial mounting.
1928: Two horizontal Bosch-Omori seismographs of 25 kg mass.
1934: Wiechert horizontal seismograph of 80 kg mass.
1936: Three component Benioff seismograph of 100 kg mass with galvanometers of 0.25 s and 60 s for short and long period recording.
1939: 12 channels seismic refraction instruments
1949: 24 channels seismic refraction instruments.
1955: Network of 6 short period seismographic stations
1962: WWSSN station with short and long period seismographs.
1987: The network was enlarged to 30 stations
1957: Ruska magnetic variometers with continuous recording of Z, H, D.
1958: Proton and rubidium magnetometers for absolute measurements

Bibliography

T. D. Barry, 1936. Weston Astronomical Observatory. *Bull. Am. Ass. Jesuit Scientists* 4, 167-168

D. Linehan, 1937. Weston Seismological Observatory. *Bull. Am. Ass. Jesuit Scientists* 14, 143-146.

D. Linehan, 1940. Field seismology at Weston College. *Bull. Am. Ass. Jesuit Scientists* 17, 200-204.

D. Linehan, 1950. Weston Seismological Observatory, Weston, Massachusetts. In J. B. Macelwane (ed.), *Jesuit Seismological Association, 1925-1950.* Saint Louis University, 248-257.

D. Linehan, 1957. Weston Observatory 1957-1958. *Bull. Am. Ass. Jesuit Scientists* 34, 101-103.

J. B. Pomeroy, 1962. A new research program at Weston Observatory. *Bull. Am. Ass. Jesuit Scientists* 39, 2-5.

UNIVERSITY OF SAN FRANCISCO SEISMOLOGICAL STATION, SAN FRANCISCO, CALIFORNIA

1950-1962
Seismology

A seismographic station was installed in 1950 in a vault under Saint Ignatius Church of the University of San Francisco. The instruments were two horizontal long period Sprengnether and a vertical short period. The director was Alexis I. Mei until he moved to Santa Clara in 1962 and the station ceased operations.

Bibliography

A. I. Mei, 1950. The Seismological Observatory at the University of San Francisco, San Francisco, California. In J. B. Macelwane (ed.). *Jesuit Seismological Association 1925-1950.* Saint Louis University 185-190.

OBSERVATOIRE DE COLLEGE JEAN DE BREBEUF, MONTREAL, CANADA

1952-
Seismology, Meteorology

A seismographic station was installed in 1952 in Collège Jean de Brébeuf in Montreal by Buist. The first instruments were a Sprengnether vertical and a Willmore vertical seismographs. In 1955 the instruments together with new short period three components Benioff seismographs were installed in an underground vault. The same year Gherzi joined the observatory and began a program on meteorology. In 1957 the observatory was moved to a new building, Lalemant Hall of the College with a new underground vault for the instruments. In 1958 a ionosonde was installed and new meteorological instrumentation. In 1961 three component Sprengnether long period seismographs were donated by the Dominion Observatory in Ottawa. A new seismograph was added in 1970. In 1975 Gouin arrived from the Observatory of Addis Ababa. Tremblay retired as the last Jesuit director in 1998. The observatory continues as part of the college under lay direction.

Directors

Maurice Buist	1952-1983
Paul Emile Tremblay	1984-1997

Collaborators

Ernesto Gherzi	1955-1976
Gabriel Leblanc	1966-1974
Florent Verreault	1966-
Conrad East 1964-	
Louis Bourgeois	
Marcel Gravel	1967-1977
Gerad Labonte	1980
Pierre Gouin	1975

Publications

Bulletin de Géophysique 1957-

Buildings and instruments

Instruments were first installed in 1955 in an underground vault. Laboratories, library and offices were installed in 1957 in Lalemant Hall with a new underground vault.

Seismology

1952: Sprengnether and Willmore short period vertical seismographs
1955: Three components Benioff short period seismographs
1961: Sprengnether long period three components seismographs (ZNE)
1970: Geotech- Helicorder short period vertical seismograph

Meteorology

1957: First meteorological station
1958: Ionospheric sounder, meteorological radar and recorder of sun radiation
1962: Thermohydrograph, anemograph

Bibliography

E. Gherzi, 1957. Foreword. *Bulletin de Geophysique, Observatoire du Collège Jean de Brébeuf* 1, 2-4.

C. East, 1965. Mémoire concernant l'avenir de l'Observatoire de Geophysique du Collége Jean de Brébeuf (manuscript) ASJCF.St Jérôme (Archives S.J. Canada Français, St. Jérôme) BO 262-3, 16 pp.

L. Bourgeois, M. Buist, C. East, G. Leblanc and F. Verreault, 1965. Mémoire sur l'Observatoire de Brébeuf (manuscript) ASJCF.St Jérôme BO 262-2, 5 pp.

M. Buist, L. Bourgeois, C. East and G. Leblanc, 1967. Mémoire sur l'Observatoire de Geophysique du Collége Jean de Brébeuf (manuscript) ASJCF.St Jérôme BO 262-3, 4 pp.

M. Buist, 1981. *L'Observatoire de Geophysique du Collège Jean de Brébeuf. Collège Jean de Brébeuf,* Montreal, 24 pp.

M. Buist, 1982. A short history of the Geophysical Station of Brebeuf College. (manuscript), 2 pp.

J. F. Gauvin, 1994. *Ernesto Gherzi et la physique du globe à Montréal, 1955-1973*. M. A. Mémoire, Université de Montreal 148 pp.

3. CENTRAL AND SOUTH AMERICA

OBSERVATORIO DEL COLEGIO DE GUATEMALA

1851-1871
Meteorology

A meteorological observatory together with a laboratory of physics were established in the Colegio de Guatemala from its foundation in 1851. From 1856 Canudas took care of the observatory and was appointed director in 1862. The observatory published an annual summary of observations, including temperature, pressure, humidity and variations of magnetic declination, with maximum, minimum and annual mean values. The observatory was closed in 1871, when Jesuits were expelled from Guatemala.

Directors

Antonio Canudas	1862-1863
José Lizarzaburu	1864-1868
Justiniano Arrubla	1869-1870

Bibliography

R. Pérez, 1898. *La Compañía de Jesús en Colombia y Centroamérica.* 3a. Parte: Desde la expulsión de la Nueva Granada en 1851 hasta la de Guatemala el año 1871. Imprenta Castellana, Valladolid, 678 pp.

OBSERVATORIO DEL COLEGIO DE BELEN, HAVANA, CUBA

1857-1961
Meteorology, Geomagnetism, Seismology

A meteorological observatory was created in 1857 in the Real Colegio de Belén in Havana with the purpose of observing tropical hurricanes and as a help for science courses. Installed on the second floor of the school the observatory had from the beginning very complete instrumentation. Regular observations began in 1858. In 1862 a magnetic station was installed which operated until 1905. In 1870 Viñes became director who gave the observatory international fame for his studies and forecasts of hurricanes. From 1877, the observatory was included in the international network of meteorological stations. In 1882 a telescope and a meridian circle were installed, but interest for astronomy decayed soon. Connection by telegraph with other stations was formalized in 1887 with the support of the shipping companies. In 1886 Viñes began his project of a network of meteorological stations in the Caribbean region with 7 stations located in Trinidad, Barbados, Martinica, Antigua,

Mayaguez (Puerto Rico), Jamaica y Santiago de Cuba. In the following years the number of stations grew to 20, but after 1891 they were reduced to only four, in Santiago, Barbados, Jamaica and Saint Thomas for daily reports and the rest only for reports in case of hurricanes. After Viñes' death in 1893 his work on observation and forecast of hurricanes was continued by his successors. In 1907 two horizontal Bosch-Omori seismographs, were installed, the first in the Caribbean region. During the American administration there were some difficulties, but the importance of the observatory was soon recognized. In 1925 the observatory was installed in the new building of the college. Since 1935 weather news from the observatory were broadcast by radio stations in Havana. In 1961 the observatory together with the college were confiscated by the communist government of Fidel Castro.

Directors

Antonio Cabré	1857-1860
Francisco Butiña	1860-1862
Felice Ciampi	1862-1863
José Reinal	1864-1865
Francisco Pons	1866-1868
Jose M. Vélez	1869-1970
Benito Viñes	1870-1893
Lorenzo Gangoiti	1893-1926
Manuel Gutierrez-Lanza	1926-1943
Simón Sarasola	1943-1947
Rafael Goberna	1947-1961

Collaborators

Tomas Ipiña	1873-1874.
Bonifacio Fernandez Valladares	1875-1877, 1883-1884
Mauricio Cid	1878-1880
Pedro Osoro	1881-1882
José Alberdi	1885-1893
Andrés Lafuente	1919
José A. Lasa	1931-1938
Gabriel González Llorente	1911
José Ma. Lasa	1939-1961

Publications

Observaciones magnéticas y meteorológicas del Colegio de Belén, 1872-1925.

Observatorio meteorológico, magnético y seísmico del Colegio de Belén, Habana, 1906 - 1920

Buildings and instruments

The first observatory was established in 1858 on the roof of the old colegio. In 1897 the observatory was moved to the new building on the two towers in the corners of the main front with a dome for the telescope in one of them . In 1907 two seismographs were installed in Luyanó outside Havana. In 1925 the observatory occupied new rooms on the roof of the new building of the college. No dome was installed.

Meteorology
1860: 4 Casella and Fortin barometers, thermometers, psicometers, hygrometers, 2 Robinson anemometers, pluviometers, nephoscopies.
1873: Secchi meteorograph and Richard barograph.

Magnetism
1862: Grubb horizontal and vertical magnetic variometers
1882: Elliot unifilar magnetometer for absolute measurements and Casella inclinometer.

Astronomy
1882: Cook telescope with equatorial mounting of 15 cm aperture, Troughton meridian circle of 4.5 cm aperture, Hilger and Browning spectroscopes, theodolites and chronometers.

Seismology
1907: Two Bosch-Omori horizontal seismographs of 28 kg mass.

Bibliography

R. H. Dana, 1859. *To Cuba and back.* (Havana, Belen and the Jesuits). Southern Illinois University Press.

Anonymous, 1885. The meteorology of Havana. *Nature* 31, 361-362.

M. Gutierrez-Lanza, 1904. *Apuntes históricos acerca del Observatorio del Colegio de Belén,* Habana. Avisador Comercial, Habana, 178 pp.

W. M. Drum, 1905. *The pioneer forecasters of hurricanes.* Stormont and Jackson, Washington, D.C., 29 pp.

P. de Vregille, 1906. Les Observatoires de la Compagnie de Jésus au début de XXe siècle. *Rev. des Questions Scientifiques* 59, 10-72, 493-579.

L. C. Kane, 1907. Belen seismograph station. A great Cuban institution. Notable achievements since its recent establishment. *World Today,* (June).

S.I. Barberena, 1912. *Informes con que se da cuenta al Ministerio de Instrucción respecto al desempeño de una misión científica a Europa*. Mis dos visitas a La Habana y el observatorio de Belén. San Salvador. 71-80.

M. Gutierrez Lanza, 1914. El Observatorio. *El Colegio de Belén en el sexuagésimo aniversario de su fundación y en el centenario del restablecimiento de la Compañía de Jesús. Albun Conmemorativo*. La Habana, 26-46.

M. Gutierrez-Lanza, 1926. Reseña histórica del Observatorio del Colegio de Belén. *Cartas de la Provincia de León*, VI-1926, 104-110.

M. Gutierrez Lanza, 1928. Observatorio del Colegio de Belén. *Cartas de la Provincia de Castilla* VII- 1927, 57-66.

M. Gutierrez Lanza, 1929. El Observatorio del Colegio de Belén. *Revista Belén* 4, 16, 855-859.

Anonymous, 1942. Noticias del Observatorio. *Noticias de la Viceprovincia de Cuba de la Compañía de Jesús,* 1, 4-7.

I. Testé, 1974. *Historia eclesiástica de Cuba*. Artes Gráficas Medinaceli, Barcelona, Tomo IV, 137-143.

OBSERVATORIO DEL COLEGIO S. IGNACIO, SANTIAGO DE CHILE

1863-1872
Meteorology, Magnetism

The observatory was founded in 1863 in the Colegio de S. Ignacio in Santiago de Chile by Enrique Cappelletti with meteorological and magnetic instruments. The magnetic instruments ceased operating in 1867 and the observatory was closed in 1872. Cappelletti left Chile and in 1886 moved to Mexico where he was director of the Observatories of Saltillo and Puebla.

OBSERVATORIO METEOROLOGICO DEL COLEGIO SAN LUIS, QUITO, ECUADOR

1864-1875
Meteorology

At a request by the President of Ecuador, Gabriel García Moreno, a meteorological observatory was established in 1864 in the Colegio de San Luis in Quito. The observatory was the central station of the National Meteorological Service of Ecuador. Federico Aguilar was its first director who was succeeded in 1870 by Johann B. Menten. The observatory was closed in 1875 when President García Moreno was killed.

OBSERVATORIO ASTRONOMICO DE QUITO, ECUADOR

1872-1876 (Under Jesuit direction)
Astronomy, meteorology

In 1872, Gabriel García Moreno, President of Ecuador, asked Johann B. Menten, a German Jesuit professor at the Polytechnic School and director of the Meteorological Observatory in the Colegio de San Luis in Quito, to take charge of the construction of an astronomical observatory. The observatory building was finished in 1875, the same year that President García Moreno was killed. Jesuits were not well accepted by the new government, and Menten, who left the Society in 1876, continued working at the observatory which was inaugurated in 1877. Its main instrument was a Merz telescope with equatorial mounting. Menten remained director until 1882.

J. B. Alenten, 1877. *Historia y descripción del Observatorio Astronómico de Quito*. Imp. Nacional, Quito.

M. Acosta Solís, 1944. *Breves anotaciones sobre la historia de la climatología del Ecuador*. Imp. Ecuador, Quito.

OBSERVATORIO DEL COLEGIO DEL SAGRADO CORAZON, PUEBLA, MEXICO

1877-1914
Meteorology, Seismology, Geomagnetism

The first meteorological observations at the Colegio del Sagrado Corazón in Puebla were made in 1875, but the observatory was established in 1877 by Spina. The observatory had very complete meteorological instrumentation. The same year three seismographs built by Heredia were installed together with magnetic instruments. In 1880 began the astronomical observations with a small Dollond telescope. In 1882 an equatorial telescope of 11.6 cm aperture and 1.8 m focal length was installed in a circular room of 3.2 m diameter with a rotating conical roof. The observatory carried out astronomical observations, mainly of the sunspots and solar faculae, seismological observations of large earthquakes in Mexico and meteorological measurements. The latter were sent to the Servicio Meteorologico of Mexico. Between 1877 and 1886 ten annual bulletins were published with observations every two hours. The observatory was closed together with the school in 1914 during the Mexican revolution.

Directors

Pedro Spina	1877-1886
Enrique Cappelletti	1886-1891
Pedro Spina	1891-1901
Gonzalo Carrasco	1901-1904
Gustavo Heredia	1905-1910
Joaquín Cordero	1912
Andrés Rongier	1913-1914

Publications

Observaciones Meteorológicas del Colegio Católico del Sagrado Corazón de Jesús en Puebla. 1877-1880.
Observaciones Astronómicas del Colegio Católico del Sagrado Corazón de Jesús en Puebla. 1880-1885

Buildings and instruments

The observatory was situated on the upper floor and roof of the college. It consisted of a room for the meteorological instruments and another for magnetometers and seismometers, a room for the meridian circle and a circular room of 3.20 m diameter with a rotating dome for the equatorial telescope and two rooms for library and offices.

1877-1906: 11 thermometers of several kinds, a thermograph, 2 Fortin barometers, aneroid barometer, evaporimeter, 2 psychrometers, 2 ozonometers, 2 anemometers, 4 pluviometers, nephoscope and nephometer.
1877: 3 horizontal pendulums designed by Heredia with recording on smoked glass plates.
1877: 2 magnetometers, and a declinometer.
1881: Dollond telescope
1882: Equatorial telescope of 11.6 cm aperture and 1.8 m focal length and a meridian circle both made in France.

Bibliography

P. Spina, 1882. *Observaciones Astronómicas del Colegio Católico del Sagrado Corazón en Puebla.* 1-27

P. Spina, 1890. Un decenio de observaciones meteorológicas en Puebla (1877-1886). *Mem. Sociedad Científica "Antonio Alzate"* (Mexico) IV, 49-66.

Anonymous, 1905. El seismógrafo inventado por el Sr. Pbro. J. Gustavo Heredia, S.J. *Bol. Soc. Astron. de México* 40, 597-599.

P. de Vregille, 1906. Les observatoires de la Compagnie de Jésus au debut du XXe siècle. *Rev. des Questions Scientifiques.* 59, 10-72, 493-579.

G. Heredia, 1906. *Perturbaciones seísmicas registradas en Puebla, 1877-1906. Observatorio Astronómico del Colegio Católico del Sagrado Corazón de Jesús, Puebla.*

G. Carrasco, 1906. Observatorio Meteorológico del Colegio Católico del Sagrado Corazón de Jesús de Puebla. In : M. E. Pastrana (ed.), *El Servicio Meteorológico de la República Mexicana.* Secretaría del Fomento, Mexico, 86-87.

G. Heredia, 1910. Catalogo de los temblores registrados en el Observatorio del Colegio del Sagrado Corazón de Jesús de Puebla desde 1877 hasta 1909. *La Espiga de Oro* II, 24, 40, 72, 88, 120. (Puebla)

J. B. Iguiniz, 1945. *Bibliografía de los escritores de la Provincia Mexicana de la Compañía de Jesús.* Buena Prensa, Mexico, 561 pp.

G. Decorme, 1960. Historia del Colegio Católico del Sagrado Corazón de Jesús de Puebla de los Angeles. 418- 424. (manuscript)

J.Gutierrez Casillas, 1971. *Jesuitas en México durante el siglo XIX.* Ed. Porrúa, Mexico.

OBSERVATORIO METEOROLOGICO DEL COLEGIO DE SAN JUAN NEPOMUCENO, SALTILLO, MEXICO.

1884-1914
Meteorology, Astronomy

In 1884 a small meteorological observatory was established at the Colegio de San Juan Nepomuceno in Saltillo, northern Mexico. The observatory had the basic meteorological instrumentation, thermometer, barometer, anemoscope, anemometer, nephoscope, nephosmeter, psychrometer, etc. The observatory had also a small telescope which was used by Spina. The observatory published a bulletin "Observaciones Meterológicas del Colegio de San Juan Nepomuceno, Saltillo" in the period 1885-1889 and "Observaciones Astronómicas" in 1888. In 1914 the observatory was closed together with the school during the Mexican revolution

Directors

Enrique Cappelletti	1884-1886
Pedro Spina	1886-1890
Enrique Cappelletti	1891-1894
Gustavo Heredia	1895-1899
Pablo Louvet	1899
Gilberto Roldad	1900
M. Guerrero	1901
M. Kubicza	1902-1903
Pedro Spina	1904-1906
Fernando Ambia	1907-1909
Santiago Maldina	1910-1911
Rafael Martinez del Campo	1912

Bibliography

E. Cappelletti, 1895. *Observaciones meteorológicas practicadas en el Colegio de Sap Juan Nepomuceno, durante los años 1891 a 1894 inclusive. Saltillo (Coahuila)* Imprenta del Colegio. 14 pp.

P. de Vregille, 1906. Les observatoires de la Compagnie de Jésus au debut du XXe siècle. *Rev. des Questions Scientifiques* 59, 10-72, 493-579.

P. Spina, 1906. Observatorio Meteorológico del Colegio de San Juan Nepomuceno en Saltillo (Coahuila). In: M. E. Pastrana (ed.), *El Servicio Meteorológico de la República Mexicana*. Secretaría del Fomento, Mexico, 92-93.

J. B. Iguiniz, 1945. *Bibliografía de los escritores de la Provincia Mexicana de la Compañía de Jesús*. Buena Prensa, Mexico, 561 pp.

J.Gutierrez Casillas, 1971. *Jesuitas en México durante el siglo XIX*. Ed. Porrúa, Mexico.

OBSERVATORIO SAN CALIXTO, LA PAZ, BOLIVIA

1892-2002
Seismology, Meteorology

The observatory began as a meteorological station installed in 1892 in the Colegio San Calixto in La Paz, Bolivia, which had been founded in 1882. The seismographic station began in 1911 in response to the recommendation of the 2nd General Assembly of the International Seismological Association in Manchester. The first seismographs were built in the observatory and installed in the crypt of the church. With the arrival of Descotes as Director in 1912, the observatory became one of the best seismographic stations in South America. Instruments were renovated continually. A small telescope was used to determine exact time by astronomical methods and from 1922 the observatory was responsible for the official time for Bolivia. Cabré, the second Director, continued the renovation of seismological instruments and established collaboration programs with United State and French institutions. International interest in the station, because of its location, assured financial support for instrumentation and research programs. Main lines of research were studies of seismicity, tectonics and crustal structure in Bolivia and the Andean region. From 1962 to the present, modern instrumentation has been installed with the support of the U.S. Geological Survey, the U.S. Air Force Technical Applications Center and the University of Paris. In 2002 the observatory was transferred to the Universidad Católica de Bolivia.

Directors

Pierre M. Descotes	1912-1964
Ramón Cabré	1965-1993
Lawrence A. Drake	1993-2001

Collaborators

Ricardo Manzanedo	1892-
Esteban Tortosa	1911-1957
Javier Lizarralde	1911-
Francisco Tortosa	1911-1959
Luís Fernández	1966-1970
Juan Enviz	1963, 1965-1968
José Marco	1962, 1968-1974
Antonio Fonoll	1966-1986
Richard Ott	1992-1994

Publications

Publicaciones del Observatorio San Calixto 1913-
Boletín del Observatorio San Calixto, 1991-

Buildings and instruments

Meteorological instruments were installed on the roof of the college. The first seismographs were installed in the crypt the church and the offices in the main building of the college. The WWSSN station (LPB) was installed in 1962 just south of La Paz. The array LRMS was installed from 1963 to 1975 in Peñas NW of La Paz. In 1972 a HGLP station (ZLP) was installed in the side of the valley of Zongo, on the flank of the mountain Huayna Potosí (inside a 70 m tunnel excavated in the granodiorite). In 1973 the offices were moved from the Colegio San Calixto to a nearby building in La Paz. In the valley of Zongo three sites were selected for short period stations. Other sites with seismographs are at Cerro Gloria, near Tianuanacu (Gloria), 52 km W of La Paz, Chanca (CNCB), and 39 km SSE of La Paz and San Ignacio de Velasco (SIV) in the Department of Santa Cruz.

1911: Horizontal seismograph of 450 kg mass of the Cartuja type.
1913: Two horizontal seismographs of same type of 500 kg mass and 15 s period and one of 1200 kg mass and 8 s period. A horizontal seismograph of the Wiechert type, but made at the observatory, for both components of 1500 kg mass and 2.5 s period.
1929: Three component Galitzin-Wilip electromagnetic seismograph
1958: Vertical Wilson-Lamison seismograph of 1 s period
1962: WWSSN station (LPB) with three component short and long period instruments.
1963-1975: LRMS array of 7 short period seismographs and 3 long period by Geotech lent by Air Force Office of Scientific Research (AFOSR).
1963: Array of 4 sensors of infrasound atmospheric pressure by University of Paris. 1972: High Gain Long Period (HGLP) station (ZLP) with seismographs of very long period with digital recording.
1980: The HGLP was transformed into an Adapted Seismic Research Observatory (ASRO) station (ZOBO)
1985-1987: Short period station in Tupiza (TPZ) by the Free University of Berlin.
1993: Three stations of short period in the valley of Zongo (ZONGO A, B, C) and a digital broad band station of three components (LPAZ) and a short period vertical component.
Seismographic stations of the Observatory in 2000 are: LPAZ (Zongo), BB-D, 1993; BOF (Collana) SPZ-D, 1993; BOA (Zongo) SP ZNE AD, 1972; BOB (Banderani) SPZ-D, 1975; LPB (Següencoma) WWSSN SPZ-A, 1962; BOE (Chanca) SPZ AD, 1982. BOD (Cerro Gloria) SP ZNE AD, 1986; SIV (San

Ignacio de Velasco) SPZ D, 1990; MOC (Mochará) SPZ D, 1993; CCH (Cochabamba) SPZ-MEQ-800 A 1985. (D = digital recording, A = analogue recording).

Meteorological instruments have been renovated periodically since the first were installed in 1892. Nearly all observations are preserved from 1892.

Bibliography

P. M. Descotes, 1913. *Le nouvel observatoire seismologique de la Compagnie de Jésus La Paz, Bolivie.* Imp. Velarde, La Paz. 15 pp.

Anonymous, 1946. El Observatorio San Calixto de La Paz. El más importante de Centro y Sudamérica. *La Razón* (17-3-1946) 3, La Paz

R. Cabré, 1982. Seismology in Bolivia - Close to the stars. *Earthquake Information Bulletin* 14, 16-18.

R. Cabré, 1985. Para la historia del Observatorio San Calixto. (manuscript) 5pp.

R. Cabré, 1985. Observatorio San Calixto, La Paz, Bolivia. (manuscript in English) 4pp.

R. Cabré, 1988. 75 Años en la vanguardia de la sismología. *Revista Geofísica* 29, 48-56.

E. Valencia, 1988. *El Observatorio San Calixto.* Tesina de grado, Lincoln Instituto, La Paz, Bolivia.

R. R. Coenraads, 1993. The San Calixto Observatory in La Paz, Bolivia. Eighty years of operation. Director Dr. Lawrence A. Drake. *Journal and Proc. Royal Society of New South Wales* 126, 191-198.

L. A. Drake, 2000. Los jesuitas y los inicios de la meteorología andina. In A. Gioda (ed.) *Actas del seminario ARCHISS*, Sucre 26-28 October 1999. Universidad Andina.

OBSERVATORIO DEL COLEGIO DE LA INMACULADA, SANTA FE, ARGENTINA

1904-1967
Meteorology

The meteorological observatory of the Colegio de la Inmaculada in Santa Fe, Argentina, was founded in 1904. In 1916 the observatory had very complete instrumentation with four Fortin barometers, a Richard anemometer and an anemoscope, pluviometers and several maximum and minimum thermometers. It was then classified as an observatory of second class. Data were sent daily and monthly to the Oficina Meteorológica in Cordoba and to the Oficina Central in Buenos Aires. Directors changed often and were science teachers in the school. The observatory was closed in 1967.

Directors

Manuel Torrents	1904-1905
Luis Teixidor	1906
Francisco Colomer	1907-1914
Juan Marzal	1915
Juan Sallaberry	1916-1920
José Doménech	1921-1925
Joaquín Añón	1926
Julián Hurley	1927-1933
Nilo Arriaga	1934
Juan van Schilt	1935-1939
Benito Reyna	1940-1954
Juan José Varas	1955-1967

Bibliography

G. Furlong, 1962. *Historia del Colegio de la Inmaculada de la Ciudad de Santa Fe y de sus irradiaciones culturales, espirituales y sociales, 1610-1962.* Ed. Soc. Exalumnos C. Inmaculada, Buenos Aires.

OBSERVATORIO DEL COLEGIO N. S. DE MONTSERRAT, CIENFUEGOS, CUBA

1910-1946

Meteorology , Astronomy

Meteorological observations in the Colegio de Nuestra Señora. de Montserrat in Cienfuegos began in 1886 and were sent to Belén Observatory in Havana. The observatory was formally inaugurated in 1910. Besides a very complete meteorological equipment it had also a Zeiss telescope of 13 cm aperture and 2.4 m focal length. The observatory worked in close collaboration with Belén and collaborated also with the U.S. Weather Bureau. Its two first directors Sarasola and Gutierrez Lanza, were afterwards directors of Belén. The observatory made 8 daily observations. The observatory was closed together with the school in 1946.

Directors

Simón Sarasola	1910-1920
Mariano Gutierrez Lanza	1920-1925
José Ma. Huidobro	1926-1930
Eulogio Vázquez	1931-1934
Luciano Estefanía	1935-1939
Eulogio Vázquez	1939-1942
José Ma. Huidobro	1943-1945

Publications

Anuario del Observatorio de Montserrat, Cienfuegos 1812-19??.

Buildings and instruments

The observatory was installed in a tower at the center of the front of the building of the colegio.

Astronomy

1910: Zeiss telescopio of 13 cm aperture and 2.4 m focal length with altazimuth mounting. A prism by Colzi was adapted for solar observations. Meridian circle by Sartorius and teodolite by Negretti-Zamba. Riefler chronometer.

Meteorology

1910: Two Newman and Fortin barometers, two anemometers, Fuess thermometers,
Pastorelli and Lambrecht psychrometers, Richard barographs, Fuess thermographs,
Richards and Friez anemometers, Fuess hydrostatic anemograph, Wilson pluviograph, Campbell heliograph, Kalocsa ceraunograph.

Bibliography

S. Sarasola, 1911. El nuevo observatorio. Descripción del observatorio y los aparatos. *Anales del Observatorio del Colegio N.S. de Montserrat*, 1. 1-11. Avisador Comercial, Habana.

M. Gutierrez-Lanza, 1922. El Observatorio, *Albún Conmemorativo del sexagésimo aniversario del Colegio de N. S. de Montserrat continuación del de Santi Spiritus 1862-1921* Madrid. 63-72.

OBSERVATORIO DEL COLEGIO DEL SAGRADO CORAZON, SUCRE, BOLIVIA

1914-1966
Meteorology, Seismology

The observatory, founded in 1914 by Cerro in the Colegio del Sagrado Corazón in Sucre, Bolivia, began as a meteorological station. It was installed on a three stories tower built for this purpose on the center of the school. It had very complete equipment: Fortin barometer, Richard barograph, Richard thermograph, anemometer and anemoscope, Fineman nephoscope, Alluard hygrometer, Augusto psychrometer, Richard heliograph, thermometers, etc. Later two mechanical seismographs built in the observatory were installed. They were a vertical seismograph with 1500 kg mass and a horizontal seismograph consisting of an inverted pendulum for the two components with a mass of 3000 kg mass, both recording on smoked paper,. The seismographs ceased operating some years later and were reinstalled in 1945 by Echevarría. They were damaged in the earthquake of 1948. From that year on the observatory continued only as a meteorological station, which in 1961was given the name of Observatorio Padre Cerro. The observatory closed in 1966.

Directors

Francisco Cerro	1914-1945
Antonio Echevarría	1945-1954
Cecilio Martín	1955-1956
Rosendo Carreras	1956-1964

Publications

Boletín mensual del Observatorio Meteorológico. 1915-1921
Observatorio del Colegio de la Compañía de Jesús en Sucre (Bolivia) fundado en 1914. Resumen anual. 1942

Bibliography

Anonymous, 1982. El Primer Observatorio Meteorológico en Sucre. *Llank'ana* (Revista del Colegio del Sagrado Corazón, Sucre) July, 1982, 3.

L. A. Drake, 2000. Los jesuitas y los inicios de la meteorología andina. In A. Gioda (ed.) *Actas del seminario ARCHISS*, Sucre 26-28 October 1999. Universidad Andina.

OBSERVATORIO METEOROLOGICO NACIONAL DE SAN BARTOLOME, BOGOTA, COLOMBIA

1922-1941

Meteorology, Seismology

In 1920 the President of the Republic requested the Jesuits to organize the meteorological service of Colombia. Sarasola who was then director of the Observatory in Cienfuegos, Cuba, was appointed for the job. The central station of the Service was established in the Jesuit Colegio de San Bartolomé in Bogota and began operating in 1922. The station was installed in new rooms added on the roof of the school. It was equipped with meteorological instruments such as thermometers, barometers, barographs, an anemocinemograph, a hygrograph and a Zeiss altazimuth telescope of 13 cm aperture and 2 m focal length with an adapted Colzi prism for solar observations. In 1924 secondary meteorological stations were installed in Bucaramanga, Ibagué, Neiva and Tunja with a project for 8 more. Collaboration was established with the U. S. Weather Bureau of Washington. In 1923 two seismographs, the first in Colombia, were installed; a Weichert horizontal of 200 kg mass and a horizontal of Cartuja type of 1000 kg. From 1923 to 1937 the observatory published periodical Anales and Noticias del Observatorio. In 1941 the Meteorological Service ceased to be directed by Jesuits when Sarasola quitted his post and founded with Ramirez the Instituto Geofísico de los Andes Colombianos in Bogota.

Bibliography

G. Fernández, 1923. El nuevo observatorio meteorológico, Colegio de S. Bartolomé, Bogota. *Cartas Edificantes de la Provincia de Castilla*, XI, 63-67.

S. Sarasola, 1924. *Noticia del nuevo observatorio con algunos datos sobre la climatología y el magnetismo de Colombia.* La Cruzada, Bogotá, 84 pp.

S. Sarasola, 1934. The meteorological service of the Republic of Colombia. *Proceedings of the Fifth Pacific Science Congress*, Ottawa, 1753-1755.

OBSERVATORIO DEL COLEGIO DOLORES, SANTIAGO DE CUBA

1934-1961

Meteorology

In 1934 a meteorological observatory was established in the Colegio Dolores in Santiago de Cuba. In 1940 a short wave radio station was installed which broadcast the weather reports. The observatory collaborated with and sent its observations to Belen. Some of its directors were also directors in

Cienfuegos. The observatory was closed when the communist government of Fidel Castro took over the school in 1961.

Directors

Santiago M. Viña	1934-1951
Luciano Estefanía	1952-1958
Eulogio Vázquez	1959-1961

Bibliography

S. M. Viña, 1941. Las trayectorias de los huracanes. *Publ. del Observatorio del Colegio Dolores, Santiago de Cuba*, 27 pp.

S.M. Viña, 1941. Ola y oleaje del huracán. *Revista Universitaria, Universidad Católica de Chile* 26, 23-55.

OBSERVATORIO DE FISICA COSMICA DE SAN MIGUEL, ARGENTINA
1935-1977
Solar Physics, Meteorology, Geomagnetism, Astrophysics

The Observatorio de Física Cósmica was established on the grounds of the Jesuit Faculty of Philosophy and Theology, Colegio Máximo de S. José, in San Miguel near Buenos Aires. It was the initiative of Mons. Devoto, Bishop of Buenos Aires and the Consejo Nacional de Observatorios. The observatory was inaugurated in 1935. Puig, from the Ebro Observatory, was its first director. Modeled on Ebro it had sections of astrophysics, geophysics and electrometeorology, with emphasis on solar-terrestrial relationship. The first instruments were for telluric currents, atmospheric electricity and meteorology. The most active period was from 1943 to 1966 with the direction of Bussolini. In 1952 the observatory began observations of sunspots and solar radiation in the H-alfa line and two radio telescopes were installed. In 1967 a heliospectrograph was installed in a new building to bolster solar physics research. In 1969, with Castex as director, the observatory was associated to the Comisión Nacional de Estudios Geo-Heliofísicos and changed its name to Observatorio Nacional de Física Cósmica. The observatory expanded in buildings and personnel, but from 1970 the Jesuit direction was only nominal. In 1977 the observatory was transferred to the Comisión Nacional de Investigaciones Espaciales of the Argentinean Air Forces and its name changed to Centro Espacial San Miguel. .

Directors

Ignacio Puig	1934-1943
José A. Bussolini	1943-1966
Albino Grassi	1966-1968
Mariano Castex	1968-1970
Hipólito Salvo	1971
Aldo Scotto	1971-1972
Hipólito Salvo	1972-1975
Victor Zorzín	1975-1977

Collaborators

Nilo Arriaga	1934-1973	
Juan Rosanas	1935-1937	
Joaquín Selles	1935-1939	
Emilio Kaiser	1939-1949	
Eugenio Gruber	1939-1964	
Oscar Dreidemie	1948-1957	
Carlos Mullin	1949-1953	
Martín González	1952-1958, 1965-1977	
Carlos Esponda	1956-1971	
Eugenio Berte	1956-1958, 1961-1966	
Tomás Paneth	1962-1977	
Arturo Yriberry	1950-1958	
Francisco Suilar	1959-1964, 1971-1974	
Ricardo Cocito	1959-1965	
Benito Reina	1960-1964	
Edmundo Benedetti	1966-1968	
Jorge Seibold	1969-1976	
Simón Yad	1970-1977	
Juan Bridarolli	1938-1949	entomology
Gregorio Williner	1940-1977	entomology

Buildings and instruments

The first central building was built in 1934 it had three stories with the third in the form of a tower. In 1970 the observatory had 13 buildings, 4 buildings of two stories for computation, ILAFIR (Instituto Latinoamericano de Fisiología de la Reproducción), physiology, biochemistry, direction and administration; a three stories building for solar physics, an eight stories building for meteorology, and other buildings for geophysics, residence and workshops. On the same ground were also the buildings of the Instituto Entomológico San

Miguel, an institute of entomology, and of the Comisión Nacional de Estudios Geo-Heliofísicos.

1935: Instruments for measurements of telluric currents with two lines, NS of 2436 m and EW of 2718 m with 10 connections and recording. Electrometer Labo-Gif (Paris), Ebert ion counter, Wulf and Gerdien electrometers. A complete meteorological station.
1952: H-alpha Lyot filter and telescope for solar observations.
1967: Ebert heliospectrograph system. Radiotelescope of 5 m diameter for 408 MHz.
1968: two Zeiss telescopes of 11/165 cm and 18/254 cm and a Heyde telescope 15/200 cm for visual observations.
1970: Radiotelescope of 1.8 m diameter for 9400 MHz.

Publications

Publicaciones del Observatorio de San Miguel, Acta Cientifica and Memorias (of irregular periodicity).

Bibliography

Anonymous, 1935. *El nuevo Observatorio de San Miguel. Lo que pretende ser.* Amorrortu, Buenos Aires, 16 pp.

I. Puig, 1935. El Observatorio de San Miguel, República Argentina. *Publ. del Observatorio de San Miguel* No. 2. 40 pp.

I. Puig, 1936. Inauguración del Observatorio de San Miguel (República Argentina). *Razón y Fe* 110, 372-386.

Anonymous, 1969. *Observatorio de Física Cósmica*, San Miguel (Bs. As.) Argentina.

Anonymous, 1970. *Observatorio Nacional de Física Cósmica.* Buenos Aires, 60 pp.

M. A. Greco, H. Grossi y J. R. Seibold, 1970. Solar physics at the Observatorio Nacional de Física Cósmica, San Miguel, Argentina. *Solar Physics* 14, 503-507.

O. Schneider, 1980. El observatorio de Física Cósmica de San Miguel. *Evolución de las ciencias en la República Argentina.* Tomo VII- *Geofísica y Geodesia.* Sociedad Científica Argentina, Buenos Aires, 85-88.

T. Paneth, 1985. 50 años del Observatorio Nacional de Física Cósmica de San Miguel. Resumen de la Historia del Centro Espacial San Miguel. (manuscript), 18 pp.

OBSERVATORIO METEOROLOGICO DEL COLEGIO DE CRISTO REY, ASUNCION, PARAGUAY

1935-1940
Meteorolog¡a

In 1935 a meteorological observatory was established in the Colegio de Cristo Rey in Asunción, Paraguay. The instruments were lent by the Servicio de Meteorología Nacional de Argentina and a Fortin barometer donated by the Bishop of Buenos Aires. The observatory was installed on the roof of the school. Its first director was Cecilio Martin. There were a project to established a network of 4 stations that was never realized. In 1940 renovation of the school building motivated its temporary closing, but the same year difficulties with the Servicio de Meteorología of Argentina which claimed its instruments led to its definitive closing.

Bibliography

L. Parola, 1973. *Historia contemporánea de la Compañía de Jesús en Paraguay*. Edición Privada, Asunción, 610 pp. Cap. 54. Observatorio, 462-466.

SEISMOGRAPHIC STATION OF ST. GEORGE'S COLLEGE, KINGSTON, JAMAICA

1940-1975
Seismology

In 1940 Linehan, Director of Weston Observatory installed a seismographic station in St. George's College in Kingston, Jamaica. The instruments were two horizontal Wood-Anderson torsion seismographs. The director was John A. Blatchford. , a teacher of science in the college. Records and readings were sent to Weston. The station was closed in 1975 after Blatchford's retirement.

Bibliography

D. Linehan, 1941. St. George's Seismographic Station. *Catholic Opinion*, January, 10.

Anonymous, 1971. A brief history of St. George's College. *The Lance 1971*. Kingston, Jamaica.

INSTITUTO GEOFISICO DE LOS ANDES COLOMBIANOS, UNIVERSIDAD JAVERIANA, BOGOTA, COLOMBIA

1941-
Seismology, Meteorology

In 1941 the Instituto Geofísico de los Andes Colombianos was inaugurated in the new Colegio de San Bartolomé, La Merced, in Bogotá. It was dedicated to seismology and meteorology. Its founders were Ramírez and Sarasola. The seismic station was excavated in the rock and its first instrument was a Beniof seismograph. The meteorological station was on the roof of the college and had very complete instrumentation. Sarasola left the Instituto in 1943 and Ramírez was its director for 40 years. In 1946 the seismographic station was completed with horizontal Sprengnether seismographs. In 1957 a new building for offices, library and a museum of minerals and fossils was built in the ground of the Jesuit Universidad Javeriana, near the college. The meteorological station was moved to the building of the Department of Engineering. In 1960 the Instituto Geofísico was officially associated to the Universidad Javeriana. Meteorological instrumentation was enlarged with the installation of a ionospheric sounder in 1957 and another in 1961 in Cartagena. In 1962 a seismological station of the network WWSSN (BOG) was installed. The Instituto was in charge of the Colombian Network of Seismological Stations and in 1972 of a network of accelerographs. In 1975 an advanced SRO (Seismic Research Observatory) seismological station was installed. In 1993 Jesuit participation ceased and a professor of engineering was named director. In 1995 the meteorological station was closed. In 1996 the name was changed to Instituto Geofísico, Universidad Javeriana and was associated to the Faculty of Engineering. At present the Instituto continues as part of the Universidad Javeriana.

Directors

Simón Sarasola	1941-1943
Jesús Emilio Ramírez	1941-1983
J. Rafael Goberna	1983-1985
Wladimiro Escobar	1985-1989
René van Hissenhoven	1989-1993

Publications

Publicaciones del Instituto Geofísico de los Andes Colombianos. Serie A Sismología; Serie B Meteorología; Serie C Geología; 1941 -

Buildings and instruments

The Instituto was first located in the building of Colegio de San Bartolomé, La Merced. Seismographs were installed in an excavated vault in the grounds of the college and meteorological equipment on its roof. In 1957 a three stories building was build on the grounds of the Universidad Javeriana for offices, library and a museum of minerals and fossils. Meteorological instruments were moved to the last floor of the Faculty of Engineering. In 1967 and 1971 two buildings were added.

Seismology

1941: Wiechert horizontal seismograph of 200 kg mass and Benioff vertical seismograph of 100 kg mass and 0.3s period.

1946: Two horizontal Sprengnether seismographs of 16 s period.

1962-1995: WWSSN (BOG) seismographic station.

1975-1995: SRO, (Seismological Research Observatory) seismographic station with digital recording and high magnification installed in a borehole at a site 14 km from Bogota.

1995, Broad-Band IRIS seismographic Station (BOCO) (Streckheisen-Quanterra)
replaced the SRO station at same site.

Seismological Network: 1946 Chinchiná (CHN) and Galerazamba (GAL), 3 comp
Sprengnether; 1957 Fúquene (FUQ) Sprengnether SP-3 components; 1971 Pasto; 1972 Bucaramanga (BMG),Benioff SP-vertical. 1995, Medellín MED Sprengnether SP-vertical, 1971 Pasto (PSO) Benioff SP-vertical.

Network of accelerographs: 1945, (Montana) Bogota; 1970, (Montana) Cartagena,
Barranquilla, Montería, Medellín (2), Manizales, Calí, Bogota .

Meteorology

1941-1995: Thermometers, thermograph, mercury barometer, barograph, microbarograph, pluviometer, pluviograph, psychrometer, heliograph, nephoscope, etc.

1957-1967, Ionospheric sounder C4 .

1961-1966, Ionospheric sounder Cossor installed in Cartagena.

Bibliography

J.E. Ramírez, 1943. The Geophysical Institute of the Colombian Andes. *Bull. Seis. Soc. Am.* 33, 81-90.

J.E. Ramírez, 1967. Bodas de Plata del Instituto Geofísico de los Andes Colombianos. *Publ. del Inst. Geofísico de los Andes Colombianos*, serie A, 27, 295-303.

J.R. Goberna, 1970. Informe sobre las actividades del Instituto Geofísico de los Andes Colombianos (Universidad Javeriana) durante los años 1967, 1968, 1969. *Publ. del Inst. Geofísico de los Andes Colombianos*, Ser. A, 29, 11 pp.

W. Escobar, 1972. Los 30 años del Instituto Geofísico de los Andes, 1941-1971. *Publ. del Inst. Geofisico de los Andes Colombianos*, Ser C, 14, 8 pp.

J.E. Ramírez, 1977. *Historia del Instituto Geofísico al conmemorar sus 35 años (Universidad Javeriana) I Antecedentes y fundación*, 31 pp. *II Los años de la guerra, 1944-1945. III Consolidación del Instituto Geofísico*, 1946-1950, 32 pp. Editora Guadalupe, Bogota

J.R. Goberna, 1984. Instituto Geofísico de los Andes Colombianos. (manuscript in English) 3 pp.

J.R. Goberna, 1983. Seismology in Colombia. *Earthquake Information Bull.* 15, 7-10.

W. Escobar, 1986. El Geofísico de los Andes. *Hoy en la Javeriana.* (Octubre)

J.R. Goberna, 1988. The historical seismograms of Colombia. In: W. Lee (ed.) *Historical Seismograms and Earthquakes of the World,* Academic Press, 467-473.

J.I. Vélez Munera and A. Gonzalez Sanchez, 1996. Nuevos horizontes en el Instituto Geofísico - Universidad Javeriana. *Noticias Colombia* 11, 43-48.

A. Amezquita, 1997. Informe estaciones sismológicas del Instituto Geofísico, Universidad Javeriana. (manuscript) 5 pp.

A. Alfaro, 2001. Instituto Geofísico Universidad Javeriana. Sesenta Años 1941-2001. *Hoy en la Javeriana*, (38) 1156, 12-13.

A. Alfaro, 2001. *60 años del Instituto Geofísico. Facultad de Ingeniería, Pontificia Universidad Javeriana*, Bogotá, 54 pp.

OBSERVATORIO SISMOLOGICO DE ANTOFAGASTA, CHILE

1949-1965
Seismology

Founded by German Saa in 1949, the seismological observatory was installed in the Colegio de San Luís in Antofagasta, Chile. The first instruments were two horizontal Bosch-Omori seismographs of 120 kg mass and recording on smoked paper. In 1957 a vertical short period Wilson-Lamison seismograph was installed by the Carnegie Institution, Washington in collaboration with the Universidad del Norte. A small telescope was also acquired and the for a time the observatory was designated as seismological and astronomical. The observatory was closed in 1962, owing to the installation of a WWSSN seismographic station in Antofagasta by the Universidad de Chile.

4. ASIA, AFRICA AND AUSTRALIA

MANILA OBSERVATORY, PHILIPPINES

1865-

Meteorology, Seismology, Geomagnetism, Astronomy, Solar Physics,

In 1865 a rudimentary meteorological station was established at the Ateneo de Manila, founded by Spanish Jesuits in 1860. In 1869 modern meteorological instruments were installed and in 1870 the first seismographs. From 1878, with Faura as director, the observatory developed sections of meteorology, magnetism, seismology and astronomy. The study and forecast of typhoons was begun in 1879 and became the main work of the observatory. In 1884 the observatory acquired official character as the headquarters of the Servicio Meteorológico de Filipinas, created by the Spanish Government. This service consisted of the observatory and of 10 secondary stations. In 1885 the observatory assumed the time-keeping service. A new building was inaugurated in 1886 and new meteorological, magnetic and seismological equipment was installed. Between 1890 and 1893 the first magnetic field survey in the Philippines was carried out. In 1891 the seismological section was reorganized with new equipment and in 1897 an astronomical section was added. In 1901 the United States administration recognized the work of the observatory and made it the headquarters of the new Philippine Weather Bureau appointing Algué Director. The Philippine Weather Bureau consisted of 9 first class stations, 25 second class and 17 third class. In 1902 new seismographs were installed in Manila, in 1909 a meteorological and seismological station in Baguio and in 1911 another seismographic station in Ambulong. Other stations were installed in Guam in 1914 and Butuam in 1915. Since 1920, American Jesuits began to replace the Spaniards. In 1926 Deppermann began studying typhoons with a modern approach. The number of secondary stations depending on the observatory increased to nearly 300 and the Philippine Weather Bureau became a complex organization with more than 200 employees. In 1942 with the occupation of Philippines by the Japanese work in the observatory ceased and in 1945 its buildings were destroyed by fire. After the war and the full independence of Philippines, reconstruction of the observatory was slow. The meteorological section was suppressed, since a new Philippine Weather Bureau was created by the Philippine Government. The observatory continued with sections of seismology, ionosphere and solar physics. In 1962 the main part of the observatory was moved to Loyola Heights, Quezon City, where a new building was built while the seismological and magnetic stations remained in Baguio. The same year two WWSSN stations were installed in Baguio and Davos. In 1963 new modern equipment for solar studies were installed in a new building, especially, a Razdow solar refractor telescope of 23 cm aperture. In 1980's work declined together with the number of Jesuits in the observatory and the continuity of the observatory was questioned. In 1997 the observatory incorporated climate and environmental studies.

Directors

Francisco Colina	1865-1867
Francisco Faura	1867-1871, 1878-1897
José Algué	1897-1926
Miguel Selga	1926-1948
Charles F. Deppermann	1948-1957
James J. Hennessey	1957-1972
Victor Badillo	1972-1992
Daniel J. McNamara	1993-

Collaborators

Martín Juan	1887-1888,	magnetism
Alphonse Renkin	1885-1886	
Miguel Saderra-Mata	1888-1938,	magnetism
Ricardo Cirera	1888-1894,	magnetism
Miguel Saderra-Masó	1890-1932,	seismology
José Coronas	1896-1931,	meteorology
Baltasar Ferrer	1901-1904,	meteorology
José Clos	1901-1902,	astronomy
John Doyle	1886-1901,	magnetism
Juan Vives	1896-1897,	seismology
Juan Comellas	1906-1926,	astronomy
William H. Stanton	1901-1904,	entomology
Robert Brown	1902-1915,	astronomy
George Zwack	1902-1913,	astronomy
James McGeary	1904-1906,	seismology
Bernard F. Doucette	1933-1966,	meteo. and seism.
Edmund Nuttall	1934-1942,	astronomy
Leo M. Welch	1930-1932,	meteorology
William C. Repetti	1928-1942,	seism and magn.
Richard A. Miller	1957-1974,	astronomy
Vicente Marasigan	1952-1968,	ionosphere
Sergio Su	1965-	seismology
Francis N. Glover	1960-1967,	ionosphere
José Dacanay	1964-1965,	seismology
Francis J. Heyden	1972-1991,	astronomy
Peter W. Walpole	1998-	environment
Jose T. Villarin	2000-	environment

Publications

Boletín mensual del Observatorio del Ateneo, 1865-1883
Boletín mensual del Observatorio Meteorológico de Manila 1884-1890
From 1890 to 1902, meteorology, seismology and magnetism.
Philippine Weather Bureau Monthly Bulletin 1902-1942
Monthly Bulletin of the Manila Observatory 1945-

Buildings and instruments

The first observatory was installed in 1865 in a tower in the NW corner of the building of the Ateneo. In 1872 new buildings were built specifically for the observatory with piers for the astronomical instruments. In 1884 the observatory was moved to a building on the grounds of the Escuela Normal in the neighborhood of La Ermita, outside Manila. A tower was built above the rest of the buildings for the instruments. The building was inaugurated in 1886. In 1894 a separate building for astronomy was built with two towers of 10 m diameter and 14 m height. In one of them a rotating dome for installed to house the equatorial telescope. These buildings were destroyed in 1945. In 1909, another observatory for meteorology and seismology was built in Baguio, Benguet at 1511 m height named Mirador Observatory. After the war the observatory was moved to Baguio where a building was finished in 1952 for seismology, magnetism, ionosphere and solar physics. In 1962 the main part of the observatory was moved again to Manila on the grounds of the Ateneo in Loyola Heights, Quezon City, where a new three stories building was built. The seismographic and magnetic stations remained in Baguio. A seismographic station was installed in Davao. Another building was added for ionosphere in Quezon City, not far from Loyola Heights. In 1963 a new three stories building was built for solar physics studies which was enlarged in 1968 for the large solar telescope.

Meteorology
1869: A barometer, thermometer, anemometer, pluviometer and Secchi meteorograph
1884: 2 Fortin barometers, Fuess maximum and minimum thermometers, Lambrecht pluviometer, Fineman nephoscopes. Sprung-Fuess and Richard barographs, Richard and Beckley anemometers, Garrigou-Lagrange clinoanemometro, Richard and Casella pluviographs; Richard and Whipple-Casella heliographs; Thompson-Mascart electrometer.

Astronomy and solar physics
1893: "Zenithal reflection telescope" designed by Algué with two lenses of 10.5 cm aperture and 64 cm focal length and several small telescopes

1897: Merz telescope with equatorial mounting of 48 cm aperture and 8 m focal length; Dondlon meridian circle of 6 cm aperture and 80 cm focal length.
1911: Angstrom pyrheliometer. Several chronometers for the time service.
1913: Repsold meridian circle of 7.5 cm aperture.
1952: Solar telescope of 9.2 cm aperture.
1963: Spectroheliograph.
1968: Razdow solar telescope of 23 cm aperture.

Seismology
1868: Two horizontal and vertical seismoscopes made by Ricart.
1881: Bertelli microseismometer, Rossi seismograph and Cecchi. seismograph
1889: Gray-Milne and Ewing seismographs.
1902: Vicentini universal microseismograph.
1906: Two horizontal pendulums of Omori type modified by Saderra-Masó made at the observatory.
1911: Wiechert horizontal seismograph of 1000 kg mass; Vicentini and Omori seismographs installed in Baguio.
1912: Vicentini and Agamemnone seismographs in Ambulong.
1914: Wiechert horizontal of 200 kg mass in the Island of Guam.
1915: Wiechert horizontal of 200 kg mass in Butuan.
1930: Three component Galtzin-Wilip seismographs in Manila.
1939: Wiechert horizontal seismograph in Tagaytay.
1951: Three component Sprengnether seismographs in Baguio.
1962: Two WWSSN stations in Baguio and Davao .

Magnetism
1887: Brunner magnetometers.
1889: Elliot (Kew) magnetometers, Dover inclinometer and two sets of Mascart-Carpentier variometers with photographic recording for D, H and Z.
1910: Instruments moved to Antipolo until 1945.
1952: Ionosphere sounder (Automatic Ionosphere Recorder C-2) at Baguio.
1955: Variometers in Baguio and Davao.
1962: Two ionosphere sounders in Balara near Loyola Heights.

Bibliography

R. Cirera and M. Juan , 1863. *El magnetismo terrestre en Filipinas*. Chofré y Cia. Manila, 42 pp.

Anonymous, 1883. *Observatorio Meteorológico de Manila bajo la dirección de los PP. de la Compañía de Jesús. Instrumentos de observación*. Manila.

Anonymous, 1890. *Observatorio Meteorológico de Manila bajo la dirección de los PP de la Compañía de Jesús*. Manila.

M. Saderra-Masó, 1895. *La Sismología en Filipinas*. Cap. 1, Reseña histórica del Servicio Seismológico establecido en el Observatorio de Manila. Manila. 2-16.

Anonymous, 1899. *El Servicio Meteorológico del Observatorio de Manila, vindicado y rehabilitado (1899)*. Imprenta del Observatorio, Manila.

Anonymous, 1900. *El Archipiélago Filipino*. Tomo II, Cap. 1. El Observatorio Meteorológico de Manila. Imprenta del Gobierno, Washington, 5-17.

J. Coronas, 1900. Climatología de Filipinas. In *El Archipiélago Filipino*, Government Printing Office, Washington, vol. 2, 1-265.

J. Algué, 1900. The Manila Observatory. *National Geographic Magazine* 11, 427-438.

J. Coronas, 1905. El observatorio de Manila en la Exposición de San Luis. *Razón y Fe* 11, 67-77.

P. de Vregille, 1906. Les Observatoires de la Compagnie de Jésus au début de XXe siècle. *Rev. des Questions Scientifiques*. 59, 426-582 (471-488).

R. E. Brown, 1906. The Hong-Kong typhoon and the Jesuit observatories. *The Month* 108, 561-570.

J. Algué, 1909. *Mirador Observatory, Baguio, Benguet*. Bureau of Printing, Manila, 11 pp.

M. Saderra-Masó, 1915. *Historia del Observatorio de Manila*. E.C. McCullough, Manila, 210 pp.

P. Pastells, 1916, 1917. *Misión de la Compañía de Jesús de Filipinas en el siglo XIX*. Ed. Barcelonesa, Barcelona, 3 vols, 1519 pp. (Vol.I, Cap. 18; Vol.II, Cap. 6 and 14; Vol.III, Cap.8 and 18)

M. Saavedra Masó, 1924. *Misiones Jesuíticas de Filipinas. 1581-1768 y 1859-1924*. Cap. 13. El Observatorio de Manila. Universidad de Santo Tomás, Manila, 110 pp.

M. Selga, 1929. *The publications of the Observatory of Manila*. Bureau of Printing, Manila. 21 pp.

C.E. Deppermann, 1931. The astronomical division of the Manila Observatory. *Bull. Am. Ass. Jesuit Scientists 8, 10-13.*

W.C. Repetti, 1946. Seventy years of seismology in the Manila Observatory. *Trans. Am. Geophys. Union* 27, 15-18.

M. Selga, 1947. La agonía del Observatorio de Manila. *Ibérica* 115, 403-407.

M. Selga, 1947. Las visitas al Observatorio de Manila durante el periodo 8 Diciembre 1941 a 1 Julio 1943. *Ibérica* 116, 13-16.

M. Selga, 1947. La labor del Observatorio de Manila durante el periodo 8 Diciembre a 1 Julio 1943. *Ibérica* 119, 119-124.

M. Selga, 1947. El Observatorio de Manila durante el periodo 1 Julio 1943-21 Septiembre 1944. *Ibérica* 121, 180-184

M. Selga, 1947. Destrucción del Observatorio de Manila. *Ibérica* 123, 259-263.

M. Selga, 1947. Servicio sísmico y magnético del Observatorio de Manila. *Ibérica,* 126, 373-375.

W. C. Repetti, 1948. *The Manila Observatory*. Washington D.C. 43 pp.

Charles E. Deppermann, 1953. The Manila Observatory rises again. *Philippine Studies* 1, 31-41.

James J. Hennesseys, 1960. The Manila Observatory. *Philippine Studies* 8, 99-120.

John N. Schumacher, 1965. One hundred years of Jesuit scientists: The Manila Observatory 1865-1965. *Philippine Studies* 13, 258-286.

Anonymous, 1966. *Manila Observatory. Geophysical Basic Research.* Manila Observatory, 14 pp.

R. A. Miller, 1968. Solar Astronomy at Manila Observatory. *Sky and Telescope*, 36, 154-155.

A. Hidalgo and V. Badillo, 1980. *Catalogue of the publications of the Manila Observatory personnel.* Manila Observatory, 100 pp.

Sergio S. Su, 1988. Historical seismograms of the Manila Observatory. In W. K. Lee, H. Meyers, K. Shimazaki (eds.) *Historical seismograms and earthquakes of the world.* Academic Press, San Diego, 490-496.

ST. XAVIER'S COLLEGE ASTRONOMICAL OBSERVATORY, CALCUTTA, INDIA

1866-1939

Meteorology, Astronomy, Solar Physics

In 1866 Lafont began meteorological observations in St. Xavier's College in Calcutta which were continued by Francotte for 34 years. In 1870 a Secchi meteorograph was installed. P. Tacchini, director of the Observatory of Palermo , who has come to India to observe the Venus transit in 1874, encouraged Lafont to create an astronomical observatory in St. Xavier's College. Telescopes and other equipment were acquired and the observatory was inaugurated in 1875. In 1877 a rotating dome for the telescope was built on the roof of the college. The observatory was moved to a separate building in 1887. The observatory was dedicated mainly to solar observations. The personnel of the observatory participated in several expeditions to observe solar eclipses in different parts of India. From 1889, the direction of the observatory was separated into two, one for meteorology and another for astronomy and solar physics. Work in the observatory began to decline at the end of 1920. The meteorological section was closed in 1925 and the solar physics section in 1939.

Directors

Eugene Lafont	1868-1879
Alphonse de Penaranda	1880-1881
Constantin de Clippeleir	1882-1884
Hermann Seitz	1885
Edward Francotte	1886
Victor de Campigneulles	1887-1888

Meteorology

Victor de Campigneulles	1889-1892
Edward Francotte	1895-1924
Emile Roeland	1925

Astronomy and solar physics

Alphonse de Penaranda	1889-1896
Constantin de Clippeleir	1902-1907
Henri Josson	1908-1910
Armand Briot	1911-1912
Medard Vermeire	1913
Leo Ameye	1914-1917
Armand Briot	1918-1919
Emile Roeland	1920

Armand Briot	1921-1922
Medard Vermeire	1923-1932
Albert Schelvis	1933-1939

Buildings and instruments

In 1866 the first meteorological instruments were installed on the roof of the college. In 1875 a rotating dome of 7 m diameter was added on the roof of the college for the telescopes. In 1878 the dome was installed on a 20 meter high separate building.

1866: first meteorological instruments.
1872: Secchi meteorograph
1875: Steinheil telescope of 22 cm aperture with equatorial mounting by Cooke. The telescope was equipped with a reversible Browning spectroscope equivalent to 10 prisms of 60 degrees.
Cooke telescope with equatorial mounting of 7.5 cm aperture. Transit telescope. Gauthier celostat with double mirror.

Bibliography

R. Meldola, 1875. Solar observation in India. *Nature* 12, 400.

V. de Campigneulles, 1899. *Observations taken at Dumraon, Behar, India during the eclipse of on the 22nd January 1998 by a party of Jesuit Fathers of the Western Bengal Mission.* Longmans, Green and Co. Bombay,

P. de Vregille, 1906. Les observatoires de la Compagnie de Jésus au début du XXe siêcle. *Rev. des Questions Scientifiques* 59 (3a série, tome 9) 10-72, 493-579.

A. Verstraeten, 1992. St. Xavier's Astronomical and meteorological observatory. Manuscrito, 2 pp.

I. Chinnici, 1996. An "Italian" observatory in India: The history of the Calcutta Observatory. *Studies in History of Medicine and Science* 14, (1-21)

OBSERVATORY OF ZIKAWEI, ZOSE AND LUKIAPANG, SHANGHAI, CHINA

1872-1949
Meteorology, Seismology, Geomagnetism, Astronomy

In 1872, a meteorological observatory was established in Zikawei near

Shanghai by French Jesuits. Since 1874 observations were made every three hours and a monthly bulletin was published. In 1876 Dechevrens was appointed director and gave the observatory true scientific standards with two sections of meteorology and magnetism. In 1879 the observatory began studying and forecasting the occurrence of typhoons which became its main work. From 1882 the observatory began its collaboration with the Maritime Customs of China and published in the newspapers daily weather forecasts, especially during typhoon season. In 1884 a system of visual signals was installed in the harbor of Shanghai, designed by Dechevrens and Froc, to warn sailors of approaching storms. The same year the observatory established the time-keeping service. A larger three stories new building was inaugurated in 1902. In 1900, Chevalier founded in the hill of Zose, 25 km South West from Zikawei, an astronomical observatory with a large dome of 10 m diameter, where a double equatorial telescope of 40 cm aperture by Gauthier was installed. In 1904 the first seismographs were installed in Zikawei, thus completing the five sections of the observatory, namely, meteorology, magnetism, astronomy, seismology and time service. In 1908, the magnetic section was moved to Lukiapang, 30 km from Zikawei, owing to the installation of electric tramways in Shanghai. In 1929 magnetic instruments were installed in Zose and the observations in Lukiapang ceased in 1934. From 1908, the observatory had a general directors and three directors of meteorology, time service and seismology (Zikawei), astronomy (Zose) and magnetism (Lukiapang), who functioned with great autonomy. In 1926 the observatory participated in the international project of revision of longitudes. In 1931 Lejay introduced several new subjects of research, such as ionospheric studies, solar physics, atmospheric physics and gravity surveys. In 1937 during the Chinese-Japanese war the observatory closed its meteorological survey which opened again in 1946. In 1949, the Communist government occupied and confiscated the observatory and expelled the Jesuits from China.

Directors

Agustine Colombel	1873-1875
Henri Le Lec	1875-1876
Marc Dechevrens	1876-1887
Bernard Oorus	1888
Stanislas Chevalier	1888-1897
Louis Froc	1896-1914
Luois Froc	1920-1926
Stanislas Chevalier	1927-1929
Louis Froc	1930
Pierre Lejay	1931-1939

Section directors

Joseph de Moidrey	1915-1920	Magnetism (Lukiapang)
Henri Gauthier	1915-1920	Meteorology (Zikawei)
Stanislas Chevalier	1902-1927	Astronomy (Zose)
Ernesto Gherzi	1940-1949	Meteorology and seismology (Zikawei)
Maurice Burgaud	1940-1949	Magnetism (Lukiapang)
Edmund de la Villemarqué	1940-1946	Astronomy (Zose)
Pierre Lapeyre	1947-1949	Astronomy (Zose)

Collaborators

Edmund Pigot	1899-1903, 1904-1907	
Paul Tsuchihashi	1906-1910	Astr.
Antoine Weckbacher	1906-1923	Astr.
Pierre Mathan	1907-1908	
George Horan	1917	
Louis Gauchet	1923-1930	Astr.
Juan Aguinagalde	1930-1949	Instr.

Publications

Bulletin des observations magnétique et météorologiques (1874-1904),
Magnétisme, Météorologie et Séismologie (1905-1907)
Météorologie et Séismologie (1908-1911)
Météorologie (1912-1949)
Annales de l'Observatoire astronomique de Zo-se (1907-1940)
Observations magnétique faites a l'Observatoire de Lu-kia-pang (1908-1934)
Calendrier-annuaire (1878, 1903-1949)
Seismological Bulletin (1912-1949)

Buildings and instruments

The first building was built in 1873 in Zikawei, near Shanghai where there were several mission institutions established by French Jesuits. The main building of 25 m length had one story except for the central part where there was a two stories tower. Ten meters from the building there was a tower of 33 meters where meteorological instruments were installed. Magnetic instruments were installed in a small octagonal building built with non-magnetic materials. A small wooden hut was used for additional magnetic measurements. In 1902 the new observatory located near the old one was inaugurated. It consisted of two buildings; the main one was a three stories building of 50 m length with a tower of 33 m in the central front. The second building, separated by 35 m, was a rectangular building (13 m by 15 m) used for magnetic instruments. The astronomical observatory on the Zose hill, 25 km South West from Zikawei, finished in 1901, consisted of a building with three parts, the central one with a dome of 10 m diameter where the equatorial telescope was installed. The magnetic observatory at Lukiapang, 30 km from Zikawei, consisted of three main buildings for the variometers, offices and living quarters and a fourth smaller one for absolute measurements. In 1932 magnetic instruments from Zikawei were installed in Zose. In 1934 Lukiapang was closed and its instruments moved to Zose.

Meteorology

1874-1877: Two Tonnelot barometers, three Fortin barometers, three aneroid barometers; four thermometers, three Baudin maximum and minimum thermometers; Montsouri actinometer, Regnault hygrometer, pluviometer, Secchi meteorograph, Robinson and Beckley anemometers

1881: A second Secchi meteorograph made by Brassart.

1886: Clino-anemometer designed by Dechevrens.

1903: a Secchi meteorograph was modified to record only atmospheric pressure.

1908: In Lukiapang; meteorological instruments, Fortin barometer, Richard barograph, Fuess thermometer, Richard thermographs, pluviometer and anemometer

1931: In Zikawei; ionospheric sounder and instruments to measure solar radiation and Ozone.

Magnetism

1877: In Zikawei; Kew-Elliott declinometer, Adie-Kew variometers, Elliot unifilar and bifilar magnetometer, two Barrow and Dover inclinometer, a magnetic balance.

1908: In Lukiapang; Kew-Elliot magnetometer, Wild-Schulzer inductor, Plath galvanometer, Dover inclinometer.

1934: Instruments moved from Lukiapang to Zose. Smith magnetic balance.

1935: In Zose; two Toepfer horizontal variometers, Godhavn vertical magnetic balance and BMZ magnetic balance.

Astronomy

1884: In Zikawei; Stackpole meridian circle of 5.5 cm aperture, Billant equatorial telescope and several chronometers

1901: In Zose; Gautier double equatorial telescope of 40 cm aperture and 6.87 m focal length.

Seismology

1904: Two Omori horizontal seismographs

1909: Wiechert horizontal seismograph of 1200 kg mass
1913: Galitzin vertical seismograph
1932: Three component Galitzin-Wilip electromagnetic seismographs

Bibliography

M. Dechevrens, 1888. L'Observatoire de Zikawei, fondé et dirigé par les missionaires de la Compagnie de Jésus, mission du Kiangnan. *Etudes* 43, 262-279

P. de Vregille, 1906. Les observatoires de la Compagnie de Jésus au début de XXe siècle. *Rev. des Questions Scientifiques* 59, 10-72, 493-579.

R. E. Brown, 1906. The Hong-Kong typhoon and the Jesuit observatories. *The Month* 108, 561-570.

J. de Moidrey, 1909. L'Observatoire magnétique de Zi-ka-wei. *Terrestrial Magnetism and Atmospheric Electricity* 14, 1-2.

H. Gauthier, 1919. L'Observatoire de Zi-ka-wei. *Relation de Chine*, Jan-Au, 2-13.

J. Bruhnes, 1924. Les travaux des Jésuites à l'Observatoire de Zi-Ka-Wei, pres Chang-hai (Chine). *Rev. d'Histoire de Missions* 1, 26-38.

P. Lejay, 1925. Trois Observatoires de mission. *Etudes* 182, 594-602.

R. Cirera, 1925. Principales observatorios de la Compañía de Jesús en las misiones. *Ibérica* 24, 41-44, 72-77.

P. Lejay, 1927. Observatoire de Zi ka wei. Situation actuelle - Avenir. (manuscript, Archive Province France S. J., APFSJ) 19 pp.

F. Canet, 1929. Le haut apostolat intellectuel en pays de Mission. La science en Chine: La determination des longitudes. (manuscript, APFSJ) 9 pp.

Anonymous, 1930. *L'Observatoire de Zi-ka-wei. Cinquante ans de travail scientifique*. Imp. d'Art G. Boüan, Paris.

L. Froc, 1930. Code de Zi-ka-wei.- Historique. Conf. des Direct. des Service Météor. d' Extréme-Oriente, *Conference des Directeurs des Services Météorologiques d'Extrême Orient,* Hong-kong. 1-4.

M. Burgaud and E. Gherzi, 1931. Annonce des typhons dans les mers de Chine par l'Observatoire de Zi-ka-wei. *Jour. de la Marine Merchande*. No. 645 (13 Août 1931) 1916-1918.

Anonymous, 1935. Observatoires de Shanghai. Chez nos savant. (based on letters by Gherzi, Burgaud and Lejay) *Relations de Chine* (Juill-Sept 1935) 165-174.

P. Lejay, 1939. Travaux executés a l'Observatoire de Zi-ka-wei de 1933 à 1939.(manuscript, APFSJ), 10 pp.

F. Maze, 1950. The Jesuits in China. Their scientific work. *Bull. Am. Ass. Jesuit Scientists* 27, 143-152. Also in *Far Eastern Economic Review*, 8, 16, 503-505.

E. Gherzi, 1950. The scientific work of the Catholic Church at the Zikawei Observatory in Shanghai. *Boletim Inst. Portugues de Hongkong*, 3, 45-57.

E. Gherzi, 1973. Un jesuite meteorologist en Chine. Le Rev Père Ernesto Gherzi S.J. par lui meme. (manuscript, ASJCF-St. Jerome)

J. Dehergne, 1976. Zi-ka-wei l'Observatoire des cyclones. *Météorologie* 4, 179-188.

L. Pyenson, 1990. Habits of mind: Geophysics at Shanghai and Algiers, 1920-1940. *Historical Studies in the Phys. and Biol. Scien.* 21, 161-196.

L. Pyenson, 1993. *Civilizing Mission: Exact sciences and French overseas expansion, 1830-1940*. Johns Hopkins University Press, Baltimore.

METEOROLOGISCHE OBSERVATORIUM BOROMA UND ZUMBO, MOZAMBIQUE

1883-1905
Meteorology

In 1883 a meteorological station was established, in Boroma at 16 km from the village of Tete, near the Zambeze River, in the present Mozambique, close to the frontier with Zambia. From 1893 to 1897 Menyhart, a Hungarian Jesuit, was in charge. He established in 1893 another meteorological station in Zumbo, a few kilometers to the north. The station of Boroma had three thermometers, a thermograph, a Richard barograph and a psycrometer. Part of the instruments were donated by Cardinal Haynald, the founder of the Kalocsa Observatory. Observations were reduced and systematized by Fenyi, who published them in the Bulletins of Kalocsa. Fenyi considered these stations as part of his observatory. Observations continued after Menyhart's death until 1905.

Bibliography

L. Menyhart, 1895. Del Afrikában tett meteorologiai megfigyelésének eredményeiröl (kiv). *Akademiai Ertesitö* 6, 644-645.

G. Fenyi, 1896. Meteorologische Beobachtungen angestellt zu Baroma in Süd Afrika von L. Menyhart. *Pub. Haynald Observatoriums, Kalocsa* 7, 76 pp.

G. Fenyi, 1905. Meteorologische Beobachtungen angestellt von Ladislau Menyhart, S.J. zu Baroma und Zumbo in Süd Afrika in den Jahren 18932-1897. *Pub. Haynald Observatoriums, Kalocsa* 9, 94 pp.

J. Fenyi, 1912. Ergebnisse der Beobachtungen der Temperatur un des Luftdruckes in Borona (Südafrika). *Kaiserl. Akademie der Wissenschaften in Wien. Mathem. Naturw.* Bd. 121, 2, 2063-2075.

G. M. Imre, 1986. *A Kalocsai Haynald Obszervatörium Története*. Orszargos Müszaki Informácios K. K, Budapest, 72-75.

OBSERVATOIRE D' AMBOHIDEMPONA, TANANARIVE, MADAGASCAR

1889-1967

Astronomy, Meteorology, Geomagnetism, Seismology

Jesuits began meteorological observations in Madagascar in 1864. The French Administration was interested in establishing an observatory in the southern hemisphere and talked Jesuits into this work. The observatory was inaugurated in 1889, built on the hill of Ambohidempona, 120 m over the plateau and 1402 m over sea level. The building had a large main dome and three smaller ones. The observatory had three sections, astronomical with two telescopes, meteorological and magnetic. It had a time service and gave support to geodetic and cartographic work. Colin was its first director. In 1895 the observatory was destroyed during the war. Rebuilt in 1898, in a more modest way, it began operating again. The observatory was the center of a meteorological network till 1920 and in charge of the observation and forecast of cyclones. From 1899 a seismological section was added. From 1902 Colin participated in geodetic and cartographic work in Madagascar. In 1930 instruments were renovated and work concentrated in meteorology and magnetism. Poisson directed the observatory for forty two years. In the 1950's the observatory began to decline. In 1967, after the independence of Madagascar, it was transferred to the University of Madagascar. Jesuits continued working at the observatory until 1969.

Directors

Elie Colin	1889-1893
Pierre Combes	1893-1894
Luois Soula	1895
Elie Colin	1896-1923
Irénée Aurand	1823
Charles Poisson	1923-1965
Louis de Laitre	1965-1967

Collaborators

Desideré Roblet	1889-1910
Louis Soula	1891-1907
Pierre Combes	1894-1895
Irénée Aurand	1908-1915
Jules Muthuen	1914-1920
Charles Rey	1923; 1931-1932
Louis Cattala	1926-1928; 1943-1949.
Jean Delpeut	1930-1932
Jean Coze,	1931-1933; 1939-1969.
Henri Gruyelle	1934-1935
André Gauvin	1936-1944
Maurice Burgaud	1956-1969

Publications

Observations météorologiques faites à Tananarive. 28 Volumes from 1889 to 1915 (includes magnetic observacions).
Etat général du temps à Madagascar (monthly). No. 1-23, 1916-1918.

Buildings and instruments

The first building was finished in 1889, built according to the plans of architect M. Lequeux (Paris). It was a large building with the form of a T, with a 32 m long front in North South direction and with a central part of octagonal plant with a circular dome of 8 m diameter. Other three smaller domes were installed on the three ends of the building. A separate small building was used for magnetic observations. After its destruction in 1895, the observatory was rebuilt in 1898 in a more modest form with a plant in the from of a L with one story and only one dome in the southern end. The magnetic building was also rebuilt.

Meteorology
1889: Three Tonnelot barometers; Richard barograph; Robinson anemometers, horizontal and vertical with mechanical and electrical recording; thermometers, August psicrometer; Richard psicrograph; Piche evaporimeter; Campbell heliograph and photographic recording by Jordan, Violle actinometer, pluvimeters, etc.

Magnetism
1889: Mascart magnetometers for the three components (declinometer, balance and bifilar); Brunner inclinometer.
1955: Proton magnetometer.

Astronomy
1889: Rigaud meridian circle from the Observatory of Paris; Eichens telescope of 20 cm aperture with equatorial mounting by Gauthier; portable telescope; Brunner solar telescope; four pendulum clocks, five chronometers and a Boutlite chronograph.
1923: lenses of Eichens telescope were renovated.

Seismology
1899: Cecchi seismograph
1927: Mainka SOM horizontal component seismographs, of 450 kg mass
1947: Grenet-Duclaux vertical seismograph

Bibliography

E. Colin, 1890. Ambohidempona Observatoire Royal de Tananarive. *Cosmos (Revue des Sciences)* 39, 320-323.

E. Colin, 1894. L'Observatoire de Tananarive. *Etudes* 62, 642-660.

E. Colin, 1897. L'Observatoire Français de Madagascar. I Pendant la guerre. II Aprés la guerre. *Etudes* 71, 308-331.

E. Colin, 1900. Mon second observatoire à Madagascar. *Etudes* 85, 213-226.

P. de Vregille, 1906. Les observatoires de la Compagnie de Jésus au début du XXe siêcle. *Rev. des Questions Scientifiques* 59 (3a série, tome 9) 10-72, 493-579.

E. Colin, 1909. *L'Observatoire de Tananarive.* Isle Maurice, Imp.à l'usage priv.

C. Poisson, 1925. *L'Observatoire de Tananarive*. Editions Spes, Paris, 77pp.

R. Cirera, 1925. Principales observatorios de la Compañía de Jesús en las misiones. *Ibérica* 24, 41-44; 72-77.

P. Lejay, 1925. Trois observatoires de Mission. *Etudes* 182, 594-602.

C. Poisson, 1930. *L'Observatoire de Tananarive et ses travaux*. Imp. Moderne de l'Emyrne, Tananarive, 38 pp.

C. Poisson, 1939. Le cinquantenaire de l'Observatoire de Tananarive. *Etudes* 240, 376-380.

C. Poisson, 1939. *Un cinquantenaire, l'Observatoire d'Abohidempona à Tananarive, Madagascar, 1889-1939*. Editions Dillen, Paris. 158 pp

C. Poisson, 1945. L'Observatoire à. la Croisée des Chemins.(manuscript, APFSJ) 6 pp.

J. Coze, 1956. Note sur l'Observatoire de Tananarive, (manuscript, APFSJ) 3 pp.

OBSERVATORY OF BULAWAYO, ZIMBABWE-RHODESIA

1903-1929
Meteorology, Geomagnetism, Astronomy

Meteorological observations began in 1897 in St. George's College, in Bulawayo. In 1901 the Jesuit Superior of the Zambeze Mission began the preparations for the establishment of an observatory. The observatory was inaugurated in 1903 with Goetz as Director, who had acquired in Europe the necessary instruments. Regular astronomical, meteorological and magnetic observations began the same year. In 1913 a telescope with equatorial mounting by Cooke was installed. Goetz collaborated with Hagen in the observations of variable stars and carried out climatological studies and magnetic field surveys. Goetz retired in 1926 and the observatory was transferred to the Colonial Government. Some of the instruments were installed in St. George's College. In his memory the observatory continues today with the name of Goetz Observatory.

Buildings and instruments

The observatory was installed in a small one story building which served also as residence. A separate small building was used for magnetic observations. In 1913 a rotating dome was added for the equatorial telescope.

The meteorological station had very complete equipment with a Tonnelot barometer, thermometers, barographs, pluviometers, etc. The magnetic station had a Nalder unifilar magnetometer and a Dover inclinometer.

The astronomical instruments were two small telescopes of 12 cm and 9 cm aperture and a Cooke telescope of 16 cm aperture with equatorial mounting installed in 1913. Auxiliary equipment was a Vion astrolabe, three theodolites and several chronographs.

Bibliography

Anonymous, 1903. Astronomical Observatory for Bulawayo. *Bulawayo Observer* (March 21st).

Anonymous, 1903. Astronomical Observatory for Bulawayo. *Zambesi Mission Record* 2, 257-258.

E. Goetz, 1903. Our Bulawayo Observatory. *Zambesi Mission Record* 2, 297-298.

P. de Vregille, 1906. Les observatoires de la Compagnie de Jésus au début du XXe siècle. *Rev. des Questions Scientifiques* 59 (3a série, tome 9) 10-72, 493-579.

Anonymous, 1908. Organizing knowledge (a report on Fr. Goetz's work on climatology). *The Bulawayo Chronicle* (May 29, 1908).

W.F. Rea, 1976. The beginnings of the Goetz Observatory. *Rhodesiana* 35, 24-32.

V. L. Bosazza, 1977. Father Goetz and the Goetz Observatory. *Rhodesiana*, 36,

J. Dach and W.F. Rea, 1979. *The Catholic Churh and Zimbabwe, 1879-1979.* Mambo Press, Gweb. 126-127.

OBSERVATOIRE DE KSARA, LEBANON

1907-1979

Meteorology, Astronomy, Seismology, Geomagnetism

The observatory was built in 1907 in Ksara, Lebanon in the Bekaa valley some 40 km from Beirut at an altitude of 960 m. The observatory was associated with the Jesuit Université Saint Joseph in Beirut and the first director was Berloty. The first astronomical instruments, installed in 1907, were a Bardoux equatorial telescope of 11 cm aperture and a meridian telescope of 4.8 cm. Magnetic instruments were installed the following year. In 1909 the building for the meteorological section was built with very complete instrumentation. In 1910 were two Mainka seismographs were installed in a separate building. In 1914 Jesuits have to leave the observatory at the beginning of the war, since Lebanon was part of the Ottoman Empire. During the war years the observatory was destroyed and most of the scientific instruments lost. In 1918 Jesuits returned to Ksara. The observatory was rebuilt with help from the French Government. And new instruments were installed. The observatory received continuous financial support from the CNRS of France. In 1921 the observatory became the headquarters of the Meteorological Service of Syria with a network of 14 stations and depended from the Office National Météorologique of France. From 1945 the Service was reduced to the territory of Lebanon. From 1966 the financial support came from the CNRS of Lebanon. Between 1967 and 1979 the observatory carried on a large work of synthesis on climatology, meteorology, graviometry and seismology. Jesuits left the observatory in 1979 and its installations were given to the CNRS of Lebanon. In 1981 the observatory was destroyed during the Lebanon war.

Directors

Bonaventure Berloty	1907-1914, 1918-1932
Charles Combier	1932-1950
Jacques Plassard	1950-1979

Collaborators

Joseph Brachet	1907	
Eugéne Monin	1908-1914	
Charles Corte	1910-1911	
Joseph Prenthaler	1912-1914	
Harold Richard	1914	
George Horan	1920-1934	meteor.
Jean Rey	1935-1975	meteor.
Henri Destezet	1936-1956	seism.
Jean Delpeut	1937-1976	magn.

Barthelemy Kogoj	1955-1979	seism.

Buildings and Instruments

The first building was built in 1907 with rooms for astronomical, magnetic and meteorological instruments, and offices. In 1910 a new building was added with larger facilities. In 1914 a 12 m high tower was built with a rotating dome for the equatorial telescope. Buildings were damaged during the war and were restored in 1920. New instruments were installed. Meteorological and seismological instruments were installed in a separate building. Magnetic equipment were placed in an underground vault.

Astronomy
1908: Gautier meridian circle of 4.8 cm aperture. Bardoux telescope with equatorial mounting of 10.8 cm aperture, another small telescope of 4.8 cm aperture, a theodolite of 4.2 cm aperture by Troughton and Simms.
1913: Dallmeyer equatorial telescope of 16 cm aperture.
1920: The equatorial telescope was renovated by Prin (Paris) with a new lens of 20.3 cm aperture, two finders of 8 cm and 2.5 cm and new clock machinery. A Borda meridian circle. Several auxiliary instruments such as Leroy, Boulite and Brillié chronographs and chronometers, and radio receptors for time signals.

Seismology
1910: Two Mainka horizontal seismographs of 135 kg mass
1921: Mainka horizontal seismograph of 460 kg mass
1933: Galitzin-Wilip vertical seismograph
1939: Short period seismographs of Ishimoto type.
1957: Grenet-Coulomb APX three component short period seismographs.

Magnetism
1908: Kew-Mascart magnetometer for absolute measurements of D and H.
Dover-Schulze inclinometer. Mascart variometers for D, H and Z with Mailat-Bouty recording.
1920: Instruments were renovated.

Meteorology
1908: Complete meteorological station with barographs, barometers, thermometers, pluviometers, etc.
1920: New instruments were installed

Publications

Annales Observations- Section de météorologique. 1921-1939.

Annales Climatologiques. 1940-1972
Climat du Liban. 1928-1966.
Annales Observations- Section de Séismologie 1921-1946
Annales Séismologiques. 1956-1972.
Bulletin Séismique Provisoire. 1935-1973.
Annales Observations de Magnetisme Terrestre. 1930-1937.
Atlas Climatique du Liban (4 vols)
Annales-Memoires (4 vols). 1924-1973. Gravimetrie, Climatologie, Séismologique.

C. Combier, 1933. La climatologie de la Syrie et du Liban. *Rev. de Geogr. Phys. et Geol. Dyn.*,6, 320-346.

C. Combier, 1945. *Les Climats de la Syrie et du Liban. Beyruth. Atlas Climatique du Liban, Tome I-IV.* Service Meteorologique du Liban, Beyruth, 1977.

J. Plassard y B. Kogoj, 1981. *Etude de la séismicité du Liban.* CNRS Libanais.

J. Rey, 1951. *Carte pluviométrique du Liban au 1/200000.* Publ. Obs. Ksara.

P. Stahl y J. Plassard, 1959. *Carte gravimétrique du Liban au 1/200000.* Beyruth.

Bibliography

B. Berloty, 1912. L'Observatoire de Ksara (Liban). *Ciel et Terre* 23, 103-108; 129-136.

B. Berloty, 1924, 1927. Notice historique sur l'observatoire fond, en Syrie par le pères de la compagnie de Jésus. *Annales de l'Observatoire de Ksara, Memoires*, Tome I, Liban, I-X, 79 pp.

P. Lejay, 1925. Trois Observatoires de Mission. *Etudes* 182, 594-602.
R. Cirera, 1925. Principales observatorios de la Compañía de Jesús en las misiones. *Ibérica* 24, 41-44, 72-77.

J. Dufay, 1932. L'observatoire de Ksara et le Service Météorologique en Syrie. *Revue Scientifique (Revue Rose)* anne 70, No. 3, 65-68.

C. Combier, 1936. Etat actuel d l'Observatoire de Ksara. Equipement, travaux, publications. (manuscript, APFSI) 4 pp.

J. Plassard, 1952. Etat actuel de l'Observatoire de Ksara. Equipement, travaux et publications.(manuscript, APFSI) 6 pp.

J. Plassard, 1995. Histoire de l'Observatoire de Ksara. (manuscript, Archives de Province du Liban S.J., APLSI) 64 pp. .

RIVERVIEW OBSERVATORY, NORTH SOUTH WALES, AUSTRALIA

1907-
Meteorology, Seismology

The Riverview Observatory was founded in 1907 in St. Ignatius College, Riverview, North South Wales, Australia. Its first director and founder was Pigot who began meteorological observations in 1907 and installed Mainka seismographs in 1909. In 1922 a telescope was installed. In 1926 a program of observations of variable stars began in collaboration with the Bosscha Observatory of Lembang, Java. O'Leary and O'Connell continued astronomical observations. In 1941 new Galitzin type seismographs were installed and seismology became the observatory main task. In 1962 a seismographic station of the WWSSN global network was installed. In 1965 Drake was appointed director who was the last formal director until 1992. In 1972 the meteorological station was closed, although rainfall measurements were continued till 1978. From 1972 support for the seismological stations came from the Bureau of Mineral Resources (from 1992, Australian Geological Survey Organization, AGSO). The seismographic station continues active today with the support of AGSO where seismograms are stored.

Directors

Edward F. Pigot	1907-1929
William J. O'Leary	1929-1939
Daniel O'Connell	1939-1952
Noel Burke-Gaffney	1952-1958
Anthony Fynn	1958-1965
Lawrence Hession	1965-1972
Lawrence A. Drake	1972-1992
Luke Tyler	1992-1996 (no S.J.)
Michael Hansen	1996-1999

Publications

Riverview Observatory Seismological Bulletin, 1909-1968
Astronomical Papers 1930-1952, 2 series

Reprints.

Buildings and instruments

The first building of the observatory was a small one story rectangular building on a small hill on the grounds of St. Ignatius College. In 1922 the building was enlarged and a rotating dome was installed for the telescope. Seismographs were installed in a separate building. The WWSSN station was installed in the basement of the college library. The meteorological instruments were installed in a separate site 6 km away.

1907: Complete meteorological station
1909: Three component (ZNE) Wiechert seismographs of 1000 kg mass
1910: Horizontal Mainka seismographs, 450 kg mass;
1922: Equatorial telescope of 17.8 cm aperture
1925: Solar radiation instrument
1931: Guiding telescope and two star cameras
1941: Three component Galitzin type seismograph built in Australia
1953: Vertical Sprengnether seismograph
1962: WWSSN seismographic station.

Bibliography

H. Doyle and R.Underwood, 1965. Seismological Stations in Australia. *The Australian Journal of Science* 28, 40-43.

L. A. Drake, 1980. Riverview Observatory. En : C. Fraser and E.L.Scarlett (eds.) *St. Ignatius Centennial 1880-1980.*O'Loughlin Bros, Lewisham, NSW. 210-214.

GEOPHYSICAL OBSERVATORY OF THE UNIVERSITY COLLEGE OF ADDIS-ABABA

1957-1971

Meteorology, Geomagnetism, Seismology

In 1955 a recommendation was passed at the Scientific Committee of the International Geophysical Year recommending the installation of a geomagnetic station near the magnetic equator, suggesting Ethiopia as a possible location. The University College of Addis Ababa, entrusted in 1945 to the Canadian Jesuits by the Government of Haile Sellassie, in response to this recommendation established in 1958 an observatory, dedicated mainly to geomagnetism, under the direction of Pierre Gouin. The observatory had very complete magnetic instrumentation: three magnetometers QHM (Quartz

Horizontal Magnetometer), Ruska Earth Inductor inclinometer, Fennel - Chasselon D-Magnetometer and three component (H,Z,D) Ruska variometers. The observatory had a complete meteorological station and a seismographic station with three components Willmore short period seismographs. In 1962 a WWSSN seismographic station (AAE) was installed. The observatory was under Gouin's direction until 1978. During this 20 years period important geomagnetic and seismological studies were carried out.

Publications

Bulletin of the Geophysical Observatory. University College of Addis Ababa. 1-16, 1959-1975 (under Gouin's direction)

P. Gouin, 1979. *Earthquake history of Ethiopia and the Horn of Africa*. IDRC, Ottawa, 259 pp.

Bibliography

P. Gouin, 1959. Magnetic Activity at Addis Ababa (January-June 1958). *Bull. Geophysical Observatory*, University College of Addis Ababa, 1, 5-15.

P. Gouin, 1994. Geophysical Observatory at the University of Addis Ababa, Ethiopia. (Manuscript) 9 pp.

PART III. JESUIT SCIENTISTS, 1814-2000

José Algué (1856-1930). Born in Manresa (Barcelona), Spain, he entered the Society of Jesus in 1871. From 1878 to 1885 he was a teacher of physics, chemistry and mathematics in the Jesuit high-school in Zaragoza. In 1889 Algué accompanied Faura in his visit to the observatory of the Collegio Romano. The same year he began his preparation for his work in Manila Observatory studying mathematics at the University of Barcelona. From 1891 to 1894 he worked on astronomy with Hagen at Georgetown Observatory where he designed a reflector zenithal telescope. In 1894 Algué arrived at Manila Observatory and became Director in 1897. The same year he published in Spanish his work on typhoons in Philippines (published in English in 1904), where he presented a complete description of these storms, their nature and paths. He also developed an instrument for detection of cyclones, named "barociclonometer", based on Faura's ideas, which was widely used by mariners in the Pacific. In 1899, after the occupation of the Philippines by the United States, Algué explained the situation of the observatory to the American authorities and, in 1901, he was appointed Director of the new Philippine Weather Bureau, organized by the United States Government which recognized the scientific work of the Manila Observatory. During his directorship the Weather Bureau developed into a large and complex organization with 248 secondary meteorological stations. In 1924 he traveled to Rome and afterwards, owing to health problems, retired from the observatory in Spain. He was member of the Royal Meteorological Society of London and of the Pontifical Academy of Sciences.

Baguios o ciclones Filipinos. Estudio teórico-práctico. Imprenta del Observatorio, Manila, 307 pp. 1897. *The cyclones of the Far East.* Bureau of Printing, Manila, 283 pp. 1904.

The Barocyclonometer. Imprenta del Observatorio, Manila, 8 pp. 1898, (2 ed. 1900, 3 ed. 1920, 4 ed. 1937).

Las nubes en el Archipiélago Filipino. Imprenta del Observatorio, Manila, 86 pp. 1899.

Atlas de Filipinas. Government Printing Office, Washington, 24 pp., 30 maps. 1900 Reprinted as : *Atlas of the Philippines* by Signal Office, U.S. War Department, 1900.

El Archipiélago Filipino (J. Algué and J. Closé, eds.) Government Printing Office, Washington, vol. I, 708 pp. vol. II, 469 pp. 1900

I. Puig, 1930. El P. José Algué. *Ibérica* 833, 392-298.

L. Rodés, 1930. El R. P. José Algué. *Razón y Fe* 91, 553-555.

C. E. Deppermann, 1951. Father Jose Algué, S.J. and microseisms. *Bull. Am. Seis. Soc.* 41, 301-302.

A.Hidalgo, 1974. El P. José Algué, S.J., científico, inventor y pacifista. Manila Observatory, Manila, 96 pp.

Bonaventure Berloty (1856-1934). Born in Lyon, France, he entered the Society of Jesus in 1874. He obtained the degree of doctor in science in the University of Paris in 1886 and taught in the colleges of Angers and Lyon and the Facultés Catholiques de Lyon from 1891 to 1901. In 1902 he was in Alexandria and in 1903 teaching at the Université Saint Joseph in Beirut. In 1906 Berloty was appointed first director of the Observatory of Ksara in Lebanon. He was in charge of the construction and installation of the first instruments of meteorology, astronomy, seismology and geomagnetism. He moved to Egypt during the World War I and returned in 1918 to reconstruct the observatory. In 1921 Berloty was named director of the Meteorological Service of Syria. He participated in several geodetic surveys carried out by the French Administration in Lebanon and Syria. He was named in 1924 corresponding member of the Académie de Sciences and was awarded the Légion d'Honneur.

Théorie des quantités complexes à n unités principales. Thèse de Doctorat . Université du Paris. 1886.

Sur la localisation des épicentres des tremblements de terre. *Comptes Rendu Acad. Science* 191, 813-816, 1930.

Annales de l' Observatoire de Ksara, 1921- 1934

C. Combier, 1934. Le Père Bonaventure Berloty. *Relations d'Orient* 15, 169-173.

M. M. Sanchez Navarro Neumann, 1935. El P. Buenaventura Berloty, S.J. *Ibérica* 32, (1076) 371-372.

Carl Braun (1831-1907). Born in Neustadt (Hesse), Germany, he studied philosophy and theology in the Gregorian University in Rome, he was professor of science in the Diocesan Seminary in Fulda for a few years and in 1861 he entered the Society of Jesus. In 1868 he completed his studies of physics and mathematics in the University of Paris and he began teaching sciences courses in the Gregorian University. During this time he worked at the observatory with Secchi. In 1878 Braun was appointed first Director of Haynald Observatory at Kalocsa, Hungary. At the observatory Braun installed a Merz telescope and a Hilger spectroscope. Following Secchi's recommendations he began observations of sunspots and solar prominences and designed a spectro-heliograph which he did not complete. He also designed an ocular for the transit telescope to diminish the observational errors. In 1884, owing to health reasons, he resigned from the direction of the observatory and retired to the college of Mariaschein in Bohemia. Affected of severe deafness, which had motivated his abandon of teaching, he continued his scientific work in his retirement. He made a long series of measurements to determine the density and mass of the earth obtaining very exact values for that time, which are quoted by Svante Arrhenius in his *Lehrbuch der kosmichen Physik*. During his last years he wrote several books and articles on cosmological subjects related to philosophy and theology

Über Kosmogonie von Standpunkt Christliche Wissenschaft nebst einer Theorie der Sonne und einigen darauf bezüglichen philosophischen Betrachtungen. Muenster, 1889, 400 pp.

Die Gravitations-Constante, die Masse und mittlere Dichte der Erde nach einer neuen experimentellem Bestimung. Vienna, 1896.

A. Carrara, 1907. L'Opera Scientifica dei P. Carlo Braun S.J. *Rivista di Fisica, Mat. e Scien. Natur. di Pavia* 16, 358-365.

A. Linsmeier, 1907. P. Karl Braun S. J. *Natur und Offenbarung* 54, 193-200; 274-285.

M. Sanchez Navarro-Neumann, 1908. El P. Carlos Braun S. J. *Razón y Fe* 20, 486-488.

Juan Antonio Bussolini (1905-1966). Born in Chivilcoy (Buenos Aires), Argentina, he entered the Society of Jesus in 1920. He was a science teacher in the Colegio de San Ignacio in Santiago de Chile. He studied astronomy at Obervatorio de la Plata, in Buenos Aires and visited several European observatories in preparation for his work in the Observatorio de Física Cósmica de San Miguel. Bussolini arrived at the observatory in 1938 and became Director in 1943. He organized scientific work and gave a great impetus to solar research. He was also a professor of science at Facultad de Filosofía de San Miguel, Universidad del Salvador and Instituto Superior del Profesorado. Under Bussolini's directorship the observatory actively participated in the programs of the International Geophysical Year. He was active in the scientific circles of Argentina. He was the first general secretary of the Asociación

Argentina de Geofísicos y Geodestas and elected member of the Academia Nacional de Ciencias of Buenos Aires.

O. Schneider, 1980. Juan Antonio Bussolini. *Evolución de las Ciencias en la República Argentina,* (vol. VIII-Geofísica y Geodesia). Sociedad Científica Argentina, Buenos Aires. 272-273.

Ramón Cabré Roige (1922-1997). Born in Barcelona, Spain, he entered the Society of Jesus in 1939. He studied physics at the University of Barcelona where he obtained his licentiate and later in 1969 his doctoral degree. In 1958 he spent some time at the Institut de Physique du Globe in Strasbourg with Prof. Rothé. In 1959 Cabré traveled to Bolivia and became subdirector of Observatorio San Calixto, La Paz. In 1964 he was appointed director and occupied this post until 1993. During the thirty year as director of the observatory he renewed continuously its seismographic instrumentation and the observatory became the most important seismological station in South America. Cabré established collaborations with American and French institutions and kept improving the capabilities of the observatory. His research was centered on the study of the seismicity of Bolivia and the Andean region. Cabré was one of the founders and first director of CERESIS (Centro Regional de Sismología de América del Sur) in 1968. He was Professor of Geophysics in the Universidad de San Andrés, La Paz. In 1971 he was elected member of the Academia Nacional de Ciencias de Bolivia and from 1971 to 1974 he was vice-president of the Comisión de Geofísica of the Instituto Panamericano de Geografía e Historia. From 1971 to 1975 he was a member of the Executive Committee of the International Association of Seismology and Physics of the Earth Interior. In 1993 he was awarded the Medalla al Mérito Científico and the Orden Nacional del Condor de los Andes by the President of Bolivia.

(With C. Lomnitz) The Peru earthquake of October 17, 1966. *Bull. Seism. Soc. Am.* 58, 645. 1968.

Ondas Lg registradas en La Paz, Bolivia. *Geofisica Panamericana* 1, 71. 1971.
(With A. Vegas) Sismicidad en Bolivia. *Publicaciones del Observatorio San Calixto,* La Paz 40, 1989.

El Clima de La Paz. *Bodas de Diamante 1913-1985. La Paz, 1988. 80 pp.*
L. A. Drake, 1997. Ramón Cabré, S.J. (1922-1997). *Journal of Seismology* 1, 111.

W.R. Ott, 1997. Rev. Ramón Cabré Roige, S.J. May 6, 1922-February 24, 1997. *Seism. Research Letters* 68, 901.

A. Gioda, 1997. Padre Ramón Cabré, s.j. De las alturas rarificadas de la ciencia a las campanas de la Parroquia de Santa Vera Cruz. *Facetas,* 9- March 1997, 5.

Enrico Cappelletti (1831-1899). Born in Aquila, Italy, he entered the Society of Jesus in Naples in 1846. In 1859 he was collaborator of Secchi in the observatory of the Collegio Romano. In 1862 he traveled to Chile where he was a science teacher in the Colegio de San Ignacio in Santiago and founded there a small meteorological and magnetic observatory. In 1880 he moved to Argentina where he taught physics and chemistry at the seminary of Santa Fe. In 1884 he traveled to Mexico where he was a science teacher at the colleges of Puebla and Saltillo and director of the observatories in both colleges. He observed the solar eclipse of 1886 and published meteorological and astronomical observations made at Puebla and Saltillo.

El eclipse de Sol del 5 de Marzo de 1886. Observaciones practicadas en el Observatorio del Colegio Católico del Sagrado Corazón de Jesus en Puebla. Imprenta el Círculo Católico, Mexico 11 pp. 1886.

Resumen de las observaciones meteorológicas ejecutadas en el Colegio Católico del Sagrado Corazón de Jesús de Puebla durante el decenio 1877 a 1886. Tipografía Secretaría de Fomento, México 20 pp. 1888.

Apuntes de astronomía elemental o cosmosgrafía. Imprenta del Colegio Pío de Artes y Oficios, Puebla, 301 pp. 1887.

J.B. Iguiniz, 1945. *Bibliografía de los escritores de la Provincia Mexicana de la Compañía de Jesús.* Buena Prensa, Mexico. 119-121.

Louis Cattala (1904-1963). Born in Bédarieux (Herault), France, he entered the Society of Jesus in 1922. He worked as an assistant at the Observatory of Tananarive, Madagascar from 1925 to 1928. In 1928 he traveled to France where he finished his studies in theology and obtained his licenciate in physics from the University of Paris. He received his formation in geophysics at the Institut de Physique du Globe of Paris and of Strasbourg and at the Institute of Fluid Mechanics of Lille. He returned to Tananarive in 1946, where he worked at the observatory and taught science in the diocesan seminary. He was an associate researcher of the French research institute CNRS (Centre National de la Researche Scientifique). Cattala was in charge of the magnetic section of the observatory, but his main work was on gravity and magnetic field surveys. He was very critical of the work being done at the observatory and had problems with its director, Poisson. In 1949 he left the observatory and continued working on magnetic and gravity field surveys. He took part in the work of the magnetic map of Madagascar and was Secretary of the Madagascar National Commission of Geodesy and Geophysics.

Etudes gravimétrique à Madagascar. *Bull. Géologique de Madagascar* 1. 75-87, 1949.
Gravimétrie à Madagascar, interprétation tectonique du Sud et de l'Ouest. *Travaux du Bureau Geologique*, no. 59, 1954.

(with H. Besairie) Interpretation tectonique de la gravimétrie. 1954

C. Poisson, 1963. L'Oeuvre scientifique du P. Louis Cattala. (manuscript, APFSJ).

Stanislas Chevalier (1852-1930). Born in St. Laurent des Autels, France, he entered the Society of Jesus in 1871. In 1883 he traveled to China and began working with Dechevrens at Zikawei Observatory. He succeeded Dechevrens as director from 1888 to 1897. He continued Dechevrens' work on typhoons. In 1897 he made a cartographic journey through the north part of the Yang-tse River with more than 1200 astronomical observations. The *Atlas* he published was awarded prizes by the Académie des Sciences and the Société de Géographie of Paris. In 1901 Chevalier founded the Astronomical Observatory of Zose, dependent of Zikawei and he was its director till 1928. He carried out many astronomical observations in particular of stars, sunspots and orbits of asteroids. His observations were published in the 19 volumes of the *Anales of the Observatory of Zose* between 1907 and 1922.

Atlas du Haut Yang-tse, de I-tchang-fou à Ping-chan-hien, levé de November 1897 à Mars 1898. Presse Orientale, Shanghai, 1899.

Le Haut Yang-tse de I-tchang-fou à P'ing-chan-hien, en 1897-1898. Voyage et description. Complement de l'Atlas de Haut Yang-tse. Presse Orientale, Shanghai, 1899.

(Editor) Coopération de l'observatoire de Zi-ka-wei à la revision international des longitudes. *Annales de l'Observatoire Astronomique de Zo-se.* Impr. Mission Catholique, Chang-hai, 156 pp., 1927.

M. Sanchez Navarro-Neumann, 1931. El P. Estanislao Chevalier, S.I. *Razón y Fe* 94, 264-266.

E. de la Villemarqué, 1933. Le Père Stanislas Chevalier (1852-1930). Un demi siècle de vie scientifique 1880-1930. *Bull. de l'Université l'Aurora* 26, 39-53.

J. de Lapparent, 1935-36. Le Père Stanislas Chevalier (1852-1930). *Lettres de Jersey* 46, 323-379.

Ricardo Cirera (1864-1932) Born in Os de Balaguer (Lerida) Spain, he entered the Society of Jesus in 1880. In 1887 he traveled to the Philippines and was named subdirector in charge of the magnetic section of Manila Observatory. In 1889 he completed the installation of the magnetic instruments and in 1892 finished the magnetic map of the Philippines Islands and the coasts of China and Vietnam. He returned to Spain in 1894 to study theology. In 1899 Cirera was put in charge of the foundation of the new Observatorio del Ebro in Roquetas, Tarragona. From 1900 to 1903 he visited the principle observatories of Europe and he stayed during 10 months at Stonyhurst, England. After these visits he proposed that the new observatory should be dedicated to the study of solar activity and its relation with terrestrial magnetic and electric phenomena. In 1914 he founded *Ibérica*, a Spanish journal dedicated to scientific and technological reviews, which he directed until 1919. In 1921 he was appointed administrator of the Jesuit mission in Bombay. In 1914 he was awarded the Gran Cruz de la Orden de Alfonso XII.

El magnetismo terrestre en Filipinas. Observatorio de Manila, 1893.

Observations magnétiques de l'Observatoire de l'Ebre à l'occasion de l'eclipse de soleil du 30 août 1905. *Comptes Rendus Académie des Sciences* 141, 285, 1905.

(with M. Balcells) Remarques sur le rapport entre l'activité solaire et les perturbations magnetiques. *Comptes Rendus Académie des Sciences* 145, 862-864, 1907.

(with J. Ubach) Observations sur le passage de la comète de Halley a l'Observatoire de l'Ebre (Espagne). *Comptes Rendus Académie des Sciences* 150, 1494-1495, 1910.

Anonymous, 1932. El R. P. Ricardo Cirera (1864-1932). *Razón y Fe* 100, 92-96.

M. Sanchez Navarro-Neumann, 1932. Le R. P. Richard Cirera S.J. *Rev. des Questions Scientifiques* 4 serie, 22, 431-436.

L. Rodés, 1932. Reverend Ricardo Cirera S.J. *Terrestrial Magnetism and Atmospheric Electricity* 37, 481-482.

F.A. Linari, 1932. R. P. Ricardo Cirera fundador y primer director de Ibérica. *Ibérica* 38, 66-70.

Elie Colin (1852-1923). Born in Grauhlet (Tarn) France, he entered the Society of Jesus in 1870. He was a science teacher at the colleges of Bordeaux, Sarlat and Monaco. In 1887, Colin was appointed Director of the projected observatory in Tananarive, Madagascar, and spent several months in preparation for this post at Stonyhurst Observatory, and the Observatoire du Bureau des Longitudes, the Bureau Central Météorologique de France and the Observatory of Paris. In 1888 he arrived at Madagascar with the instruments for the observatory, some of them donated by these institutions. The observatory was completed in 1889. From that year on Colin carried out astronomical, meteorological and magnetic observations. In 1893 he began observations and studies of cyclones which crossed Madagascar. In 1893 Colin returned to France for health reasons. On the following years the observatory was destroyed in the French-Magadascan war. Colin returned to Tananarive in 1896 and rebuilt the observatory. From 1896 he participated in several geodetic and cartographic surveys made by the French administration. He measured the base line and made the triangulations from Tananarive to the coast and in the Imerina region. Colin together with D. Roblet received in 1898 the prize Herbert-Fournet of the Société de Géographie de Paris, in 1903 the prize Gay of the Académie des Sciences and the Golden Medal of the prize Louis Bourbannaud for his geodetic work. He was a corresponding member of the Académie des Science and received the Légion d'Honneur in 1921.

Observations météorologique faites à Tananarive. vol I - XXVIII, Imp. de la Mission Catholique, Tananarive, 1891-1918.

Levées géodésique, astronomique et magnétique à Madagascar. *Comptes Rendus Académie des Sciences* 127, 708-711, 1898.

Le régime des vents à Madagascar. *Bull. Académie Malgache* 9, 173-177, 1911.

Les Cyclones à Madagascar. Imp. Offic. de Tananarive, 1913.

Le magnétism terrestre à Madagascar. *Bull. Economique de Madagascar*, 18. 167-174. 1921.

P. de la Devèze, 1923. *Le père Elie Colin. Directeur de l'Observatoire de Tannarive (1852-1923)*. With an appendix by C. Poisson, L'Oeuvre scientifique du R.P. Colin. Imp. Tardy Pigelet, Burges.

Aloysius L. Cortie (1859-1925). Born in London, England, he studied at Stonyhurst College and entered the Society in 1878. He studied science at the

University of London. He was a teacher of mathematics and sciences from 1885 to 1889 and from 1895 to 1925 at Stonyhurst. In 1895 he began working at Stonyhurst Observatory and became Director in 1919. He took part in several scientific expeditions to observe solar eclipses in Spain in 1905, in the island of Tonga in 1911 and in the north of Sweden in 1914. His scientific work covered the relationship between solar activity and terrestrial magnetism, stellar spectrography and solar eclipses. In particular he studied the relationship between sunspots and magnetic storms, establishing that the latter are caused by the jets of influence from sunspots. His work on stellar spectroscopy was centered on the study of spectra of novas. He served in the Council of the Royal Astronomical Society in 1904-1905, 1908-1909, 1912-1916 and 1918.

Solar prominences and terrestrial magnetism. *Astrophysical Jour.* 18, 287-293, 1903.

Report on the total solar eclipse of 1911, April 28. *Proceed. Royal Society, London.* ser A, 87, 293-301, 1912.

Sun spots and terrestrial magnetic phenomena, 1898-1911. *Monthly Notices Roy. Astr. Soc.* 73, 52-60, 148-155, 431-436, 1912.

The spectrum of Nova Cygni III. *Monthly Notices Roy. Astr. Soc.* 81, 57-61, 1920.

Solar and terrestrial magnetic phenomena 1913-1921. *Monthly Notices Roy. Astr. Soc.* 83, 204-215, 1923.

Anonymous, 1925. Father Aloysius Cortie. *Letters and Notices* 60, 210-218.
H. H. Turner, 1926. The Rev. Aloysius Laurence Cortie, S.J. *Monthly Notices Roy. Astr. Soc.* 86, 175-177.

Marc Dechevrens (1845-1923). Born in Chène-Bourg (Genéve), Switzerland, he studied at the Seminary of Fribourg and entered the Society of Jesus in 1862. From 1869 to 1872 he was a teacher of physics in the colleges of Vannes and Vaugirard. In 1873 he was appointed to Zikawei Observatory, Shanghai (China) and spent two months of preparation with Perry in Stonyhurst. In 1876 Dechevrens became formally Director of Zikawei. He was the first to give the observatory its true scientific character. His main work was in meteorology, centered on the study of typhoons. Between 1879 and 1885 he published reports on 62 typhoons. In 1883 Dechevrens was asked to establish a service of warnings to sailors about storms. He established a network of meteorological stations at the Maritime Customs offices on the coast of China which sent data to Zikawei and a system of visual signs at Shanghai harbor. He studied the vertical component of winds and presented one of the first models for the generation of cyclones as formed by atmospheric turbulence with vortices at a variable height. He pointed out the importance of the observation of high altitude clouds, such as cirrus and fracto-cumulus, and their relationship with the changes in direction of cyclones. He observed also the vertical changes in temperature and humidity. In 1887 for health reasons Dechevrens returned to Europe and in 1893 founded an observatory in the Island of Jersey, where he continued his research on the vertical component of winds. He published 153 scientific articles. He was elected in 1887 member of the Pontificia Accademia dei Nuovi Lincei.

Le typhon du 31 Julliet 1879. Imp. Mission Catholique, Zikawei, 50 pp., 1879.

L'inclinaison des vents sur l'horizon. Imp. Mission Catholique Zikawei, , 32pp. 1881.

Courant verticaux dans le cyclones, Imp. Mission Catholique, Zikawei, 35pp. 1887.

Les tourbillons atmosphérique, leur formation, leur constitution.Variation de température et d'humidité. *Atti della Accademia Pontificia* IV, Rome, 1888.

H. Gauthier, 1924. Un maitre en physique du globe (M. Dechevrens). *Etudes* 178, 272-291.

L. Palazzo, 1924. Marco Dechevrens. *Atti della Accademia dei Nuovi Lincei* 77, 6, 175-158.

Charles E. Deppermann (1889-1957). Born in New York, he entered the Society of Jesus in 1910. From 1923 to 1925 he studied physics at Johns Hopkins University, Baltimore, where he obtained his doctoral degree. In 1925-1926 he worked on astronomy at Lick Observatory, California. In 1926 Deppermann arrived at Manila Observatory where he took charge of the astronomy section and the time service and participated in a program of measurements of longitude. From 1929 to 1931 he studied variable stars. In 1931 he took charge of the meteorological section and worked on atmospheric electricity and the ionosphere. In 1932 Deppermann traveled to the U.S. Weather Bureau in Washington and to Norway where he visited the Geophysical Institute in Bergen and the Meteorological Service in Oslo. In Norway he learned the new theories of weather fronts proposed by Bjerknes and Petersen. On his return to Manila he applied, for the first time, the theories of frontgenesis and air mass analysis to the storms in Philippines, and published his most original work which he continued until the occupation of the observatory by the Japanese. In 1944 he was imprisoned by the Japanese at

Los Baños concentration camp. After World War II, in 1945, he began plans for the reconstruction of the observatory and spent in 1947 some time at Saint Louis University studying seismology with Macelwane. In 1948 he became director of the new Manila Observatory dedicated now to seismology, magnetism, ionospheric studies and solar physics. Deppermann was one of the most renowned Jesuit meteorologists.

Outline of Philippine frontology. Bureau of Printing, Manila, 27 pp.,1936.

Typhoons originating in the China Sea. Bureau of Printing, Manila 51 pp., 1938.

Some characteristics of Philippine typhoons. Bureau of Printing, Manila, 143 pp., 1939.

Upper air circulation (1-6 km) over the Philippines and adjacent regions. Bureau of Printing, Manila, 64 pp. 1940.

Notes on the origin and structure of Philippines typhoons. Bull. Am. Meteor. Soc. 28, 399-404, 1947.

J. J. Hennesseys, 1957. Charles E. Deppermann S.J.: Philippine Scientist. *Philippine Studies* 5, 311-335.

J. J. Hennesseys, 1958. Father Charles E. Deppermann. *Woodstock Letters* 87, 123-134

Pierre M. Descotes (1877-1964). Born in St. Genis sur Guiers (Savoy), France, he joined the Society of Jesus in 1895 in Granada, Spain. In 1912 he arrived at La Paz, Bolivia, after spending some time with Navarro-Neumann at Cartuja Observatory, Granada. He was appointed Director of Observatorio San Calixto and science teacher at Colegio de San Calixto. He installed new seismographs built by himself, similar to those of Cartuja. His careful and exact analysis of seismograms made the seismographic station one of the most reliable of South America. In 1931 Descotes discovered that there was an error in the longitudes determined in 1879 by the Green and Davies Commission and in consequence all geographic coordinates in South America were displaced 400 meter to the west. With the installation in 1929 of three components Galitzin-Wilip electromagnetic seismographs, the observatory was equipped with the most modern instruments. His scientific work was recognized with the award of the Orden del Condor de los Andes by the Government of Bolivia and the Légion d'Honneur by France.

Una rectificación de la longitud de La Paz y de las demas longitudes del continente Sudamericano. *Anales de la Soc. Cientifica de Bolivia, y El Diario, (15-1-1931) 4, La Paz,* 1931.

Courbes isodiastématiques. Nouvelle méthode pour les calculer. Impr. Daupeley Gouverneur. 1933.

With R. Cabré. Historia sísmica de Bolivia. *Boletín del Instituto. Boliviano del Petróleo* 5. 1965.

Anonymous, 1964. P. Pedro Descotes, R.I.P. Diaspora. Noticias de la Viceprovincia Boliviana (November) 16-20.

F. Gun Bayer, 1971. *Erinnerungen an Chile.* (Father Descotes). Canisius Haus, Vienna. 15 pp.

P. Menacho, 1981. Los jesuitas en Bolivia 1575-1767, 1881-1981. Cochabamba.

Antonio Due Rojo (1998-1975). Born in Málaga, Spain, he entered the Society of Jesus in 1915. He was professor of physics and chemistry in the Faculty of Philosophy of Granada from 1926 to 1929 and worked at Cartuja Observatory as an assistant to Navarro Neumann until 1931. In 1940 he was appointed Director of the observatory until 1966. He continued the seismological work of Navarro Neumann about the earthquakes in Spain and published periodically reports on the seismicity of Spain under the title of "Notas Sismológicas" and annual summaries "Movimientos Sísmicos en España durante el año". He published several popular books on science and on the relationship between science and religion.

Dios y la ciencia. Facultad Teológica de Granada, 1941, 232 pp.

Periodicidad sísmica en la provincia de Granada. *Revista de Geofísica* 1, 362-368. 1941.

El problema de los microsismos. *Revista de Geofísica* 10, 320-331. 1951.

Die südspanischen Erdbeben von März bis August 1951. *Neues Jahrbuch der Geol. und Paläont.* 1, 4-5. 1952.

Anonymous, 1975. El Padre Due, S.J.. *Ideal* (8 Enero 1975), Granada

Michael A. Esch (1869-1938). Born in Eupen (Rheinland), Germany, he entered the Society of Jesus in 1891. He participated in the installation of the observatory of Ignatiuskolleg in Valkenburg. From 1896 to 1898 he traveled to Georgetown and collaborated with Hagen in the observation of variable stars. In 1898 he was appointed Director of the Observatory of Valkenburg. In 1904 he spent a year with Fenyi in Kalocsa Observatory and in 1905 traveled to Spain to observe the solar eclipse of 30 of August and studied the solar crown. From 1905 to 1909 he was in charge of the Observatory of Valkenburg. In 1909 Esch studied astronomy at the University of Vienna where he obtained his doctoral degree in 1916. From 1916 to 1918 he taught astronomy at the University of Innsbruck. In 1918, Esch returned to Valkenburg where he was again director of the observatory until his death. During this time he collaborated with Hagen, now director of the Vatican Observatory, in the observation of variable stars and in the publication of Hagens' *Atlas*. He was named Astronomo Onorario of the Vatican Observatory.

M. Esch, 1930-1937. *Beobachtungen veränderlichen Sterne*. Sternwarte des Ignatiuskollegs, Valkenburg, vol. 1-7.

J. Stein, 1938. P. Michael August Esch, S.J. 1869-1938. *Astronomische Nachrichten* 266, 47.

H. Schmitz, 1939-1941. P. Michael Esch, gestorben am 28 April 1938 zu Valkenburg. *Mitteilungen aus dem deutschen Provinzen der Gesellschaft Jesu* 15, 319-329.

Federico Faura (1840-1897) Born in Artés (Barcelona), Spain he entered the Society of Jesus in 1859. As a young Jesuit he was sent to Manila in 1866 as a teacher of physics and mathematics at the Ateneo and took charge of the observatory which had been created in 1865. In 1868 he made his first scientific expedition to observe a solar eclipse in the Celebes Islands. In 1871 he returned to Spain to complete his theological studies and in 1877 he visited the observatories of the Collegio Romano and Stonyhurst where he worked with Secchi and Perry. In 1878 Faura returned to Manila and became again director of the observatory. He began studies of typhoons stating, some hypotheses on their nature and precursory phenomena and made the first forecasts of these storms in 1879. He studied the earthquakes of 1880 and their damage and renovated the seismological instrumentation of the observatory. In 1882 he made the design of a barometer for the detection of typhoon occurrences. In 1884 Faura was named Director of the official Servicio Meteorológico Filipino, created by the Spanish Colonial Government and established a network of meteorological stations in the archipelago. In 1888

Faura traveled to Barcelona to presents the work of the observatory in the Exposición Universal and returned to Manila in 1894.

Resumen de los ciclones de 1879. Manila, 1880.
Observaciones sismométricas de los terremotos del mes de julio de 1880. La Oceanía, Manila, 1880.

Señales precursoras de temporal en el Archipiélago Filipino. Manila ,14 pp., 1882.

El barómetro aneroide aplicado a la previsión del tiempo en el Archipiélago Filipino. C. Valdezco, Manila, 13 pp., 1886.

B. F. Doucette, 1929. Father Frederic Faura. *Bull. Am. Ass. Jesuit Scientists* 7, 4, 24-25.

L. Rodés, 1930. Rdo. P. Federico Faura, creador del Observatorio de Manila. *Memorias Real Academia de Ciencias y Artes de Barcelona*, 22, No. 5, 71-77.

E. Fontseré, 1930. Rdo. P. Federico Faura. *Memorias Real Academia de Ciencias y Artes de Barcelona,* 22, No. 5, 78-84.

E. Alcobé, 1930. Rdo. P. Federico Faura, creador del Observatorio de Manila. *Memorias Real Academia de Ciencias y Artes de Barcelona*, 22, No. 5, 85-92.

A. Hidalgo, 1974. *El P. Federico Faura S.J. y el Observatorio de Manila.* Manila Observatory 72 pp.

J. O. Cardús, 1997. El P. Frederic Faura S. J. meteoròleg a Filipines. *III Jornades de Meteorologia Eduard Fontseré, Asoc. Catalana de Meteorologia.* Barcelona, 63-67.

Gyula Munkai Fényi (1845-1927) Born in Sopron, Hungary, he entered the Society of Jesus in 1864. He began working at Haynald Observatory in Kalocsa, in 1871 when he was a teacher of mathematics and physics at the Jesuit College (Stephaneum Gimnasium). In 1885, after finishing his studies of theology in Innsbruck, where he also attended lectures in physics and mathematics, Fényi was appointed Director of the observatory. During the 28 years he occupied this post the observatory experienced its most active time. He continued as Director Emeritus until his death in 1927. He worked on solar physics and during 32 years he made more than 40000 observations of sunspots, solar prominences and faculae, studying their characteristics and spectra. He published about 6000 of these observations. He also proposed theories to explain these solar phenomena. According to him solar prominences and sunspots have the same origin in the motion of the currents of incandescent material of the solar crown. He also applied this explanation to the characteristics of solar spectra. Fényi studied the relationship between solar activity and magnetic storms, a controversial subject at that time. His observations and interpretations were published in the 154 articles published in English (Astrophysical Journal), French (Comptes Rendus Academie de Science, Paris) in German (Meteorologische Zeitschrift, and Astronomische Nachrichten), in Italian (Memorie Societa degli Spectroscopisti Italiani) and in Hungarian in the publications of the observatory and local journals. In 1902 he was named member of the Accademia Pontificia dei Nuovi Lincei and in 1916 corresponding member of the Academy of Sciences in Budapest.

Protuberances solaires extraordinaires observées à l'Observatoire Haynald. *Comptes Rendus Acad. des Sciences*, Paris 20,63-68, 1891

Ueber einen neuen Gesichtspunkt und neue Erklärungen der Erscheinungen auf der Sonne. *Astronomische Nachrichten* 140, 290-302, 1896.

A new point of view for regarding Solar phenomena and a new explanation of the appearances on the surface of the Sun. *Astrophysical Jour.* 4, 18-37, 1896.

Zur Theorie des Gewitterregistrator. *Meteorologische Zeitschrift* 36, 536-537, 1901.

Protuberances beobachtet in den Jahren: *Publicationen des Haynald Observatoriums, Kalocsa* 1887, 6(1892) 78 pp., 1888-1890, 8(1902) 132 pp., 1891-1892, 10(1911) 144 pp., 1893-1895, 12(1923) 105 pp., 1896-1903, 13(1923) 126 pp., 1904-1910, 14(1924) 127 pp., 1911-1917, 15(1924) 106 pp.

T. Angehrn, 1928. Todesanzeige von P. Julius Fényi S.J. *Astronomische Nachrichten* 232, 127-128.

J. Hagen, 1928. Necrologio del P. Fényi. *Atti della Pontificia Accademia delle Scienze* 81, 122.

I. Mojzes. 1986. *A. Kalocsai Haynald Observatórium története.* Orzágos Müszaki I. K. K., Budapest, 114-135.

T. Horvath, 1995. Hundred fiftieth Anniversary of the Hungarian Jesuit scientist, Gyula Fényi's birthday (1845-1927). (manuscript) 3 pp.

Louis Froc (1859-1932). Born in Brest, France, he entered the Society of Jesus in 1875. He arrived in China in 1883 where he began working with Dechevrens at Zikawei Observatory. In 1887 he returned to France where studied and obtained the licentiate in sciences at the University of Paris. He became director of Zikawei in 1896. Froc developed a system of visual signs for the harbor of Shanghai consisting of flags and other signs which was used to warn mariners of the changes in the weather and the approach of severe storms. Accepted by the Chinese Maritime Customs the system was used in most harbors of China from 1898. For his ability in forecasting typhoons, Froc was known among sailors as the "Père des Typhons". He studied the path of 620 typhoons in the coast of China from 1893 to 1918. He participated in the division of China into four time zones. In 1899 Froc was called by the French General Governor of Indochina to organized the network of meteorological stations. In 1931 he received the Croix de la Légion d'Honneur.

L'atmosphère en Extrême-Orient. *Annales Hidrographiques*, 1900. 180pp.

Signaux aux Marins. Zikawei, 1909, 23pp.

La pluie en Chine, 1900-1910. Onze années d'observation. Zikawei, 1912, 63pp.

Atlas of the tracks of 620 typhoons (1893-1918). Zikawei, 1920.

Code de Zi-ka-wei . Historique. *Conference des Directeurs des Services Meteor.d'Extreme-Orient, Hong-Kong*, 1930.

H. Gauthier, 1932. Au service de tous et de chacun. Le père Louis Froc (1859-1932). *Etudes* 213, 373-286.

Anonymous, 1933. El P. Luis Froc. *Ibérica* 963, 90-91.

J. Dorronsoro, 1933. La obra de un meteorólogo. El R. P. Luis Froc, S. J. *Ibérica* 966, 136-138.

Telsos, 1933. El padre Luis Froc S. J. *Estudios* 22, 279-295

Ernesto Gherzi (1886-1976). Born in Sanremo (Imperia), Italy, he entered the Society of Jesus in 1903 in the French novitiate with the idea of going to the China mission. In 1910 he arrived in China where he was a professor of physics in the Jesuit University of Aurora. After his studies of theology in France he returned to China in 1920 and was appointed to Zikawei Observatory in charge of the sections of seismology and meteorology. In meteorology Gherzi studied the nature of typhoons, the climatology of China and tried to relate weather forecasts with changes in the ionospheric layers. In seismology he was a pioneer in assigning the origin of microseisms to the oscillation of atmospheric pressure in storms. He continued this research during many years. In 1949 after the Communist Government confiscated Zikawei Observatory, he was asked by the Portuguese Colonial Government to organize the Meteorological Service of Macao. In 1955 after a brief stay at St. Louis University, Gherzi went to the Observatory of Jean de Brebeuf in Montreal, Canada, where he continued his work on atmospheric electricity, radio propagation, solar radiation and seismology. He was a member of the Pontifical Academy of Science and of the academies of Portugal and New York

Remarks on some new researches on the nature of tropical cyclones. *Gerlands Beiträge zur Geophysik* 53, 316-322, 1938.

Microsisms associated with storms. *Gerlands Beiträge zur Geophysik* 25, 145-147, 1930.

Climatological atlas of East Asia. Shanghai, 1944, 175 pp.

Ionospheric reflections and weather forecasting for eastern China. *Bull. Amer. Meteor. Soc.* 27, 114-116, 1946.

The meteorology of China. *Serviço Meteorológico Provincia de Macau, Imp. Nacional de Macau.* vol. 1, 388 pp. vol. 2, tables y maps, 1951.

E. Gherzi, 1970?. Un Jesuite meteorologiste en Chine. Le Rev. Pere Ernesto Gherzi S.J. par lui-meme. (manuscript, ASJCF-St. Jerome) 84 pp.

M. Buist, 1974. Ernesto Gherzi, S. J., 1886-1973, Memorial. *Bull. Seis. Soc. Am.* 64, 507.

G.L. Belt, 1974. Father Gherzi S.J., 1886-1973. An appreciation. *Jour. Hong Kong Branch Royal Asiatic Soc.* 14, 85-91.

P. J. H., 1974. Ernesto Gherzi, 1886-1973. *Bull. American Meteor. Soc.*, 55, 344

V. Cantú, 1974. Note commemorative. Padre Ernesto Gherzi, S. J. *Rivista di Meteorologia Aeronautica* 34, 357-365.

J. F. Gauvin, 1994. *Ernesto Gherzi et la physique du globe à Montréal, 1955-1973*. M.A. Thesis, Université de Montréal, 148 pp.

Edmund Antoine Goetz (1865-1933). Born in Ste. Marie-aux Mines (Alsace), France he moved to England in 1880 and entered the Society of Jesus in 1883. Between 1887 and 1889 he was a teacher of physics and mathematics in Aidan's College, Grahamstown and in the Island of Jersey. En 1894 he studied sciences in the University of Paris and obtained his Licentiate in Science in 1900 with certificates of astronomy, geography, physics, mineralogy and geology. He prepared for his work as the first director of the new Observatory of Bulawayo, Zimbabwe-Rhodesia, spending some time at Stonyhurst Observatory and eighteen months with Hagen at Georgetown Observatory, working on the observation of variable stars. Goetz arrived at Bulawayo in 1903 and began his work in the observatory with regular astronomical, meteorological and magnetic observations. He continued his collaboration with Hagen on variable stars and carried out studies of climatology of the region. Between 1909 and 1914 he carried out magnetic field surveys with two lines in Zimbabwe and Congo. He was an active member of the Rhodesian Scientific Association and a fellow of the Royal Astronomical Society.

Notes on the climate of Bulawayo. *Zambesi Mission Record*, 2, 390-392, 587-589. 1904.

Notes on the climate of Southern Rhodesia. *Zambesi Mission Record* 4, 570-573, 1913; 5, 24-28, 1914.

The Rainfall of Rhodesia. *Proc. Rhodesia Scientific Association*. vol. 8, part 3. 128 pp. 1909.

Magnetic Observations in Rhodesia. *Transactions Royal Society of South Africa*, 8, 4, 297-302, 1920.

Anonymous, 1933. Obituary, Fr. Edmund Goetz. *Zambesi Mission Record* 9, 438-440.

Anonymous, 1934. The Reverend Father Edmund Goetz, S.J. *Monthly Notices Roy. Astron. Soc.* 95, 323-324.

V. L. Bosazza, 1977. Father Goetz and the Goetz Observatory. *Rhodesiana*, 36, 10-11.

Mariano Gutierrez-Lanza (1865-1943). Born in Pardave (Leon), Spain, he joined the Society of Jesus in 1883. He began teaching physics and chemistry at Colegio de Belen in Havana, Cuba in 1891. In 1900 he studied meteorology and astronomy at Georgetown University, Washington. In 1902, he became Subdirector of Belén Observatory and in 1920 was appointed Director of the

Observatory of Montserrat at Cienfuegos. Gutierrez-Lanza returned to Belén Observatory in 1924 and in 1928 became its Director. He published three books about the climatology of Cuba and the tropical hurricanes in the Caribbean region. In the observatories of Monserrat and Belén he worked on the observation and forecast of hurricanes. In 1935 he was awarded by the President of Cuba with the Cruz de la Orden Nacional del Mérito and the same year by the Government of Spain with the Orden de la República Española and in 1943 with the Cruz of Isabel la Católica.

Climatología de Cuba, Habana, 1913.

Conferencias de seismología. Lloredo y Ca., Habana, 1917.

Génesis y evolución del huracán del 20 de Octubre de 1926 y catálogo de ciclones en la isla de Cuba 1865-1926. A. Dorrbecker, Habana, 1927.
Los ciclones de las Antillas, Habana, 1935.

J. García Moll , 1945. P. Mariano Gutierrez-Lanza, S. J. (1865-1943). *Ibérica,* 19, 461-463.

Johann Georg Hagen (1847-1930). Born in Bregenz, Austria, he entered the Society of Jesus in 1863. In 1870-1872 he studied science at the universities of Münster and Bonn. In 1880 he traveled to America where he was a teacher of mathematics at Sacred Heart College in Prairie du Chien, Wisconsin, where he installed a small astronomical observatory and began the study of variable stars. In 1888 he was appointed director of the Observatory of Georgetown University, Washington. There he continued his studies of variable stars and published the three first volumes of his *Atlas*. In 1906 Hagen was appointed director of the Specola Vaticana by Pope Pius X, the first Jesuit who occupied this post. He gave a great impetus to the observatory building new observation domes and installed new astronomical equipment. He gave the observatory its definite astronomical character and suppressed the sections of meteorology and magnetism. He continued the work in the international program of the Carte du Ciel, advancing the part assigned to the observatory. He completed and published 5 more volumes of his Atlas of variable stars. From 1922 to 1927 Hagen published a preliminary catalogue of the luminosity of nebulae and began the study of dark nebulae. His last research project was the study of the of the earth's rotation using Atwood's machine, Foucault's pendulum and an instrument of his own design, consisting of a torsion balance with moveable masses. He published a treatise on higher mathematics in four volumes and 158 scientific articles in astronomical and astrophysical journals. He was member of the Pontifical Academy of Sciences and other scientific societies, received doctoral degrees honoris causa from the universities of Bonn (1926) and

Münster (1927) and the golden medal of astronomy from the Accademia delle Scienze (1926).

Synopsis Höherer Mathematik, (4 vol.).1. Arithmetische und algebraichen Analyse. 2. Geometrie der algebraichen Gebilde. 3. Differential und Integralrechnung; 4. Metrische Differentialgeometrie der Ebene und des Raumes. F. L. Dames Steglitz, Berlin. 1891-1931

Atlas Stellarum Variabilium. (7 vol.), F.L. Dames Steglitz, Berlin. 1905, 1906-1908.

La rotation de la terre, ses preuves mechaniques anciènne et nouvelles. Specola Vaticana, Roma, 1911.

Die veränderlichen Sterne, geschichtlich-techniser Teil. Herder, Freiburg, 1921.

J. Stein, 1930. El P. Juan Jorge Hagen S.J. *Ibérica* 34, 296-300.

J. Stein, 1931. P. Giovanni Giorgio Hagen S.J. *Atti Pont. Acad. delle Scienze, Nuovo Lincei* 84, II.

H. Dopp, 1931. Johann Georg Hagen S.J., 1847- 1930. *Rev. des Questions Scientifiques* 99, 5-37.

A.C.D.C., 1931. Johann Georg Hagen. Monthly Notices Royal Astr. Soc. 91, 337-338.

S. Maffeo, 2001. *La Specola Vaticana. Nove Papi, una missione*. Publicazioni della Specola Vaticana, Vatican City, 67-70, 72-89, 305-308.

Gustavo Heredia (1869-1926) Born near Mexico City, Mexico, he entered the Society of Jesus in 1887. He studied theology in Oña, Spain and in 1903 spent a year with Sidgreaves in Stonyhurst Observatory. From 1905 he was a teacher of science and director of the observatory at the Colegio del Sagrado Corazón de Jesús in Puebla and from 1908 a science teacher at the Seminario Palafoxiano, also at Puebla until 1914 when in the Mexican revolution both institutions were closed. He carried out astronomical observations and studies of spectra of stars. He designed the seismographs that were installed in Puebla from 1877 to 1914 and published the seismological observations. In 1904 he was elected a fellow of the Royal Astronomical Society of London.

La rayas de emisión en el espectro de Lirae durante el periodo de mínima principal. *Memorias Sociedad Científica "Antonio Alzate"* XXIII, 5-8. 1905.

Oposición del planeta Marte en el mes de Julio de 1907. *Memorias Sociedad Científica "Antonio Alzate"* XXVI 14-15. 1907.

Catálogo de los temblores registrados en el Observatorio del Colegio del Sagrado Corazón de Puebla desde 1877 hasta 1909. *La Espiga de Oro* II, 24, 40, 72, 88, 120. 1910.

J. B. Iguiñiz, 1945. *Bibliografía de los escritores de la Provincia Mexicana de la Compañía de Jesús*. Buena Prensa, Mexico, 296-298.

Francis J Heyden (1907-1991). Born in Buffalo, New York, he entered the Society of Jesus in 1924. In 1945, he finished his doctoral studies at Harvard University, where he was a teaching fellow in astronomy from 1942 to 1944. From 1950 to 1972 Heyden was Director of Georgetown Observatory. He was also Professor at Georgetown University where he taught courses in descriptive astronomy, stellar statistics and galactic structure. His principal research was devoted to the study of planetary and solar spectra and he built several large-scale spectrographs. Between 1947 and 1963 he took part in seven expeditions to different parts of the world to observe solar eclipses. In 1972 after Georgetown Observatory was closed, Heyden went to Manila Observatory, Philippines where he continued working on solar spectroscopy and taught at the Ateneo de Manila. He was fellow of the Royal Astronomical Society and member of the International Astronomical Union.

The complete astronomer. Ateneo de Manila, Quezon City, 253 pp. 1975
M. F. McCarthy, 1992. Francis J. Heyden S.J. (1907-1991). *Quart. J. Roy. Astr. Soc.* 33, 265-267.

Richard E. Ingram (1916-1967). Born in Belfast, Ulster, he studied at the Belvedere College and joined the Society of Jesus in 1933. He got the degrees of B.Sc. and M.Sc. in mathematics at the University College in Dublin. From 1934 he was in charge of the Seismological Observatory of Rathfarnham Castle, Ireland. In 1946 he studied mathematics at Johns Hopkins University, Baltimore, and obtained his Ph.D. in 1948. In 1949 he spent a year at California Institute of Technology, Pasadena studying mathematics and seismology. In 1949 Ingram returned to Rathfarnham Castle where he was in charge of the seismological station and taught mathematics at the University College of Dublin where in 1966 he became Associate Professor of Mathematics. In 1961-1962 he was professor of mathematics at Georgetown University, Washington and worked in seismology at the U.S. Coast and Geodetic Survey and at the Dominion Observatory in Canada. He published works in mathematics and theoretical aspects of seismology. Ingram collaborated in the edition of the mathematical works of W.R Hamilton and was a member of the council of the Royal Irish Academy.

Character of symmetric group. Ph.D. Thesis. Johns Hopkins University, Baltimore.

Phase change of pP y PP on reflection at a free surface. *Bull. Seis. Soc. Am.* 46, 203-214.

An integral solution of the electromagnetic seismograph equation. *Bull. Seis. Soc. Am.* 50, 461-466.

Focal mechanism of double couples without moment. *Bull. Seis. Soc. Am.* 53, 817-820.

T. Murphy, 1968, The Rev. R. E. Ingram. *Quart. Jour. Roy. Astr. Soc.* 9, 226-227.

Anonymous, 1968. Reverend Professor Richard E. Ingram, S.J., M.Sc., Ph. D. *Report of the President of the University College, Dublin*. 98-99.

Anonymous, 1968. Father Richard Ingram, 1916-1967. *Irish Province News* 47-53.

J. R. Timoney, 1968. Reverend R. E. Ingram- A tribute. *Irish Province News*, 53-54.

Eugene Lafont (1837-1908). Born in Mons, Belgium, he studied at St Barbara College in Ghent and entered the Society of Jesus in 1854. He studied philosophy and sciences in Namur and taught science at the Jesuit colleges of Gand, Liege and Anvers. In 1865 he traveled to India and the following year began teaching natural sciences at St. Xavier's College, Calcutta. He was an excellent professor and besides his classes at the college he began in 1868 a series of public lectures on scientific topics which he held for 19 years. These lectures were very popular and attracted large audiences. In 1867 Lafont began meteorological observations in the college. In 1870 he installed a Secchi Meteorograph and began publishing a meteorological bulletin In 1874 P. Tacchini, director of the Observatory of Palermo, encouraged Lafont to create in St. Xavier's College an astronomical observatory dedicated mainly to solar observations. The Observatory was inaugurated in 1875. In 1877 he was named

Fellow of Calcutta University and in 1904 President of the faculty of Arts. Lafont was twice rector of St. Xavier's College from 1871 to 1877 and from 1901 to 1904. He was a strong promoter of scientific education in India and founder with Mahendralar Sarkar of the Indian Association for the Cultivation of Science. Lafont received many awards, in particular, the Companion of Indian Empire and the Medal of Indian Star. In 1898 King Leopold of Belgium named him Chevalier de l'Ordre de Léopold.

Anonymous, 1908. Rev. Father Eugene Lafont, S.J. *Nature* 78, 35

A.K. Biswas, 1969. *Science in India*. (Rev. Father Lafont of St. Xavier's College) K. L. Muhopadhyay, Calcutta, 67-89.

A. Verstraeten, 1991. A physicist and astronomer, Fr. Eugene Lafont S. J. (1837-1908) In V. L. Sundaran (ed.) *Jesuit profiles. Some eminent Jesuits of South Asia*. Gujarat Sahita Prakash, Gujarat, 217-222.

A. Verstraeten, 1991. Excellence, Eugene Lafont (1837-1908), Alphonse de Penaranda (1834-1896). *Jivan*, February, 13-14.

Pierre Lejay (1898-1958). Born in La Seyne, near Toulons, France, he joined the Society of Jesus in 1915. In 1921 he obtained the Licenciate in physics and mathematics and in 1926 the doctor's degree from the University of Paris. His doctoral dissertation was on the propagation of electromagnetic waves in the atmosphere. From 1922 to 1926 Lejay worked at the Observatory of Paris and collaborated with the Service Méridien and the Service de l'Heure. In 1926 he traveled to China and in 1930 was appointed director of Zikawei Observatory. In 1933 he participated in the International Program of Longitudes for which Zikawei was selected as one of the three main participant observatories. Lejay introduced in Zikawei research in physical meteorology, in particular, atmospheric water vapor spectrography, solar absorption and ozone contents. In 1938 he installed a ionospheric sounder and introduced improvements in its design. In 1933, he developed a gravity meter, in collaboration with Fernand Holweck, based on an inverted pendulum, with a sensitivity 200 times greater than the conventional instruments and of small size (Holweck-Lejay gravimeter). With this instruments he carried out several gravity surveys in the south of France, Near East, Indochina, Philippines, China and Japan. In 1939 Lejay returned to France where he continued his research in ionosphere and atmospheric electricity. He published more than 130 scientific articles in the fields of atmospheric physics, ionosphere and gravimetry, most in the Comptes Rendus de la Académie de Science and a textbook on gravimetry. In 1945 he was appointed Research Director of the Centre National de la Recherche Scientifique and in 1946 Director of the recently created Bureau

Ionospherique Français and President of the la Societé des Radioélectriciens. He was an active collaborator of the International Union of Geodesy and Geophysics (IUGG) and in 1927 president of the section of gravimetry, president of URSI (Union Radioélectrique Scientifique International) between 1952 and 1957 and of other scientific associations. In 1935 Lejay was elected a corresponding member of the Académie de Sciences and in 1946 a non-resident member. In 1953 he was named officer of the Légion d'Honneur.

Sur les variations de la quantité d'ozone contenue dans l'atmosphère au voisinage de Shanghai. *Comptes Rendus Académie des Sciences*. 205, 307, 1937.

Etude gravimétrique des isles Philippines. Observatoire de Zikawei, Shanghai, 1938

Recherche sur l'ionosphère à Shanghai. *Onde Electrique* 212, 8-35, 1940.
Mensures de pesanteur executé le sud de la France en 1941. L'atraction des couches minces. *Comision National de Géodésie*, 1942.

Devéloppements modernes de la gravimétrie. Gauthiers-Villars, Paris, 1947.

(with M.D. Lepechinsky) Formation de couches ionisée. Influence de la temperature. *Comptes Rendus Académie des Sciences* 232, 2058. 1951

Anonymous, 1946. *Notice sur le travaux scientifique du R. P. Pierre Lejay, directeur de l'Observatoire de Zi Ka Wei, Correspondant de l'Académie des Sciences et du Bureau des Longitudes*. Soc. Gen. d'Imprimmerie et d'Edition, Paris, 6 pp.

F. Russo, 1958. Le Père Pierre Lejay, S.J. (1898-1958). *Etudes* 299, 416.

Anonymous, 1958 R.P. Pierre Lejay (1898-1958) Necrología. *Rev. de Geofisica* 17, 537-539.

M. P. Tardi, 1959. Le R.P. Pierre Lejay. Notice nécrologique. *Rev. des Questions Scientifiques* 130, 1-12.

B. Decaux, 1959. Le R. P. Pierre Lejay, 1898-1958. *Onde Electrique*, 384, 1-4.

H. Milloux, 1961. *Notice sur la vie et les travaux de Pierre Lejay (1898-1958).* Institut de France. Académie des Sciences, Palais de l'Institut, Paris. 7 pp.

Levallois, J.J., 1988. *Mesurer la terre. 300 ans de geodésie Français*. A.F.T., Paris (P. Lejay) pp. 193-194.

L. Pyenson, 1990. Habits of mind: Geophysics at Shanghai and Algiers, 1920-1940. *Historical Studies in the Phys. and Biol. Scien.* 21, 161-196.

Daniel Linehan (1904-1987). Born in Beverly, Massachusetts he entered the Society of Jesus in 1922. He studied physics at Boston College and geology at Harvard University. He was professor of physics and geology in Holy Cross College and Boston College. In 1935 Linehan began working at Weston Observatory and was appointed director from 1955 to 1974; he remained as Director Emeritus until his death. In 1948 he founded the Department of Geophysics at Boston College and was its Director till 1963 and Professor of Geophysics till 1972. Among his early seismological work was the study of the T phase in earthquakes with marine epicenter. His main work was on shallow seismic exploration, specially of the glacial deposits in New England. In 1951 he applied this methodology to archeological findings in the Vatican. Linehan was best known for his scientific expeditions to the Antarctica and the Arctic, between 1954 and 1958. There he measured the thickness of the ice by seismic methods. He participated between 1961 and 1958 in several UNESCO mission to Africa, Asia and South America. He was a prolific writer and a popular lecturer. In 1954 was chairman of the Eastern Section of the Seismological Society of America, from 1970 to 1986 president of the Jesuit Seismological Association and in 1958 received the Distinguished Public Service Award of the U.S. Navy.

Earthquakes in the West Indian Region. *Trans. Am. Geophys. Union*, 229-232, 1940.

Earthquakes of the Northeastern United States and Eastern Canada, 1938, 1939 and 1940. *Bull. Seis. Soc. Am.* 32, 11-17, 1942.
Seismic prospecting in New England. *Trans. Am. Geophys. Union*, 227-278, 1942.

J. F. Devane and J. P. McCaffrey, 1988. Reverend Daniel Linehan (1904-1987). Memorial. *Bull. Seis. Soc. Am.* 78, 1618-1620.

C. C. Bates, T. F. Gaskell and R. B. Rice, 1982. *Geophysics in the affairs of man*. Pergamon Press, New York, 116-117, 183, 205.

John Joseph Lynch (1894-1987). Born in London he traveled with his family to Philadelphia and in 1914 he joined the Society of Jesus in New York. He studied theology at Valkenburg, Holland and between 1923 and 1927 worked during the Summers as an assistant at the Observatory of Oxford University. He studied physics at Fordham University where he obtained the doctoral degree in 1939. From 1928 to its closing in 1977, he was Director of the Seismological Observatory of Fordham University. In 1946, in response to a request of the Government of the Dominican Republic, Lynch carried out a field study of the damage of the earthquake of 4 August. He helped to install there the first seismographic station. In 1950 he accompanied Linehan in his work of archeological seismic exploration in the Vatican. Lynch published several articles on the nature of earthquakes and the propagation of seismic waves. From 1957 to 1970 he was president of the Jesuit Seismological Association and in 1930 chairman of the Eastern Section of the Seismological Society of America.

Geographical distribution of deep focus quakes. *Bull. Seis. Soc. Am.* 26, 195-199. 1936

What is the state of the Earth's core? *Bull. Seis. Soc. Am.* 30, 337-341. 1940.

Dominican earthquake of August 1946. *Bull. Seis. Soc. Am.* 38, 1-17. 1948.

Watching our trembling earth for 50 years. Dodd, Mead and Co. New York, 110 pp. 1970.

E. Norton, 1964. Rev. J. Joseph Lynch, S.J. *News from Fordham University,* 6 pp.

D.M.C., 1965. Bene Merenti Citation, Reverend J. Joseph Lynch, S.J. (manuscript), 3 pp.

James Bernard Macelwane (1883-1956). Born in Port Clinton, Ohio, he entered the Society of Jesus in 1903. He study sciences at Saint Louis University and obtained the M. Sc. in 1912. That year he began teaching physics at the same university and collaborating with the seismographic station. In 1920 he studied at the University of California, Berkeley where he graduated in physics in 1923 with a doctoral thesis on the dispersion of seismic waves under Prof. A. C. Lawson. From 1923 to 1925 he was a professor in the Department of Geology of the University of California. In 1925 he returned to Saint Louis University where he founded the Department of Geophysics, one of the first in North America to offer doctor degrees in geophysics and meteorology. He became also Director of the Seismological Observatory. The same year he organized the Jesuit Seismological Association of which he was president until his death. In 1936 Macelwane published the first textbook of seismology in America. Under his directorship Saint Louis University became a worldwide known center of geophysical research. He published a total of 133 articles in scientific journals and was without doubt the most renowned Jesuit seismologist. Travel times of seismic waves, the constitution of the interior of the earth and the nature of microseisms and their relation to atmospheric storms were a few of the topics of his research papers. Macelwane was in 1928 president of the Seismological Society of America, founder and first president in 1928 of its Eastern Section and in 1953 president of the American Geophysical Union which had awarded him in 1948 the Bowie Medal. In 1944 he was elected member of the American National Academy of Science. The American Geophysical Union and the American Meteorological Society have established in his honor medals awarded each year to young scientists of outstanding ability.

A study in the relation between the periods of elastic waves and distances traveled by them, based on the seismographic records of the California earthquake of January 31, 1922. *Bull. Seis. Soc. Am.* 13, 13-69. 1923.
A new table of observed travel times of earthquake waves for distances between 10 and 180 at one degree intervals, applicable only to earthquakes of shallow focus. *Trans. Am. Geophys. Union* 307-309, 1933.

Introduction to Theoretical Seismology, part I, Geodynamics. J. Wiley, New York, 366 pp. 1936.

Evidence of the interior of the Earth derived from seismic sources. In B. Gutenberg, *Internal Constitution of the Earth*, McGraw-Hill, New York, 219-290. 1939.

Storms and the origin microseisms. *Annales de Geophys*. 2, 281-289. 1946.

When the Earth Quakes. Bruce, Milwaukee, 288 pp. 1947.

V. J. Blum, 1956. Sketch of the life of James Bernard Macelwane, S.J. *Earthquakes Notes* 27, 9-11

E.A. Hodgson, 1956. The contribution of Father Macelwane to the founding of the Eastern Section, Seismological Society of America. *Earthquakes Notes* 27, 11-12.

H.F. Birkenhauer, 1956. Father Macelwane and the Jesuit Seismological Association. *Earthquakes Notes* 27, 12-13.

R. R. Heinrich, 1956. James B. Macelwane, S.J., scholar. *Earthquakes Notes* 27, 13-15.

E. J. Walter, 1956. Reverend James B. Macelwane. *Earthquakes Notes* 27, 15-16

V. T. Allen, 1956. James Bernard Macelwane (S.J.) (1883-1956). *Bull. Am. Ass. Petrol. Geol.* 40, 2038-2039.

V. J. Blum, 1956. James Bernard Macelwane, S. J. *Trans. Am. Geophys. Union* 37, 135-136.

W. V. Stauder and P. Byerly, 1957. Memorial to James Bernard Macelwane (1883-1856). *Proc. Geological Society of America* (Annual Report 1956) 159-164.

P. Byerly and W. V. Stauder, 1958. James B. Macelwane, September 28, 1883-February 1956. *Biographical Memoirs National Academy of Sciences*, 31, 254-281.
C. C. Bates, T. F. Gaskell and R. B. Rice, 1982. *Geophysics in the affairs of man.* Pergamon Press, New York, 9, 35-37, 52, 87, 103, 131.

Daniel J. K. O'Connell (1896-1982). Born in Rugby, England, he entered the Society of Jesus in Ireland in 1913. He studied physics and mathematics in the University of Dublin and obtained his doctorate in science in 1930 from the Irish National University. From 1931 to 1933 he did research at the observatory of Harvard University, Cambridge, Massachusetts, in the subject of variable stars with professor Harlow Shapley. Due to health problems he was sent to Australia and was assigned to Riverview College Observatory in Sidney, where he became director in 1938. There he continued his work on variables stars and got interested in seismology. In 1952 O'Connell was appointed director of the Specola Vaticana and occupied this post till 1970. He continued working on variable and binary stars. He discovered an effect, since known as the "O'Connell Effect" concerning the rotation of the line of the apsides, that is, the major axis of the double star's elliptical orbit. He studied and photographed the phenomenon present at sunsets, known as the green ray. From 1955 to 1961 he was President of the Commission on Variable Stars of the International Astronomical Union. He enjoyed a close relationship with Pope Pius XII. He was member of the Pontifical Academy of Science and its president from 1968 to 1972.

Das Schmidt Teleskop der Vatikanische Sternwarte. *Mitt. Astronomische Gessellschaft Hamburg* 24, 1956.

(with C. Treusch) The green flash and other low sun phenomena. *Ricerche Astronomiche della Specola Vaticana* 4. 1958.

G. V. Coyne and M. F. McCarthy, 1982. *Daniel Joseph Kelly O'Connell, S.J.: In Memoriam.* Vatican Observatory, 4 pp.

G. V. Coyne and M. F. McCarthy, 1982. Daniel Joseph Kelly O'Connell, S.J. *Quart. Jour. Roy. Astr. Soc.* 24, 363-364.

S. Maffeo, 2001. La Specola Vaticana. Nove Papi una missione. Pubblicazioni della Specola Vaticana.

Frederick L. Odenbach (1857-1933). Born in Rochester, New York, he entered the Society of Jesus in 1881. From 1885 he taught science at Canisius College, Buffalo. In 1891 he went to England to study theology. In 1896 he was appointed to John Carroll University, Cleveland, where he began meteorological observations. He studied atmospheric electricity and designed an instrument to detect distant electric discharges in the atmosphere. In 1900 he became interested in seismology and build a horizontal pendulum with Hengler-Zöllner suspension. This was the first seismograph installed by a Jesuit in North America. In 1909, he founded the Jesuit Seismological Service (JSS), a network of sixteen seismographic stations in North America with uniform instruments. Although the JSS did not operate for long, Odenbach interested many Jesuit colleges and universities in seismology. He can be considered as the pioneer of Jesuit seismology in the United States. He was one of the founder members of the Seismological Society of America.

The International Seismological Association. *Bull. Seis. Soc.* Am. 1, 103-106.

A small library on modern seismology. *Bull. Seis. Soc. Am.* 1, 108-109.

J. B. Macelwane, 1933. Memorial to the reverend Frederick L. Odenbach, S.J. *Trans. American Geophysical Union, 14 Annual meeting*, 317-318.

J. G. Doherty, 1933. Frederick L. Odenbach, S.J. *Bull. Am. Ass. Jesuit Scientists* 11, 104-106.

William J. O'Leary (1869-1939). Born in Dublin, Ireland, he entered the Society of Jesus in 1886. He studied mathematics, physics and astronomy at the University of Louvain, Belgium and was a science teacher at several Jesuit colleges in Ireland. In 1909 O'Leary founded an observatory at Mungret College in Limerick with meteorological and seismological instruments. He designed a seismograph consisting of an inverted pendulum with a threefold suspension. In 1915 he founded another observatory in Rathfarnham Castle, Dublin. He built a seismograph with a mass of 1.5 Tn and several types of

clocks, in particular of the free pendulum type. In 1929 O'Leary traveled to Australia and became director of Riverview Observatory. There he worked in the observation of variable stars in collaboration with Bosscha Observatory in Lembang, Java. He designed and built several scientific instruments of different kinds, in particular a blink comparator and a new type of vertical seismograph. His lectures on subjects of astronomy and seismology were very popular. He was a fellow of the Royal Astronomical Society, the Royal Irish Academy and of the Australian National Committee on Astronomy.
A new Cepheid variable in Puppis. *Riverview Coll. Obs. Publ.* 2, 17-18, 1936

D.J.K. O'Connell, 1939. Rev. William J. O'Leary, 1869-1939. *Bull. Am. Assoc. Jesuit Scientists* 17, 60-62.

Anonymous, 1939. William J. O'Leary. *Monthly Notices Roy. Astr. Soc.* 100, 255-256.

Alphonse de Penaranda (1834-1896). Born in Brussels, Belgium, he entered the Society of Jesus in 1851. He traveled to India and became a professor of mathematics and astronomy at St. Xavier's College in 1874. He collaborated with Lafont in the observatory and was its director from 1880 to 1881 and director of the astronomical section from 1889 to 1896. Penaranda was very active on astronomy and a frequent lecturer in the public lectures organized by the Association for the Cultivation of Science. In 1890 he participated in the expeditions to observe solar eclipses in Bagalpore (Bihar), and in Meghna (East Bengal). Penaranda was highly esteemed by Rabindranath Tagore who had been his student.

A. Verstraeten, 1991. Excellence, Eugene Lafont (1837-1908), Alphonse de Penaranda (1834-1896). *Jivan*, (February), 13-14.

Stephen J. Perry (1833-1889). Born in London, England, he entered the Society of Jesus in 1853. In 1858 he studied mathematics in London and in 1859 in Paris where he attended lectures by Bertrand, Liouville, Cauchy and Serret. From 1860 to 1863, he was appointed Director of the observatory of Stonyhurst College and professor of mathematics and physics. In 1868 after finishing his theological studies, Perry returned to Stonyhurst where he was appointed again Director of the observatory. In 1868 with Walter Sidgreaves he carried out a magnetic survey in the west part of France with 15 stations and in 1869 in the east part with 21 stations. In 1871, with the help of William Carlisle, he did a third magnetic survey in Belgium with 20 stations. These were the first magnetic surveys done in Europe, after those of Sabine in England in 1858. In 1870 he was put in charge of the scientific expedition to observe the solar eclipse in San Antonio (Cadiz), Spain. In 1874 he was appointed chief of the British expedition to the Kerguelen Islands in the South Indian Ocean, organized by the Royal Society to observe the solar transit of Venus. During this expedition, which lasted five months, he carried out many magnetic, meteorological and astronomical observations during the journey and in the Kerguelen Islands. In 1882, he took part in a similar expedition to Madagascar. In 1887 he participated in two other expeditions to observe solar eclipses in Pogost, Russia, and Carriacou, Caribbean. His last expedition was in 1889 to the Salut Isles, French Guyana, where he died of severe fever. Besides his work in geomagnetism he worked on solar spectrography and physics. From 1880 Perry made detailed images of sunspots and studied their spectra. He confirmed the 11 years cycle discovered by Schwabe and studied the relation between solar activity and magnetic storms. Other activities were the application of photography to astronomy and the observation of Jupiter satellites, comets and meteorites. Besides his scientific work he taught physics and mathematics at Stonyhurst College and gave lectures on scientific subjects, which were very popular. He published a total of 129 scientific articles on astronomy, solar physics, meteorology and geomagnetism. Perry was a member of many scientific societies, and was elected fellow of the Royal Society in 1874. He served in the Council of the Royal Astronomical Society in 1873-1878 and 1887-1890.

Magnetic survey of the west of France in 1868. *Phil. Trans. Royal Soc. London*, 160, 33-50, 1870.

Magnetic survey of Belgium in 1871. *Phil. Trans. Royal Soc. London*, 163, 341-357, 1873.

Magnetic observations taken during the transit of Venus expedition to and from Kerguelen. *Proc. Royal Soc. London*, 27, 1-11, 1878.

Observations of Sun-spot spectra in 1883. *Monthly Notices Roy. Astr.* Soc. 44, 244-248, 1884.

The surface of the Sun in 1888. *Monthly Notices Roy. Astr. Soc.* 49, 121-122, 1889.

Anonymous, 1890. Father Perry. *Letters and Notices* 20, 131-146.

A. A. C., 1890. Stephen Joseph Perry. *Proc. Roy. Soc.* 48, xii-xv.

Anonymous, 1890. Stephen Joseph Perry. *Monthly Notices Roy. Astr. Soc.* 50, 168-170.

A. L. Cortie, 1890. *Father Perry, F.R.S. The Jesuit astronomer.* Catholic Truth Society, Londres. 113 pp.

A. L. Cortie, 1890. Father Stephen Joseph Perry S.J. *The Month* 68, 305-323.

A. L. Cortie, 1890. The scientific work of Father Perry. *The Month* 68, 474-488.

J. Thirion, 1890. Le R. P. Perry. *Rev. des Questions Scientifiques* 27, 201-208.

A. L. Cortie, 1892. *P. Perry, F.R.S. Jesuit Astronom, sein Lebend, sein Wirken und sein Tod.* Pustet, Regensburg, xiv-136 pp.

J. Ashbrook, 1966. Father Perry's expedition to Kerguelen island. *Sky and Telescope*, (June) 340-341.

G. D. Bishop, 1977. Stephen Joseph Perry (1833-1889) Priest, scientist, educator. Master Sci. Thesis, University of Manchester.

G. D. Bishop, 1979. Stephen Perry (1833-1889): Forgotten Jesuit scientist and educator. *J. Brit. Astron. Assoc.* 89, 473- 484.

G. D. Bishop, 1979. Stephen Perry (1833-1889): Stonyhurst astronomer and educator. *The Stonyhurst Magazine* 41, 7-15.

Edward F. Pigot (1858-1929). Born in Dundrum, near Dublin, Ireland, he obtained the degree of doctor of medicine by Trinity College, Dublin in 1882 and entered the Society of Jesus in 1885. He arrived in Australia in 1888 and taught science at Xavier College, Melbourne and at St. Ignatius College, Riverview. In 1899 he was appointed to Zikawei Observatory where he stayed from 1899 to 1907. In 1907, owing to health reasons, he returned to Australia. The same year Pigot founded Riverview Observatory at St. Ignatius College, dedicated to meteorology, seismology and astronomy. In 1910, 1911 and 1922 he observed solar eclipses in Brunei, Tonga and Queensland. From 1915 to 1917 he built and experimented with a Foucault's pendulum, the first in the southern hemisphere. In 1926 Pigot collaborated with Bosscha Observatory in Lembang, Java, on the observation of variable stars and stellar photography. Between 1925 and 1929 he made observations of solar radiation and made observations of earth tides by means of a pendulum in a mine in Combar, NSW. Pigot was member of the Australian National Research Council and represented his country in several international scientific congresses in particular at the first general assembly of the IUGG in Roma in 1922.

Note on the new Wiechert seismometers at Riverview College, Sidney. *Roy. Soc. of New South Wales. Journal and Proceedings* 43, 388-393, 1909.

(with L.A. Cotton) Seismology in Australia. *Proceed. 1st PanPacific Science Conference, Hawaii,* 2, 409-410, 1920.

D.J.K. O'Connell, 1952. Father Edward Francis Pigot, S.J. *Studies* 41, 189-196, 323-332.

L. A. Drake, 1988. Edward F. Pigot (1858-1929) Jesuit priest, astronomer and seismologist. In: G. Serle (ed.), *Australian Dictionary of Biography*, Melbourne University Press, 230-231.

Charles Poisson (1882-1965). Born in Rion des Landes, France, he was an officer of the French Navy. In 1905 he was in charge of the astronomical and geophysical observations on the navy ship Lavoisier. He entered the Society of Jesus in 1910. During the First World War, Poisson was mobilized as a Navy Lieutenant and participated in the front of Dixmude, where he was awarded the Croix de la Légion d'Honneur and later commanded a battery of heavy artillery in the fronts of Champagne and Somme. After the war, Poisson was ordained priest in 1923 and went to Madagascar to succeed Colin as director of the Observatory of Tananarive. Poisson directed the observatory for forty one years. During this time he never forgot completely his years as a Navy commandant. His main scientific work was on meteorology specially on the study the cyclones of the Indian Ocean. In 1934 he received the Prize Victor Raulin of the Académie des Sciences and in 1935 was named corresponding member of the Institut de France in the section of Geography and Navigation. He was a member of the Madagascan Academy.

Journal de route d'un officier de fusiliers marins. 1.Le mois d'October 1914: Gans, Dixmude. 2. Le fin de Dixmude. 3. Dunkerque, Lizerne la grand dune. 4. Nieuport, Saint Georges. *Etudes* 145, 108-127; 210-235; 354-375; 530-543. 1915.

Quarante-cinq trajectoires de cyclones à Madagascar. *Bull. Econom. de Madagascar* 23, 1-12. 1926.

Les cyclones tropicaux. *Annales de l'Academie de Sciences Coloniales*, 5, 51 pp., 1931.

Météorologie de Madagascar. Tananarive 1945.

J. Trefouel, 1965. Charles Poisson (Note nécrologique) *Comptes Rendus Acad. Science, Paris,* 261 (Groupe O) 5258.

J. C., 1965. Le Père Charles Poisson, 1882-1965. *Madure, Madagascar,* 116, 218-223.

J. C., 1965. Le Pere Charles Poisson S.J. *Courrier des Isles (Viceprovince de Madagascar),* 75, 1-2.

Ignacio Puig Simón (1887-1961). Born in Manresa (Barcelona), Spain, he joined the Society of Jesus in 1903. From 1912 to 1916 he was a science teacher at the Colegio de Santo Domingo in Orihuela and from 1920 to 1924 he taught chemistry in the Instituto Químico de Sarriá in Barcelona. In 1925 he became subdirector of the Observatorio del Ebro and worked in astronomy and cosmic physics. In 1934 he was asked by the Consejo Nacional de Observatorios of Argentina to found and direct the new Observatorio de Física Cósmica de San Miguel, near Buenos Aires. He was its director during ten years and began the different scientific programs of the observatory. He was a prolific writer of scientific and popular books on science, with more than 80 titles. He returned to Spain in 1943 and the Holy See entrusted him with the job of founding an observatory in Addis Ababa, Ethiopia which could not be carried out due to the opposition of the Ethiopian government. In 1944 he became the editor of the Spanish scientific journal *Ibérica*. Puig was a member of the Academia de Ciencias Argentina and honorary member of the Academy of Science of Río de Janeiro.

Curso general de química. Manuel Martín, Barcelona, 592 pp. 1927.

Los últimos adelantos de la técnica. Revista Ibérica, Barcelona 475 pp. 1948

La astronomía contada con sencillez. Escelicer, Cadiz, 361 pp. 1951.

A. G. Galmarini, 1934. *El Padre Ignacio Puig.* Sociedad de Estudios Geográficos GAEA, Buenos Aires

C. Eguía Ruiz, 1942. El Padre Ignacio Puig S.J., Director del Observatorio de San Miguel. *El Pueblo, Buenos Aires,* 7 Junio 1942.

I. Sala de Castellarnau, 1961. R.P. Ignacio Puig, S.I. (1887-1961). *Ibérica* 34, No. 461, 392-394 y 399.

Jesús Emilio Ramírez (1904-1981). Born in Yolombó (Antioquia), Colombia, he joined the Society of Jesus in 1919. He studied geophysics in Saint Louis

University, Missouri, where he received the degrees of M. S. (1931) and Ph. D. (1939). His doctoral thesis, directed by Macelwane, was on the detection of cyclones using microseismic waves observed in three seismological stations. In 1941, together with Sarasola, Ramírez founded the Instituto Geofísico de los Andes Colombianos at Bogota. He was its director for 38 years. In 1949 he was during one year director of the Servicio Meteorológico Nacional of Colombia and from 1947 to 1960 he was professor of geophysics in the state Universidad Nacional of Bogota. In this context he published in 1957 a textbook in applied geophysics. From 1960 to 1966 Ramírez was Rector of the Jesuit Universidad Javeriana in Bogota where he was Professor of Geophysics from 1939. His main subject of research was the seismicity and tectonics of Colombia and the Andean region. Ramírez published more than 140 scientific articles in particular studies of many important Colombian earthquakes. From 1973 to 1979 he was in charge of the organization of the Nariño I-IV Project for the study of the seismicity and crustal structure under Colombia, with the participation of several national and foreign institutions. Ramírez was the most prominent figure in the field of seismology in Colombia and contributed to the promotion of science in his country. He was a member and president of the Academia Colombiana de Ciencias Exactas and a corresponding member of Real Academia de Ciencias Exactas, Físicas y Naturales of Spain. He was president of CERESIS (Centro Regional de Sismología de América del Sur) and member of many scientific societies.

Earthquake history of Colombia. *Bull. Seis. Soc. Am.* 23, 13-22.

An experimental investigation on the nature and origin of microseisms at St. Louis, Missouri. *Bull. Seis. Soc. Am.* 30, 35-84; 139-178, 1940.

(with L.G. Durán), *Nociones de Prospección Geofísica.* Artes Gráficas, Madrid, 241 pp. 1957.

Historia de los terremotos en Colombia. Instituto Geográfico Agustín Codazzi. Editorial Agra, Bogota, 218 pp. 1969.

(with L.T. Aldrich) (eds.), *Nariño Proyecto Cooperativo Internacional-1973. La transición océano-continente en el suroeste colombiano.* Editora Guadalupe, 313 pp. 1977.

Anonymous, 1971. Jesús Emilio Ramírez, S.J. *Anales de Ingeniería*, vol. 10, 20-25.

J. E. Ramírez, 1978. Jesús Emilio Ramírez, S.J., Curriculum Vitae y bibliografía (manuscript) 18 pp.

J. R. Goberna, 1984. Rev. P. Jesús Emilio Ramírez, S.J. (1904-1981). (manuscript) 21 pp.

J. R. Goberna, 1984. P. Jesús Emilio Ramírez Gonzalez. (manuscript) 3 pp.

A. Alfaro, 2001. Instituto Geofísico Universidad Javeriana, sesenta años 1941-2001. Padre Jesús Emilio Ramírez Gonzalez S.J. *Hoy en la Javeriana* 1156, 12-13.

Willian C. Repetti (1884-1966) Born in Washington, D.C., he entered the Society of Jesus in 1907. In 1910 he was director of the Seismological Station of Canisius College, Buffalo, New York. In 1914 he taught physics and mathematics in Fordham University, New York, and was director of its seismological observatory until 1918. In 1926 he studied geophysics in Saint Louis University with Macelwane and obtained his doctoral degree with a thesis on the discontinuities in the structure of the earth interior. In 1928 he traveled to Philippines and was put in charge of the seismological section of the Manila Observatory. There he completed a new catalogue of earthquakes in Philippines. After the World War II, he returned to the United States and worked at Georgetown University.

New values for some of the discontinuities in the earth. *Seismological Bulletin. Manila Central Observatory* 75-89, 1930.

Tectonic lines of the Philippines. *Seismological Bulletin, Manila Observatory*, 57-71, 1934.

Catalogue of earthquakes felt in Guam 1925-1938. *Seismological Bulletin . Manila Observatory*, 27-43, 1939.

Catalogue of Philippine earthquakes. *Bull. Seis. Soc. Am.* 36, 133-322, 1946.

J. B. Macelwane, 1943. The Rev. William Charles Repetti, S.J. *Bull. Seis. Soc. Am.* 33, 83-84.

Jerome S. Ricard (1850-1930) Born in Plaisians, Buis les Baonies (Drôme), France he enter the Society of Jesus in 1871 and traveled as a Jesuit novice to California in 1872. He began teaching mathematics at Santa Clara University in 1891, after a short period of studies at Johns Hopkins University and Creighton University. In 1895 he installed a small meteorological and astronomical observatory at Santa Clara University where he began to observe sunspots and their correlation with weather. In 1907 Ricard began his weather forecasts at medium and long terms. From 1915 he published a monthly bulletin with weather forecasts and news of meteorological interest. He held constant debates and controversies regarding his theory about the relationship between sunspots and weather. His answers to his opponents were full of humor and ironies and his opinions were readily accepted by newspapers and the public who called him "Padre of the Rains". He published his ideas and observations in *The Sunspot*, a journal he founded. His last years were shadowed by his project to build a large reflecting telescope and the deception of the optician who took his money and never delivered the mirror.

Latest advances in weather forecasting at a long range by sunspots and planetary positions. *Popular Astronomy* 21, 131-142, 1913.

The Sunspot, Santa Clara College, vol 1-16, 1915-1930.
R. Buckley, 1931. Rev. Jerome Sixtus Ricard, S.J., Meteorologist. *Bull. Am. Ass. Jesuit Scientists* 8, 3, 6-8.

G. McKevitt, 1979. *The University of Santa Clara, a history 1851-1977.* Stanford University Press, (The Padre of the Rains), 221-229.

William F. Rigge (1857-1927). Born in Cincinnati, Ohio, he entered the Society of Jesus in 1875. In 1895-1996 he worked at Georgetown Observatory as an assistant of Hagen and obtained his doctoral degree in astronomy by Georgetown University. In 1896 he went to Creighton University as a professor of mathematics and astronomy and director of the observatory that had been founded ten years before. He did continuous observations of solar eclipses, stellar occultations, comets and planets. He published numerous articles in scientific journals and also of more popular character. He wrote two books on eclipses and occultations and harmonic curves. The latter described a "harmonic motion machine" he built which could trace a large number of harmonic curves. In 1910 he solved a criminal case by determining the hour a photograph was taken by the length of a shadow. He remained in his post in Creighton during 31 years. He was the first American Jesuit to be elected a fellow of the Royal Astronomical Society and was member of the American Astronomical Society.

Jesuit astronomy. The restored Society, 1814-1904. *Popular Astronomy* 12, 230-239, 303-310, 375-385. 1904.

The graphic construction of eclipses and occultations. Chicago, 1924.

Harmonic curves. Omaha, 1926

J. McC., 1927. Father Rigge, S.J. *The Province News-Letter, Missouri Province of the Society of Jesus* 8, 73-74.

Luís Rodés (1881-1939). Born in Santa Coloma de Farnes (Gerona), Spain he entered the Society of Jesus in 1897. He taught physics at the Colegio San Ignacio at Barcelona from 1906 to 1910 and studied physics at the University of Barcelona. From 1912 to 1914 he collaborated at the observatory at Valkenburg. In 1914 he began working at the Ebro Observatory and represented it in the observation of a solar eclipse in Hernosand, Sweden. From 1916 to 1918, Rodés studied astronomy and astrophysics at the universities of Harvard and Chicago and at the observatories of Yerkes and Mount Wilson in the United States. During this time he worked on the spectroscopic determination of the rotation of the interior planets. In 1919 he became Director of the Ebro Observatory. During the 20 years in this post he worked on the influence of the solar activity in the magnetic and electric field of the earth, publishing many articles about this subject. His work about sunspots and their relation with magnetic storms was especially important. He studied the influence of the sun and moon on the occurrence of earthquakes, though this subject has never been demonstrated. During the Spanish Civil War he made a great effort to continue the work of the observatory in spite of the enormous difficulties.

A discussion of the spectroscopic method of determining the rotation periods of the inner planets. *The Observatory* 42, 308-315, 1919.

La Terre exerce-t-elle une influence sur la formation de taches du Soleil? *Comptes Rendus Académie des Sciences* 173, 550-551, 1921.

Some new remarks on the cause and propagation of magnetic storms. *Terrestrial Magnetism and Atmospheric Electricity* 32, 127-131, 1927.

Sur une nouvelle méthode pour mesurer la parallaxe solaire. *Comptes Rendus Académie des Sciences*, 187, 527-529, 1928.

El Firmamento. Salvat, Barcelona, 586 pp. 348 figs. 1927. 2nd edition 1939.

I. Puig, 1939, 1940. El astrónomo jesuita Padre Luís Rodés, *El Pueblo* (Buenos Aires) 22-6-1939. *Actualidades Científicas* 17, 127-139.

J. B. Macelwane, 1940. Padre Luís Rodés S. J., 1881-1939. *Terrestrial Magnetism and Atmospheric Electricity* 45, 87-91.

I. Puig, 1940. El astrónomo Padre Luís Rodés, S. J. *Revista Astronómica*, 260-266.

I. Puig, 1945. R. P. Luis Rodés S. I. (1881-1939). *Ibérica* (2a época), 7, 169-172.

Antonio Romañá (1900-1981). Born in Barcelona, Spain he entered the Society of Jesus in 1917. From 1922 to 1927 he studied mathematics and physics at the University of Barcelona and obtained his doctoral degree in 1930 from the University of Madrid. He spent some time in 1932 at the Observatory of the University of Vienna, working on astronomy with the Profs. Graff, Berheimer and Krumpholz. The same year he began his work at Ebro Observatory. In 1934 he was in charge of the sections of magnetism and atmospheric and terrestrial electricity. In 1939 Romañá was appointed Director of the Ebro Observatory. His first task was to restore the observatory to working conditions after the damage suffered during the last days of the Spanish Civil War. From 1950 to 1970 he made a complete renewal of the instrumentation of the observatory and introduced ionospheric studies with the installation of the first ionosonde in Spain. Romañá's main scientific work was the study of rapid magnetic variations and storms. He had an active presence and participation in national and international institutions in several of which he held office. Especially he was for many years president of the Commission for the Study of Magnetic Rapid Variations of the International Association of Geomagnetism and Aeronomy. He had positions of responsibility in the Consejo Superior de Investigaciones Científica, the Spanish Research Council created after the Civil War. In 1966 he was elected a member of the Real Academia de Ciencias Exactas, Físicas y Naturales and was also a member of many scientific societies. He received several honors, in particular the Gran Cruz de Isabel la Católica and was a member of the French Légion d' Honneur.

Le monde, son origine et sa structure aux regards de la science et de la foi. In I. Kologrivof (ed.) *Essai d'une Somme Catholique contre les sans-Dieu.* Spes, Paris, 45-95, 1936. English translation 1937 and 1938, Spanish translation 1943. Also in: J. de Bivort (ed.) *Essai sur Dieu, l'homme et l'univers*, Castermann, Tournai, Paris 113-172, 1950. 2nd ed. 1951, 3rd ed. 1953, 4th. ed. 1957. Translations: Italian 1953, German 1956, English 1953, Spanish 1959 and 1961.

A propos de la caractérisation électrique des jours dans l'étude du gradient du potentiel atmosphérique. *Trans. IATME meeting Washington*, 245-254, 1939.

(With J. Princep) Contribution à l'étude des effects géomagnétique des differentes manifestations de l'activité solaire. Trans. *IATME meeting*, Oslo, 13, 355-357, 1950.

(With J. Bartels, J, Veldkamp, and D. van Sabben) Geomagnetic data. Rapid Variations 1958 a 1978, *IAGA Bulletin*, 12 m2 a 12 x2, 1962 a 1971, 32b a 32j, 1972 a 1980.

Provisional atlas of rapid variations. *IAGA Commission on Rapid Variations and Earth Currents.* Tortosa 20 pp. 1958.

J.M. Torroja, 1981. Antonio Romañá , sacerdote y científico. *Arbor* 110, 307-310.

A. Blanch, 1981. Antonio Romañá, lección y ejemplo de un jesuita científico. *Razón y Fe* 204, 500-505.

J. O. Cardús, 1983. Antonio Romañá Pujó (1900-1981). *Vol. Conmemorativo del 75 aniversario del Observatorio del Ebro,* Roquetas, 9-19.

P. M. Lamet, 1986. Antonio Romañá, científico y amigo. *Jesuitas* 9, 32.

J. O. Cardús, 1995. Antoni Romañá i Pujó, Barcelona 1900 - Sant Cugat del Vallés 1981. In: J. M. Camarasa y A. Roca (Eds.) *Ciéncia i Técnica als Països Catalans: Una aproximació biográfica.* Fundacio Catalana per la Recerca, Barcelona, 1403-1442

Miguel Saderra-Masó, 1865-1939. Born in Olot (Gerona), Spain, he joined the Society of Jesus in 1882. In 1890 he arrived at Manila Observatory and was put in charge of the seismological section. In 1895 he published the first work on the seismology in the Philippines. He returned to Spain to complete his studies and returned to Manila in 1901. He continued working on seismology and also in magnetism at Manila Observatory. He installed the seismographic stations of Guam, Butuam, Ambulong and Baguio. He traveled to different islands of the archipelago to study earthquakes, volcanic eruptions and to establish meteorological and magnetic stations. In 1910 he published the first catalogue of earthquakes of Philippines and in 1913 the first work on seismotectonics in the Philippines. In 1932 he returned to Spain after suffering a stroke.

La sismología en Filipinas. Ramirez y Cia. Manila, 130 pp., 1985

Catalogue of violent and dectructive earthquakes in Philippines: 1599-1909. Bureau of Printing, Manila, 27 pp. 1910.

The relation of seismic disturbances in the Philippines to geological structures. *Bull. Seis. Soc. Am.* 3, 151-186, 1913.

Historia del observatorio de Manila. E. C. McCullough, Manila, 1915.

W. C. Repetti, 1939. Rev. Miguel Saderra-Masó, 1865-1939. *Bull. Am. Ass. Jesuit Scientists* 17, 66-67

M. Selga, 1939. El P. Miguel Saderra. *Cultura Social* 27, 207-209.

Manuel Sanchez Navarro-Neumann (1867-1941) born in Malaga, Spain, he obtained the doctoral degree in medicine in 1893 and entered the Society of Jesus in 1900. In 1908 he was put in charge of the seismological section of the Observatorio de Cartuja, Granada and in 1915 appointed Director of the observatory until 1931, when Jesuits were expelled from Spain. He returned to the observatory after the Civil War in 1939. From 1908 Navarro-Neumann modernized the seismological instruments, building most of them. He began building mechanical seismographs of the Mainka type and of Wiechert type with a large mass of 3000 kg and a magnification of 1000, a large value for mechanical instrument. In 1924 he built horizontal and vertical electromagnetic seismographs with photographic recording of Galitzin type with 7.5 kg mass and periods of 30 and 12 seconds connected to a Deprezt-D'Arsonval galvanometer of 19 seconds period. In these instruments he actually improved the original designs by Galitzin. Navarro-Neumann besides his observational work carried out very detail studies of the seismicity of Spain and some works on theoretical seismology. His most important contributions were the first modern catalogues of earthquakes in Spain published in 1917 and 1920. He published about 300 articles on seismological topics and in 1916 the first textbook of seismology in Spanish.

Travail produit par un tremblement de terre. *Boll. della Società Sismologica Italiana* 19, 351-353, 1915

Ensayo sobre la sismicidad del suelo español. *Bol. Real Sociedad de Historia Natural* 13, 83-109, 1917.

Bosquejo sísmico de la península Ibérica. *La Estación Sismológica y el Observatorio Astronómico y Meteorológico de Cartuja*, Granada, 3-68, 1921.

Terremotos, sismógrafos y edificios. Gabriel López del Horno, Madrid, 243 pp. 1916.

Les séismographes de la station séismologique de Cartuja (Granada). *Bull. Union Geod. et Geophys. Int. Sismologie*, ser. A, 4, 119-131, 1927.

A. Due Rojo, 1941. El R.P. Manuel Ma. Sánchez Navarro S.J. *Bol. Real Sociedad. de Historia Natural* 39, 117-120.

A. Due Rojo, 1940. *Labor científica del R. P. Manuel Sánchez Navarro S.I.* Observatorio de Cartuja.

A. Due Rojo, 1941. Um grande sismologo espanhol. *Broteria* 22, 284-287.

A. Due Rojo, 1945. El P. Manuel Ma. Sánchez Navarro S.I. (1867-1941). *Ibérica* 10, 239-243.

Simón Sarasola (1871-1947). Born in Beliarrain (Guipúzcoa), Spain, he entered the Society of Jesus in 1887. In 1897 he was a science teacher at the Colegio de Belén at Havana, Cuba and began working at the observatory with Gangoiti. In 1900 he studied theology at Woodstock College, Maryland and worked in meteorology for some time at the Observatory of Georgetown University and at the U. S. Weather Bureau. In 1905 he returned to Belen Observatory and in 1909 founded an observatory in Cienfuegos, where he was director until 1920. That year, called by the President of Colombia, he was put in charge of the Servicio Meteorológico Nacional of Colombia and founded the Observatorio Meteorológico de San Bartolomé in Bogota. In 1940 he resigned from this post and together with Ramirez founded in 1941 the Instituto Geofísico de los Andes Colombianos at Bogota. He became its first director, in charge of the meteorological section until 1943. In 1943 he returned to Belen Observatory and was its director until his death. There he continued the studies of tropical hurricanes carried out by Viñes, Gangoiti and Gutierrez Lanza. He was awarded by the Government of Colombia with the Cross of Bocayá, in recognition of his work in meteorology in Colombia.

Investigaciones sobre los centros de acción de la atmósfera. *Ibérica* 27, 137-139, 1915.

Los huracanes de las Antillas (2a edición). Bruno del Amo, Madrid, 254 pp. 1928.

La obra de los católicos y creyentes en las ciencias exactas, físicas y naturales. Imp. del Corazón de Jesús, Bogota, 196 pp., 1933.

La meteorología en Colombia. *Revista Javeriana* 123, 155-163, 1946

.

La predicción de los ciclones en Cuba. Algunas notas sobre perturbaciones atmosféricas en el Cantábrico. *Revista de Geofísica* VII, 27, 309-327, 1947.

J.E. Ramirez, 1948. The Rev. Simón Sarasola S.J. 1871-1947. *Bull. Seis. Soc. Am.* 38, 229-231.

Anonymous, 1948. R. P. Simón Sarasola, S.J. *Revista de Geofísica* VII, 27, 307-308.

W. Escobar, 1971. El Padre Simón Sarasola, eminente científico jesuita. *Publ. del Inst. Geofisico de los Andes Colombianos,* serie B, No, 8, 8 pp.

Pietro Angelo Secchi (1818-1878). Born in Reggio, (near Parma), Italy, he joined the Society of Jesus in 1833 in Rome. He studied philosophy and science in the Collegio Romano. In 1840 he taught physics at the College of Loreto. In 1848 during the expulsion of the Jesuits from Italy, he went to Stonyhurst College in England to continue his studies of theology. There he got the first contact with astronomy as an assistant of Weld at the observatory. In 1849 he traveled together with de Vico to Washington, where he worked at the observatory of Georgetwon University. In 1850 Secchi returned to Rome where he was appointed Director of the Osservatorio del Collegio Romano. He remained director for 28 years. In 1853 he built a new observatory on top of the roof of Saint Ignatius Church dedicated to astronomy, meteorology and geomagnetism. His earliest research dealt with the observation of the planets, the Moon, stellar nebula and several comets. Soon he became aware of the importance of physical study of the stars and planets. In 1851 he begun his studies about the characteristics of the sun, its spots and protuberances and the distribution of temperature on its disk and since 1868 its spectra. In 1860 he observed the solar eclipse of 18 July in the Desierto de las Palmas, Alicante, Spain. In 1863 he begun what will be his most important contribution, the spectroscopic study of stars, observing more than 4000 spectra. In 1867 established the classification in 4 spectral classes that mark an important temperature sequence. Secchi's classification later extended and modified has become one of the basic tools of astrophysicists. He founded in 1872 the Societá degli Spettroscopisti Italiani. He is considered rightly as one of the pioneers of stellar spectroscopy. Besides his work in astronomy he began the measurement of a degree of latitude along the via Apia, which had been measured before by Boscovič in 1751. He built an instrument he called the meteorograph for the simultaneous recording of atmospheric pressure, temperature, rainfall, and intensity and direction of the wind which was awarded a gold medal in the World Fair of Paris in 1867. He published about

500 scientific articles in Italian, French and German in particular in *Astronomiche Nachrichten, Comptes Rendus de l'Academie des Sciences* and *Atti dell' Accademia Pontificia de Nuovi Lincei.* In 1864 he published his book on the unity of physical forces that was translated into French and German. His book on the sun was published in French in 1870, and then translated into German and Spanish. He published a book on the stars and after his death a book was published on terrestrial physics. He received many honorific awards among them the Cross of the Légion d'Honneur of France and was a member of many scientific academies and societies in particular of the Royal Society, elected in 1856. He is the only Jesuit mentioned in the Biographical Encyclopedia of Scientists published by the American Institute of Physics.

L'Unitá delle forze fisiche, saggio di filosofia naturale. Tip. Forense, Roma, 1864. 2nd. Ed. Milan, 1874. French translation: *L'Unité de forces physique, essai de philosophie naturelle.* Savy, Paris, 1869; 2nd. Ed. 1874. German translation: *Die Einheit der Naturkräfte. Ein Beitrag zur Naturphilosophie.* Frohberg, Leipzig, 1876. 2nd. Ed. O.Salle, Braunschweig, 1892.

Le Soleil. Gauthier-Villars, Paris, 1870, 921 pp. 2nd Ed. 1875. German translation: *Die Sonne*, H. Schellen, Braunschweigh, 1872. Spanish Translation: *El Sol*, Victoriano Suarez, Madrid, 930 pp.

Le stelle. Saggio di Astronomia siderale. Dumolar, Milano, 1877, 432 pp. German translation: *Die Sterne, Grundzüge der Astronomie der Fixsterne.* Brockhaus, Leipzig, 1878, 413 pp. French translation: *Les Etoiles. Essai d'astronomie sidérale.* Germer Baillière, Paris, 1879, 443 pp.

Lezioni elementari di Fisica terrestri. Ermanno Loescher, Roma, 1879. German translation: *Die Grosse der Schöpfung.* E. Bidder, Leipzig, 1882, 60 pp. 2nd Ed. 1883, 3rd Ed. 1885. Spanish translation: *Lecciones elementales de Física terrestre*, Suces. de Rivadeneira, Madrid, 1886.

Memorie dell'osservatorio del Collegio Romano Vol. 1-6, Roma, 1851-63

Bulletino Meteorologico dell'Osservatorio del Collegio Romano, Vol. 1-14, 1862-1877

Catalogo delle stelle di cui si è determinato lo spettro luminoso. Rome, 1867

Sugli spettri prismatici delle stelle fisse. Rome, 1868.

V. van Tricht, 1878. Le Père Secchi. *Revue des Questions Scientifiques* 4, 358-402.

Anonymous, 1879. Angelo Secchi. *Monthly Notices Roy. Astr. Soc.* 39, 238-241.

C. Bricarelli, 1888. *Della vita e delle opere del P. Angelo Secchi D.C.D.G. con un elenco de suoi scritti.* Rome, 65 pp.

T. Pépin, 1889. Le Père Secchi. *Études* 46, 456-475.

Anonymous, 1903. *Al P. A. Secchi nel XXV della morte il comitato romano.* (Colection of 26 articles of different authors) Desclée, Rome, 88 pp.

J. Pohle, 1904. *P. A. Secchi, ein Lebens und Kulturbild aus den neunzehnten Jahrhundert.* J.P. Bachen, Colonia, 288 pp.

A. L. Cortie, 1923. Angelo Secchi, S. J. (1818-1878). In B. Windle (ed.) *Twelve catholic men of science.* Catholic Truth Society, London, 149-164.

G. Abetti, 1928. *Padre Angelo Secchi. Il pionieri dell'astrofisica.* G. Agnelli, Milano, 116 pp.

M. McCarthy, 1950. Fr. Secchi and stellar spectra. Popular Astronomy 58, 153-169.

R. Finzi, 1970. *P. Angelo Secchi (1818-1878).* Lombardini Motori, Reggio Emilia, 94 pp.

M. Vigano, 1978. Cento anni di astrofisica. Nel centenario della morte del padre Angelo Secchi. *Civiltà Cattolica* 129, 4, 538-549.

Anonymous, 1979. *Padre Angelo Secchi nel centenario della morte (1878-1978)* (Collection of 11 articles of different authors) Accademia Nazionale delle Scienze. Scritti e documenti II, 116 pp.

L. Houziaux, 1979. Angelo Scchi et les origines de la clasification spectral. *Ciel et Terre* 95, 349-353.

J. Casanovas, 1983. Fr. Angelo Secchi, a pioneer in the study of solar terrestrial relationships. *Volumen Conmemorativo del 75 aniversario del Observatorio del Ebro*, Roquetas, 59-62.

J. Daintith, S. Mitchell, E. Tootill y D. Gjertsen, 1994. *Biographical Encyclopedia of Scientists* (2 ed). Institute of Physics Publ. Philadelphia, (P.A. Sechi, 808-809).

Miguel Selga (1879-1956). Born in Rajadell (Barcelona), Spain, he entered the Society of Jesus in 1895. He studied sciences in the University of Zaragoza and obtained his licentiate in 1908. In 1913 and 1914 he visited the American observatories of Georgetown (Washington), Harvard, Boston, Yerkes (Wisconsin) and Lowell (Arizona) where he worked on stellar photography and photometry and Lick Observatory (California) where he worked on stellar spectra and binary stars. In 1915 traveled to Manila and in 1926 was appointed director of the Manila Observatory and of the Philippine Weather Bureau, succeeding Algué. From 1925 to 1927 Selga was professor of astronomy and meteorology at the University of Philippines. His main fields of research were astronomy and meteorology. He studied binary stars and stellar spectrography, and the occurrence of typhoons in the Philippines. In 1945 he witnessed the destruction of the observatory by the Japanese Army. He retired as Director in 1948.

Observaciones fotométricas de Nova Aquilae 3 en Manila. *Rev. de la Sociedad Astronómica de España y America*, 8, 90-91 (1918); 9, 12-13; 31-34 (1919).

Astronomical observations made in the Philippines prior to 1927. *Manila Observatory Publ.* vol. I, no. 8, 74 pp.

The velocity of wind at Manila, Baguio, Iliilo and Cebu. Bureau of Printing, Manila 35 pp. 1931.

Observations of rainfall in the Philippines. Bureau of Printing, Manila, 289 pp. 1935.

Charts of remarkable typhoons in the Philippines 1902-1934. Catalogue of Typhoons 1848-1934. Bureau of Printing, Manila, 55 pp. 1935.

Anonymous, 1957. Father Miguel Selga. *Woodstock Letters* 86, 71-75.

A. Hidalgo, 1967. Miguel Selga, 1879-1956: Priest and scientist. *Philippine Studies* 15, 307-347.

Walter Sidgreaves, 1837-1919. Born in Grimsargh (Lancashire), England, he studied from 1848 at Stonyhurst College and entered the Society of Jesus in 1855. He was a teacher of mathematics from 1862 to 1868 at Stonyhurst College and Director of the observatory from 1863 to 1868. During this period the observatory was established as a first class meteorological station and magnetographs were installed. Sidgreaves accompanied Perry in the magnetic surveys in France and the scientific expeditions to Kerguelen Islands and Madagascar. Sidgreaves became Director of the observatory again after Perry's death from 1890 to 1919 and taught physics at the college from 1883 to 1905. He collaborated with Perry and continued Perry's work about the relation between sunspots and magnetic storms. His principal work was on astronomy about the characteristics of the spectra of stars. He was a member of the Royal Astronomical Society and served in its Council from 1892 to 1897.

Spectrum of Nova Aurigae. *Memoirs of the Roy. Astr. Soc.* 51, 29-43, 1892-1895.

On the connection between Solar spots and Earth magnetic storms. *Memoirs of the Roy. Astr. Soc.* 54, 85-96, 1899-1901.
Spectrum of Nova Persei from 1901 Feb. 28 to Apr. 26. *Monthly Notices of Roy. Astr. Soc.* 62, 137-156, 1901-1902.

Spectrographic study of beta-Lyrae. *Monthly Notices Roy. Astr. Soc.* 64, 168-182, 1903-1904.

Anonymous, 1919. A Jesuit astronomer. Father Walter Sidgreaves, S.J. F.R.A.S. *The Tablet* (June 21) 768.

Anonymous, 1919. Father Walter Sidgreaves. *Letters and Notices* 35, 118-123.

A. L. Cortie, 1919. Astronomer and Jesuit. *America* 21, 612-613, 636-637.

A. L. Cortie, 1920. Father Walter Sidgreaves, S.J. *Monthly Notices Roy. Astr. Soc.* 80, 355-359.

Pedro Spina (1839-1925) Born in Rimini, Italy, he entered the Society of Jesus in 1863 in Rome. He traveled to Mexico in 1872. He was a science teacher at the colleges of Puebla and Saltillo until 1906. In 1877 he founded an observatory at the Jesuit college in Puebla and was its director until 1886 and again from 1891 to 1901. He was director of the observatory of Saltillo from 1886 to 1890 and from 1904 to 1906. He carried out in Puebla and Saltillo astronomical and meteorological observations. Spina developed a curious cosmological theory based on a cosmic process of universal rotation and impulsion. He published two textbooks on analytic geometry and calculus and the meteorological observations in Puebla from 1877 to 1896 and the astronomical observations from 1881 to 1885. He returned to Italy in 1906. He was a member of the Academia Mexicana and the Sociedad Científica Antonio Alzate.

Clima de Pluebla. *Rev. Mensual de Climatología* I, 42-44. 1881.

Teoría de la rotación universal. Imp. Colegio Pío de Artes y Oficios, Puebla, 21 pp.

El clima, nociones generales. Tip. Sec. del Fomento, M,xico. 1890.

Un decenio de observaciones meteorológicas en Puebla (1877-1886). *Memorias de la Sociedad Científica Antonio Alzate*, IV, 49-66. 1890.

J. B. Iguiniz, 1945. *Bibliografía de los escritores de la Provincia Mexicana de la Compañía de Jesús*. Buena Prensa, Mexico, 484-488.

Johan Stein (1871-1951). Born in Grave, Holland, he entered the Society of Jesus in 1888. He studied mathematics and physics at the University of Leiden where he obtained the doctoral degree. From 1906 to 1910 he was an assistant of Hagen at the Specola Vaticana. Returned to Holland and was a teacher of physics at Saint Ignatius College in Amsterdam from 1910 to 1930. In 1930 he was commissioned by Pope Pius XI to create a section of the Specola Vaticana in Ethiopia, project which could not be carried out. The same year he was appointed Director of Specola Vaticana where in 1933 he directed the transfer and reorganization of the observatory in Castelgandolfo. Stein completed the work of Hagen on variable stars, continued the work of the observatory in the Carte du Ciel program and of the observation of double stars. He was member of the Pontifical Academy of Sciences and of the Royal Academy of Sciences of Holland.

(with J. Hagen) *Die Veränderlichen Sterne. Band II, Mathematisch-Physikalischer Teil*. Herder, Freiburg im Breisgau, 1924.

Le stelle dopie nel catalogo astrographico della Specola Vaticana. *Atti Pontificia Accademia Nuovi Lincei* 79, 125, 1926.

Lo stato attuale del problema delle nubi cosmiche oscure. *Atti Pontificia Accademia Nuovi Lincei* 87, 125, 1934.

J. de Kort, 1952. In memoriam Father Johan Stein S.J. 1871-1951. *Ricerche Astronomiche* 2, 371-374.

S. Maffeo, 2001. *La Specola Vaticana. Nove Papi, una missione*. Pubblicazioni della Specola Vaticana, 117-120, 354-356.

Francis A. Tondorf (1870-1929). Born in Boston, Massachusetts, he entered the Society of Jesus in 1888. In 1897-1899 he studied physics at Johns Hopkins

University. In 1905 he joined Georgetown University, where he worked on astronomy with Hagen and where in 1913 he became Professor of Physics. In 1911 he installed the first seismological station at Georgetown and began his research on seismology. Tondorf was an active member of the Jesuit Seismological Service and afterwards of the Jesuit Seismological Association. He has been described as the father of seismology in the Jesuit colleges and universities in the United States. He was a keen observational seismologists and held to the belief that earthquakes could be foretold, although he was aware that a dependable forecasting was many years away.

Seismological Report of Georgetown University. *Monthly Weather Review* 1915-1923.

The registration of earthquakes and press dispatches on earthquakes. Georgetown University, 1916-1929.

J. P. Merrick, 1929. Father Francis A. Tondorf, S.J., Seismologist. *Bull. Am. Ass. Jesuit Scientists.* 7, 3-4.

Anonymous, 1929. Francis A. Tondorf. *Bull. Seis. Soc. Am.* 19, 245-246.

S. Sarasola, 1930. Un eminente sismólogo. En memoria del Reverendo Padre Tondorf, S.J. *Noticias Científicas. Observatorio Nacional de S. Bartolomé*, vol 4, N. 32, 69-70.

Anonymous, 1931. Rev. Francis Anthony Tondorf, S.J. *Monthly Notices Roy. Astr. Soc.* 91, 321.

Patrick Joseph Treanor (1920-1978). Born in London, England, he studied at Saint Ignatius College and entered the Society of Jesus in 1937. In 1943 studied physics at Oxford University where he obtained his doctoral degree in 1950 with a thesis on new techniques for measurements of interference phenomena. In 1954 he was accepted as fellow of Balliol College, Oxford. From 1954 to 1959 he did research in the Observatory of Oxford University and Balliol College where he worked on the solar spectrum under Prof. Plaskett. From 1949 to 1953 he was editor of the journal "Observatory". In 1960 Treanor spent two years at the observatories of Lick and Yerkes in the United States. At the end of 1961 he became a member of the staff of the Specola Vaticana and in 1970 he was appointed director. His main research was on the study of heavily polarized stars and the characteristics of the spectrum of fainted and reddened stars. He designed a rotating polarimeter to study the light from stars and other instruments. He died the same year of the celebration of the centennial of Angelo Secchi's death.

A new technique for stellar polarimetry. *Monthly Notices Roy. Astr. Soc.* 138, 325-336, 1968.

An image-tube spectrograph for small telescopes. *The Observatory* 90, 251, 1970.

(with L. B. Otten) A new spectral broadening technique for objective prism spectra. *Vatican Observatory Publication* 13, 1977.

(with G. Coyne) The Vatican Spectral Atlas Project (1933-1977) *Vatican Observatory Publication* 14, 1978.

M. F. McCarthy, 1979. Patrick J. Treanor, S.J. *Quart. Jour. Roy. Astr. Soc.* 20, 333-334.

S. Maffeo, 2001. *La Specola Vaticana. Nove Papi, una missione*. Pubblicazioni della Specola Vaticana, 144-148, 206-209, 217-218, 360-362.

Francesco de Vico (1805-1849) born in Macerata, Italy, he entered the Society of Jesus in 1829. In 1832 he was Professor of Mathematics in the Collegio Romano and in 1833 he became an assistant of Dumouchel, director of the observatory. His first astronomical work was the calculus of the position for the return of the comet Halley in 1835, which enabled the observatory to make the first observations. In 1838 he was appointed Director of the observatory while he continued with his classes of mathematics and astronomy at the college. He began his astronomical research with the observation of the satellites of Saturn, together with other characteristics of this planet. In 1842 he completed four years of observations of Venus, determining its period of rotation and the inclination of its equatorial plane with respect to the ecliptic. From 1844 to 1847 de Vico observed seven comets and for this work he received an award from the King of Denmark. He studied the Andromeda and Orion nebulae and began a project of a catalogue of stars down to magnitude 10, which he could not complete. He was also and excellent musician, director of the College Choir and composed some religious music. In May 1848, he was obliged to leave Rome owing to the revolts and the banishment of Jesuits from Italy. He traveled to the United States where he was appointed Director of the Astronomical Observatory of Georgetown University, Washington. In 1849 he died in London where he had gone to acquire new equipment for the observatory. De Vico was a founding member of the Pontificia Accademia de Nuovo Lincei and a member of several other scientific societies.

Beobachtungen von Flecken auf der Venus in Collegio Romano. *Astr. Nachrichten* 18, 307, 1840

Determination de la rotation de Vénus sur son axe. *Comptes Rendus Académie des Sciences,* 1840.

Diverse decouvertes sur l'atmosphére de Saturne. *Comptes Rendus Académie des Sciences*, 15, 748, 1842.

Anonymous, 1849. Francis De Vico. *Monthly Notices Roy. Astr. Soc.* 9, 65-66.

A. Secchi, 1850. Ragguaglio intorno alla vita e ai lavori del P. Francisco de Vico. *Memorie dell Oservatorio del Collegio Romano.* 133-149.

J. Stein, 1949. Francesco de Vico e i suoi contributi alle scienze astronomiche. *Civiltà Cattolica* (100) II, 190-200; 314-324.

S. Maffeo, 2001. *La Specola Vaticana. Nove Papi, una missione*. Pubblicazioni della Specola Vaticana, Vatican City, 11-14, 27, 81.

Benito Viñes Martorell (1837-1893). Born in Poboleda (Tarragona), Spain, he joined the Society of Jesus in 1856. During seven years he taught physics, chemistry and mathematics in the seminary of Salamanca. In 1870 he traveled to Cuba and became director of the observatory of the Colegio de Belén at Havana. In 1875 Viñes began the study of tropical hurricanes and the publication of their forecasts. Commissioned by the Academy of Sciences of Havana he made a field study of the effects of the 1877 hurricane in Cuba and Puerto Rico. In 1877 Viñes published the first treatise on tropical hurricanes where he proposed his empirical laws of the circulation and translation of cyclones. In 1880 he studied the damages produced by earthquakes in the west part of Cuba. En 1882 he traveled to Europe where he visited several observatories and spent sometime in Stonyhurst with Perry. In 1886 he began to organize a network of meteorological stations in the Caribbean region which in 1890 was formed by 20 stations. He made timely and accurate forecasts of hurricanes which save many lives and loss of property and were greatly appreciated specially by the maritime community. A second book about the nature and properties of hurricanes was published posthumously in 1895. Viñes was a member of the Real Academia de Ciencias Médicas, Físicas y Naturales of Cuba and of several other scientific societies.

Apuntes relativos a los huracanes de las Antillas en Septiembre y Octubre 1875 y 1876. El Iris, Habana, 1877

El ciclonoscopio de las Antillas. Havana, 1880.

Investigaciones relativas a la circulación y traslación ciclónica en los huracanes de las Antillas. Avisador Comercial, Habana, 1895.

M. Gutierrez Lanza, 1936. *El P. Viñes y su obra científica humanitaria al frente del Observatorio de Belen.* Molina, Habana.

M. Gutierrez Lanza, 1942. El P. Benito Viñes S.J.. In: *Figuras cubanas de la investigación científica*. Habana, 1942.

A. Lopez de Santa Ana, 1958. *Contribución a una biografía completa del R.P. Benito Viñes Martorell, S.J., célebre meteorólogo de las Antillas*. Bedia, Santander.
J. R. Goberna, 1954. La personalidad científica del Padre Viñes y su aporte al estudio y conocimiento de los huracanes de las Antillas. *Ecos de Belén*, 1954.

R.A. de Angelis, 1968. The hurricane priest. *ESSA World*. October, 16-17.

J. Fernandez-Partagás, 1993. El papel del P. Benito Viñes S.J. en el conocimiento del huracán visto a cien años de su muerte. (manuscript), 19 pp.

L. E. Ramos Guadalupe, 1996. *Benito Viñes, S. J. Estudio biográfico.* Academia, La Habana, 128 pp.

A. Linés Escardó, 1996. La figura de Benet Viñes. *Treballs de la Societat Catalana de Geografía* 39, 153-166.

Alfred Weld (1823-1890). Born in Leagram, England, he studied at Stonyhurst college and entered the Society of Jesus in 1842. In 1844 he obtained his B.A. from London University and began teaching science at Stonyhurst College. He was director of the observatory from 1846 to 1851 and from 1856 to 1860. He began the true scientific work of the observatory with observations of sunspots, stars positions and occultations. He was responsible for Secchi's interest in astronomy while a student of theology in Stonyhurst. In 1858 Weld contacted Sabine and installed a temporary and afterwards a permanent magnetic station at the observatory. In Weld's time the observatory was acknowledged as one of the first order meteorological stations of England. He was the first English Jesuit to be elected fellow of the Royal Astronomical Society and one of the first members of the British Meteorological Society. His scientific carrier was cut short by his appointment as Novice Master in 1864 and afterward Provincial of England and other important positions in the Jesuit Order in Africa and India.

Anonymous, 1890. Father Alfred Weld. *Letters and Notices* 20, 317-325.

Anonymous, 1890. Rev Alfred Weld. *Monthly Notices Roy. Astr.* Soc. 51, 198.

F. Edwards, 1985. *The Jesuits in England.* Tunbridge Wells, 333.

J. Bommer, 1992. Fr. Alfred Weld S.J., the "Father" of Stonyhurst Observatory. (Manuscrito) 1 p.

BIBLIOGRAPHY OF PART I, CHAPTERS 1 TO 3

Chapter 1

Albers, P., 1914. *Liber saecularis Socistatis Jesu ab anno 1814 ad annum 1914.* Curia Generalis Societatis Iesu, Roma.

André, C., G. Rayet and A. Angot, 1878. *La astronomie practique et les observatoires en Europe et en Amerique.* Gauthier-Villars, Paris.

Baldini, U., 1992. *Legem impone subactis. Studi su filosofia e scienza dei gesuiti in Italia 1540-1632.* Bulzoni Editori, Roma, 601 pp.

Dear P., 1987. Jesuit mathematical science and the reconstruction of experience in the early seventeenth century. *Studies in Hist. and Phil. of Science* 18, 133-175.

Draper, J. W., 1874. *History of the conflict between religion and science.* Appleton, New York.

Harris, S. J., 1989. Transposing the Merton thesis: apostolic spirituality and the establishment of the Jesuit scientific tradition. *Science in Context* 3, 29-65.

Harris, S. J., 1995. Les chaires de mathématiques. In: L. Giard, *Les Jésuites a la renaissance.* Presses Universitaires de France, Paris, 239-261.

Heilbron, J. L., 1982. *The elements of early modern physics.*(II.1. Jesuits. Their magisterium. Their eclecticism. Their constraint, 93-106) University of California Press, Berkeley

Lancaster, A., 1890. *Liste général des observatoires et des astronomes, des sociétés et de revues astronomiques.* Hayez, Brussels.

Merlin, E. and Somville, 1910. *Liste de observatoires magnétiques et des observaoires séismologiques.* Hayez, Brussels.

O'Connell, D., 1956. Jesuit men of science. *Studies* 45, 307-318.

O'Neill, C. E. and J. M. Domínguez, 2001. *Diccionario Histórico de la Compañía de Jesús. Biográfico-Temático.* Universidad Pontificia Comillas, Madrid. Vol. 1-4.

Polgar, L., 1981-1990. *Bibliographie sur l'histoire de la Compagnie de Jésus 1901-1980.* IHSI, Rome. Vol. 1-5.

Sommervogel, C., 1890-1909. *Bibliotheque de la Compagnie de Jésus. Bibliographie*. Brussels, Vol. 1-10.

Stein, J., 1941. La Compagnia di Gesú e le scienze fisiche e matematiche. *Publ. Universitá Cattolica del S. Cuore, Milan. (Vita e Pensiero)* serie 5, **19**, 1-23.

Stroobart, P., J. Delvosal, H. Phillippot, E. Delport and E. Merlin, 1904. *Les observatoires astronomique et les astronomes*. Hayez, Brussels.

Udías, A., 1996. Jesuit contribution to meteorology. *Bull. Am. Meteor. Soc.* **77**, 2307-2315

Udías, A., 2000. Observatories of the Society of Jesus, 1814-1998. *Arch. Hist. Soc. Iesu* **69**, 151-178.

Udías, A. 2001. Serving God and Science. *Astronomy and Geophysics* 42, 2, 23-24.

Udías, A. and W. Stauder, 1991. Jesuit geophysical observatories. *EOS, Trans. Am. Geophys. Union* **72**, 185-187.

Udías, A. and W. Stauder, 1996. Jesuit contribution to seismology. *Seismological Research Letters* **67**, 10-19.

Udías, A. and W. Stauder, 1998. Jesuits and the Earth. In: G. A. Good (ed.) *Sciences of the Earth. An encyclopedia of events, people and phenomena.* Garland Pubs. New York, 493-498

Vrégille, P. de, 1906. Les observatoires de la Compagnie de Jésus au début du XXe siècle. *Rev. des Questions Scientifiques* **59**, 10-72, 493-579.

Wallace, W. A., 1989. The problem of apodictic proof in early seventeenth century mechanics, Galileo, Guevara and the Jesuits. *Science in Context* 3, 67-87.

Chapter 2

Aoust, M. l'abbé, 1870-1871. Etude sur le P. Pézenas, astronome marseillais. *Mémoires de l'Académie de Marseille*, tome XX, 15 pp.

Bigourdan, G., 1918. La station astronomique du Collége de Clermont (première pèriode) et la mission astronomique de Siam. *Comptes Rendus Académie des Sciences*

(Paris) **166**, 833-844.

Bigourdan, G., 1918. L'observatoire du Collége Louis-le-Grand (derniere pèriode) et les travaux astronomiques de la mission française de Pékin. *Comptes Rendus Académie des Sciences* (Paris) **166,** 871-877.

Crétinau-Joly, J. 1846. *Histoire de la Compagnie de Jésus*. Mellier Fréres, Lyon, vol. **4**, 253-254.

Dainville, F. de, 1978. *L'education des jesuites (XVI-XVIII siècles)*. Les Edition de Minuit, Paris.

Delattre, P., 1953. *Les établissements de Jesuites en France depuis quattre siècles*. Meester Frères, Wetteren (4 vols.)

Duhr, B., 1921. *Geschichte der Jesuiten in den Länder deutcher Zunge*. G. J. Manz, Munchen-Regensburg, 3 vols.

Fischer, K. A. F., 1978. Jesuiten-Mathematiker in der Deutschen Assistenze bis 1773. *Arch. Hist. Soc. Iesu* **47**, 159-224.

Fischer, K. A. F., 1983. Jesuiten Mathematiker in der französischen und itelianischen Assistenz bis 1762 bzw 1773. *Arch. Hist. Soc. Iesu* 52, 52-92.

García-Villoslada, R., 1954. *Storia del Collegio Romano*.(Analecta Gregoriana Vol. 66) Universitat Gregoriana, Rome, 209 pp.

Gehlert, O., 1990. Jesuiten als Naturwissenschafter (Maximilian Hell, Joseph Liesganig and Karl Braun). *Kalsburger Korrespondenz* **218**, 28-36.

Harris, S. J., 1995. Les chaires de mathématiques, In: L. Giard (ed.), *Les jésuites à la Renaisssance*. Presses Universitaires de France, Paris, 239-261.

Heilbron, J. L., 1982. *The elements of early modern physics*. (II.1. Jesuits. Their magisterium. Their eclecticism. Their constraint, 93-106) University of California Press, Berkeley

Heilbron, J.L., 1999. *The Sun in the Church. Cathedrals as solar observatories*. Harvard University Press, Cambridge, Mass. 366 pp.

Hopmann, J. 1957. 200 Jahren Astronomie in Wien. In: *1756-1956 Jubiläumfeier der Universitätsternwarte in Wien*. Vienna, 9-21.

Knobloch, E., 1988. Sur la vie et l'oeuvre de Ch. Clavius. *Rev. d'Histoire des Sciences* 41, 331-356.

Lackner, F., 1976. *Die Jesuitenprofessoren an der philosophischen Fakultät der Wiener Universität 1712-1773*. Dissertation der Universität Wien 128/11, Vienna, 346 pp.

Lalande, J. J., 1803. *Histoire de Mathématique de Montucla*, H. Agasse, Paris, Part 5, Lib. 7, 347-367.

Lattis, J.M., 1994. *Between Copernicus and Galileo. Christopher Clavius and the collapse of Ptolemaic cosmology*. The University of Chicago Press, Chicago, 293 pp.

Lerner, M. P., 1995. L'entrée de Tycho Brahe chez les jesuites ou le chant du cygne de Clavius. In: L. Giard (ed.), *Les jesuites a la Renaisssance*. Presses Universitaires de France, Paris, 145-185.

MacDonnell, 1989. *Jesuit Geometers. A study of fifty-six prominent Jesuit geometers during the first two centuries of Jesuit history*. Vatican Observatory Publications, Vatican City, 80 pp.

Maynard, M. 1853. *Des études et de l'enseignement des Jésuites à l'epoque de leur suppression (1750-1773)*. Poussielgue-Rusand, Paris, 259-279.

Naux, C., 1983. Le pére Christophore Clavius (1537-1612) Sa vie et son oeuvre. *Rev. des Questions Scientifiques* **154**, 55-67, 181-193, 325-347.

Oñate-Gillén, C. 1991. Los jesuitas y la ciencia moderna (siglos XVI al XVIII*). Letras de Deusto* **21**, 297-324.

Randles, W. G. L., 1999. *The unmaking of the medieval Christian cosmos, 1500-1760. From solid heavens to boundless aether*. Ashgate, Aldershot (U.K.), 274 pp.

Rodrigues, F., 1950. *História da Companhia de Jesus na Assistencia de Port*ugal. Apostolado da Imprenta, Porto, 3 Vols.

Romano, A., 1999. *La contre-réforme mathématique. Constitution et diffusion d'une culture mathématique jésuite à la Renaisssance (1540-160)*. École Française de Rome. 691 pp.

Sanchez Navarro-Neumann, M.M., 1937. Jesuitas astronómos. *Broteria* **24**, 423-236.

Sarton, G., 1944. Vindication of Father Hell. *Isis* 35, 97-105

Schreiber, J., 1903. Die Jesuiten des 17 and 18 Jahrhunderts und ihr Verhälnis zur Astronomie. *Natur und Offenbarung* **49**, 129-143.

Seydl, O., 1938. Z nejstarsich dejin prazske hvezdarny (From the oldest history of the Observatory in Prague) *Cesky casopis historicky* (Czech Historical Magazine) **44**, 486-502.

Simón-Díaz, J., 1952. *Historia del Colegio Imperial de Madrid.* Instituto de Estudios Madrileños, Madrid , 2 vols.

Thirion, J., 1880. Les jesuites astronomes jugés par le Baron de Zach. *Precis Historique.* A. Vromant, Bruxelles.

Udías, A., 1996. Meteorology in the observatories of the Society of Jesus. *Arch. Hist. Soc. Jesu,* **65**, 157-170.

Udías, A., 2000. Contribución de los jesuitas a la ciencia en los siglos XVI al XVIII. *Arbor* **167**, 207-228.

Volk, O., 1982. 400 Jahre Mathematik und Astronomie an der Universität Würzburg: Alma Julia Herbipolensis, 1582-1982. *Celestial Mechanics* **28**, 243-250.

Vrégille, P. de, 1906. L'Observatoire du Collège de la Trinité à Lyon, 1556-1794. *Relations d'Orient* (supl. 3). Polleunis, Brussels, 21 pp.

Vrégille, P. de, 1906. *L'Observatoire de Sainte Croix à Marseille* (1702-1763). Victor Pétaux, Paris, 15 pp.

Wrba, J., 1985. Der Orden der Gessellschaft Jesu im alten Universitätsviertel von Wien. In: G. Hamann, K. Mühlberger and F. Skacel (eds.), *Das alte Universitätsviertel in Wien, 1385-1985*, Vienna, 47-74.

Chapter 3

Anonymous, 1810. Observations géographique faites en 1734 par des Pères Jésuites pendant leur voyage de Chandernagpore à Dehli et à Jaëpur. *Lettres Edifiantes et Curieuses* (N.E. Sens, Toulouse), vol. 15, 269-290.

Bernard, H. 1935. Galilée et les jésuites des missions d'Orient. *Rev. des Questions Scientifiques* 108, 357-382.

Bernard-Maitre, H., 1951. *La science européen au tribunal astronomique de Péquin, XVII-XIX siècles.* Palais de la Découverte, serie D, No. 9, Paris.

Bigourdan, G., 1918. La station astronomique du Collége de Clermont (première pèriode) et la mission astronomique de Siam. *Comptes Rendus Académie des Sciences*
(Paris) **166**, 833-844.

Bigourdan, G., 1918. L'observatoire du Collége Louis-le-Grand (derniere pèriode) et les travaux astronomiques de la mission française de Pékin. *Comptes Rendus Académie des Sciences*(Paris) **166**, 871-877.

Bossman, H., 1912. Ferdinand Verbiest, directeur de l'Observatoire de Peking (1623-1688).*Rev. des Questions Scientifiques* **71**, 195-375.

V. Cronin, 1959. *The Wise Man from the West.* Rupert Hart-Davies, London, 300 pp.

Dehergne, J., 1973. *Répertoire des jésuites de Chine de 1552 à 1800.* Institutum Historicum Societatis Iesu, Rome, 430 pp.

Dunn, G.H., 1962. *Generation of Giants.* Burns and Oates, London,

Furlong Cardiff, G., 1929. *Glorias Santefesinas* (Buenaventura Suárez, 1679-1750). Editorial Surgo, Buenos Aires, 79-140.

Golvers, N., 1993. *The Astronomia Europaea of Ferdinad Verbiest, S.J. (Dillingen, 1687). Text, translation and commentaries.* Monumenta Serica Monograph Series XXVIII, Steyler, Nettetal.

Ho Peng Yoke, 1985. *Li Qi and Shu: An introduction to science and civilization in China.* Ch. 7. The decline of Chinese mathematics and the coming of the Jesuits. Hong Kong University Press.

Hsia, F., 1999. Jesuits, Jupiter's satellites and the Académie des Sciences. In: J.W. O'Malley, G. A. Bailey, S. T. Harris and F. Kennedy. *The Jesuits. Cultures, sciences and the arts, 1540-1773.* University of Toronto Press, Toronto, 241-257.

Kochhar, R. K., 1994. Secondary tools of empire: Jesuit men of science in India. In: T. R. de Souza (ed.) *Discoveries, missionary expansion and Asian cultures.* Concept Pub., New Delhi, 175-183.

Lichtenstern, A., 1977. Ignaz Kögler Mandarin aus Landberg. *Ignaz Kögler Gymnasium, Jahrebericht 1977*. Landberg am Lech, 1-10.

Needham, J., 1959. *Science and Civilization in China*. Vol III, Mathematics and the science of heavens and earth. The time of the Jesuits. Cambridge University Press, 437-461.

Prahlad Singh, 1986. *Les observatoires de Pierre (Jantar-Mantars)*. Holiday Publications, Jaipur. 94 pp.

Rochemonteix, C de, 1903. Le pére Amiot et la mission française de Pékin a la fin du XVIIIe siècle. *Études* 94, 26-47, 175-196, 338-355.

Rodrigues, F., 1925. *Jesuitas portugueses astronomos na China*. Tip. Porto Medico, Porto.

Rowbotham, A.H., 1942. *Missionary and Mandarin. The Jesuits at the Peking court of China*. University of California Press, Berkeley.

Sharma, V. N., 1982. The impact of the eighteenth century Jesuit astronomers on the astronomy of India and China. *Indian Jour. of History of Science* 17, 345-352.

Sharma, V. N., 1982. Jai Singh, his European astronomers and the Copernican revolution. *Indian Jour. of History of Science* 18, 333-344.

Sharma, V. N. and L. Huberty, 1984. Jesuit astronomers in eighteenth century India. *Arch. Int. D'Histoire des Science* 34, 99-107.

Udías, A., 1994. Jesuit astronomers in Beijing, 1601-1805. *Quart. Jour. Roy. Astr. Soc.*, **34**, 463-478.

Vaeth, A. 1991. *Johan Adam Schall von Bell S.J., Missionar in China keiserlicher Astronom und Ratgeber am Hofe von Peking (1592-1666). Ein Leben und Zeitbild* (Unter Mitwirkung von L. Van Hee S.J.) Monumenta Serica Monograph Series XXV. Steyler Verlag, Nettetal.

Witek, J. W. (ed.), 1994. *Ferdinand Verbiest (1623-1688) Jesuit Missionary, Scientist, Engineer and Diplomat*. Monumenta Serica Monograph Series XXX, Steyler Verlag, Nettetal.

INDEX BY OBSERVATORIES

INDEX BY NAMES

Previously published in Astrophysics and Space Science Library book series:

- **Volume 285**: **Information Handling in Astronomy - Historical Vistas**
  Editor: André Heck
  Hardbound, ISBN 1-4020-1178-4, March 2003
- **Volume 283**: **Mass-Losing Pulsating Stars and Their Circumstellar Matter**
  Editors: Y. Nakada, M. Honma, M. Seki
  Hardbound, ISBN 1-4020-1162-8, March 2003
- **Volume 282**: **Radio Recombination Lines**
  Authors: M.A. Gordon, R.L. Sorochenko
  Hardbound, ISBN 1-4020-1016-8, November 2002
- **Volume 280**: **Organizations and Strategies in Astronomy III**
  Editor: André Heck
  Hardbound, ISBN 1-4020-0812-0, September 2002
- **Volume 279**: **Plasma Astrophysics , Second Edition**
  Author: Arnold O. Benz
  Hardbound, ISBN 1-4020-0695-0, July 2002
- **Volume 278**: **Exploring the Secrets of the Aurora**
  Author: Syun-Ichi Akasofu
  Hardbound, ISBN 1-4020-0685-3, August 2002
- **Volume 277**: **The Sun and Space Weather**
  Author: Arnold Hanslmeier
  Hardbound, ISBN 1-4020-0684-5, July 2002
- **Volume 276**: **Modern Theoretical and Observational Cosmology**
  Editors: Manolis Plionis, Spiros Cotsakis
  Hardbound, ISBN 1-4020-0808-2, September 2002
- **Volume 275**: **History of Oriental Astronomy**
  Editor: S.M. Razaullah Ansari
  Hardbound, ISBN 1-4020-0657-8, December 2002
- **Volume 274**:**New Quests in Stellar Astrophysics: The Link Between Stars and Cosmology**
  Editors: Miguel Chávez, Alessandro Bressan, Alberto Buzzoni,Divakara Mayya
  Hardbound, ISBN 1-4020-0644-6, June 2002
- **Volume 273**: **Lunar Gravimetry**
  Author: Rune Floberghagen
  Hardbound, ISBN 1-4020-0544-X, May 2002
- **Volume 272**:**Merging Processes in Galaxy Clusters**
  Editors: L. Feretti, I.M. Gioia, G. Giovannini
  Hardbound, ISBN 1-4020-0531-8, May 2002
- **Volume 271**: **Astronomy-inspired Atomic and Molecular Physics**
  Author: A.R.P. Rau
  Hardbound, ISBN 1-4020-0467-2, March 2002
- **Volume 270**: **Dayside and Polar Cap Aurora**
  Authors: Per Even Sandholt, Herbert C. Carlson, Alv Egeland
  Hardbound, ISBN 1-4020-0447-8, July 2002
- **Volume 269**:**Mechanics of Turbulence of Multicomponent Gases**
  Authors: Mikhail Ya. Marov, Aleksander V. Kolesnichenko
  Hardbound, ISBN 1-4020-0103-7, December 2001

- **Volume 268:** **Multielement System Design in Astronomy and Radio Science**
  Authors: Lazarus E. Kopilovich, Leonid G. Sodin
  Hardbound, ISBN 1-4020-0069-3, November 2001
- **Volume 267:** **The Nature of Unidentified Galactic High-Energy Gamma-Ray Sources**
  Editors: Alberto Carramiñana, Olaf Reimer, David J. Thompson
  Hardbound, ISBN 1-4020-0010-3, October 2001
- **Volume 266:** **Organizations and Strategies in Astronomy II**
  Editor: André Heck
  Hardbound, ISBN 0-7923-7172-0, October 2001
- **Volume 265:** **Post-AGB Objects as a Phase of Stellar Evolution**
  Editors: R. Szczerba, S.K. Górny
  Hardbound, ISBN 0-7923-7145-3, July 2001
- **Volume 264:** **The Influence of Binaries on Stellar Population Studies**
  Editor: Dany Vanbeveren
  Hardbound, ISBN 0-7923-7104-6, July 2001
- **Volume 262:** **Whistler Phenomena**
  Short Impulse Propagation
  Authors: Csaba Ferencz, Orsolya E. Ferencz, Dániel Hamar, János Lichtenberger
  Hardbound, ISBN 0-7923-6995-5, June 2001
- **Volume 261:** **Collisional Processes in the Solar System**
  Editors: Mikhail Ya. Marov, Hans Rickman
  Hardbound, ISBN 0-7923-6946-7, May 2001
- **Volume 260:** **Solar Cosmic Rays**
  Author: Leonty I. Miroshnichenko
  Hardbound, ISBN 0-7923-6928-9, May 2001
- **Volume 259:** **The Dynamic Sun**
  Editors: Arnold Hanslmeier, Mauro Messerotti, Astrid Veronig
  Hardbound, ISBN 0-7923-6915-7, May 2001
- **Volume 258:** **Electrohydrodynamics in Dusty and Dirty Plasmas**
  Gravito-Electrodynamics and EHD
  Author: Hiroshi Kikuchi
  Hardbound, ISBN 0-7923-6822-3, June 2001
- **Volume 257:** **Stellar Pulsation - Nonlinear Studies**
  Editors: Mine Takeuti, Dimitar D. Sasselov
  Hardbound, ISBN 0-7923-6818-5, March 2001
- **Volume 256:** **Organizations and Strategies in Astronomy**
  Editor: André Heck
  Hardbound, ISBN 0-7923-6671-9, November 2000
- **Volume 255:** **The Evolution of the Milky Way**
  Stars versus Clusters
  Editors: Francesca Matteucci, Franco Giovannelli
  Hardbound, ISBN 0-7923-6679-4, January 2001
- **Volume 254:** **Stellar Astrophysics**
  Editors: K.S. Cheng, Hoi Fung Chau, Kwing Lam Chan, Kam Ching Leung
  Hardbound, ISBN 0-7923-6659-X, November 2000
- **Volume 253:** **The Chemical Evolution of the Galaxy**
  Author: Francesca Matteucci
  Hardbound, ISBN 0-7923-6552-6, May 2001

- **Volume 252: Optical Detectors for Astronomy II State-of-the-art at the Turn of the Millennium**
  Editors: Paola Amico, James W. Beletic
  Hardbound, ISBN 0-7923-6536-4, December 2000
- **Volume 251: Cosmic Plasma Physics**
  Author: Boris V. Somov
  Hardbound, ISBN 0-7923-6512-7, September 2000
- **Volume 250: Information Handling in Astronomy**
  Editor: André Heck
  Hardbound, ISBN 0-7923-6494-5, October 2000
- **Volume 249: The Neutral Upper Atmosphere**
  Author: S.N. Ghosh
  Hardbound, ISBN 0-7923-6434-1, July 2002
- **Volume 247: Large Scale Structure Formation**
  Editors: Reza Mansouri, Robert Brandenberger
  Hardbound, ISBN 0-7923-6411-2, August 2000
- **Volume 246: The Legacy of J.C. Kapteyn Studies on Kapteyn and the Development of Modern Astronomy**
  Editors: Piet C. van der Kruit, Klaas van Berkel
  Hardbound, ISBN 0-7923-6393-0, August 2000
- **Volume 245: Waves in Dusty Space Plasmas**
  Author: Frank Verheest
  Hardbound, ISBN 0-7923-6232-2, April 2000
- **Volume 244: The Universe Visions and Perspectives**
  Editors: Naresh Dadhich, Ajit Kembhavi
  Hardbound, ISBN 0-7923-6210-1, August 2000
- **Volume 243: Solar Polarization**
  Editors: K.N. Nagendra, Jan Olof Stenflo
  Hardbound, ISBN 0-7923-5814-7, July 1999
- **Volume 242: Cosmic Perspectives in Space Physics**
  Author: Sukumar Biswas
  Hardbound, ISBN 0-7923-5813-9, June 2000
- **Volume 241: Millimeter-Wave Astronomy: Molecular Chemistry & Physics in Space**
  Editors: W.F. Wall, Alberto Carramiñana, Luis Carrasco, P.F. Goldsmith
  Hardbound, ISBN 0-7923-5581-4, May 1999
- **Volume 240: Numerical Astrophysics**
  Editors: Shoken M. Miyama, Kohji Tomisaka, Tomoyuki Hanawa
  Hardbound, ISBN 0-7923-5566-0, March 1999
- **Volume 239: Motions in the Solar Atmosphere**
  Editors: Arnold Hanslmeier, Mauro Messerotti
  Hardbound, ISBN 0-7923-5507-5, February 1999
- **Volume 238: Substorms-4**
  Editors: S. Kokubun, Y. Kamide
  Hardbound, ISBN 0-7923-5465-6, March 1999
- **Volume 237: Post-Hipparcos Cosmic Candles**
  Editors: André Heck, Filippina Caputo
  Hardbound, ISBN 0-7923-5348-X, December 1998

- **Volume 236:** **Laboratory Astrophysics and Space Research**
  Editors: P. Ehrenfreund, C. Krafft, H. Kochan, V. Pirronello
  Hardbound, ISBN 0-7923-5338-2, December 1998
- **Volume 235:** **Astrophysical Plasmas and Fluids**
  Author: Vinod Krishan
  Hardbound, ISBN 0-7923-5312-9, January 1999
  Paperback, ISBN 0-7923-5490-7, January 1999
- **Volume 234:** **Observational Evidence for Black Holes in the Universe**
  Editor: Sandip K. Chakrabarti
  Hardbound, ISBN 0-7923-5298-X, November 1998
- **Volume 233:** **B[e] Stars**
  Editors: Anne Marie Hubert, Carlos Jaschek
  Hardbound, ISBN 0-7923-5208-4, September 1998
- **Volume 232:** **The Brightest Binaries**
  Authors: Dany Vanbeveren, W. van Rensbergen, C.W.H. de Loore
  Hardbound, ISBN 0-7923-5155-X, July 1998
- **Volume 231:** **The Evolving Universe**
  Selected Topics on Large-Scale Structure and on the Properties of Galaxies
  Editor: Donald Hamilton
  Hardbound, ISBN 0-7923-5074-X, July 1998
- **Volume 230:** **The Impact of Near-Infrared Sky Surveys on Galactic and Extragalactic Astronomy**
  Editor: N. Epchtein
  Hardbound, ISBN 0-7923-5025-1, June 1998
- **Volume 229:** **Observational Plasma Astrophysics: Five Years of Yohkoh and Beyond**
  Editors: Tetsuya Watanabe, Takeo Kosugi, Alphonse C. Sterling
  Hardbound, ISBN 0-7923-4985-7, March 1998
- **Volume 228:** **Optical Detectors for Astronomy**
  Editors: James W. Beletic, Paola Amico
  Hardbound, ISBN 0-7923-4925-3, April 1998
- **Volume 227:** **Solar System Ices**
  Editors: B. Schmitt, C. de Bergh, M. Festou
  Hardbound, ISBN 0-7923-4902-4, January 1998
- **Volume 226:** **Observational Cosmology with the New Radio Surveys**
  Editors: M.N. Bremer, N. Jackson, I. Pérez-Fournon
  Hardbound, ISBN 0-7923-4885-0, February 1998

Missing volume numbers have not yet been published.
For further information about this book series we refer you to the following web site: http://www.wkap.nl/prod/s/ASSL

To contact the Publishing Editor for new book proposals:
Dr. Harry (J.J.) Blom: harry.blom@wkap.nl

GPSR Compliance

The European Union's (EU) General Product Safety Regulation (GPSR) is a set of rules that requires consumer products to be safe and our obligations to ensure this.

If you have any concerns about our products, you can contact us on ProductSafety@springernature.com

In case Publisher is established outside the EU, the EU authorized representative is:

Springer Nature Customer Service Center GmbH
Europaplatz 3
69115 Heidelberg, Germany

Batch number: 10607204

Printed by Printforce, the Netherlands